Laws and

Laws and Symmetry

Bas C. van Fraassen

CLARENDON PRESS · OXFORD

Oxford University Press, Walton Street, Oxford OX2 6DP
Oxford New York Toronto
Delhi Bombay Calcutta Madras Karachi
Petaling Jaya Singapore Hong Kong Tokyo
Melbourne Auckland
and associated companies in
Berlin Ibadan

Oxford is a trade mark of Oxford University Press

Published in the United States
by Oxford University Press, New York

First published 1989
Reprinted 1990, 1991

British Library Cataloguing in Publication Data
Van Fraassen, Bas C. 1941–
Laws and symmetry
1. Science. Laws. Philosophical aspects
I. Title
501

ISBN 0–19–824811–3
ISBN 0–19–824860–1 (Pbk.)

Library of Congress Cataloging in Publication Data
Van Fraassen, Bas C., 1941–
Laws and symmetry / Bas C. van Fraassen.
Bibliography: p. Includes index.
1. Nature. Philosophy of. 2. Necessity (Philosophy)–
–Controversial literature. 3. Symmetry. 4. Philosophy and science.
I. Title.
BD581.V27 1989 113—dc19 89–30366

ISBN 0–19–824811–3
ISBN 0–19–824860–1 (Pbk.)

Printed in Great Britain by
Courier International Ltd., Tiptree, Essex

To Etruscan places

PREFACE

THE idea that there are laws of nature is by no means the sole property of philosophers. That this idea is the key to what science is, often appears as self-evident in popular thought: in *The History of Landscape Painting*, John Constable writes 'Painting is a science' and he adds, unconscious of any leap in thought, 'and should be pursued as an inquiry into the laws of nature.' That these laws constitute also walls of impossibility that border the course of events, is a constant preoccupation of Dostoevsky's narrator in *Notes from Underground*: 'all the same', he says, 'the laws of nature have mistreated me constantly, more than anything else in my life'.

But the idea that there are such laws has also played a major role in philosophy concerned with science. We find this idea perhaps most prominently in the philosopher-scientists of the seventeenth century, but it survived in ours even the logical positivists' radical rejection of history. Reichenbach, Hempel, and Goodman themselves invoked laws to shed light on issues prominently associated with their names. Their invocation was perhaps critical and tentative, but it seeped into other areas of philosophy as certainty. When moral philosophers discuss free will, for example, should they not be allowed to lean on such certainties as philosophy of science has established concerning cause, necessity, counterfactuals, and nomological explanation? And similarly, given those supposed certainties, isn't it quite proper too for philosophy of mind to approach uncritically the question whether there are psycho-physical laws?

In the first part of this book I shall argue that no philosophical account of laws of nature does or can succeed. In the second I shall rebut the argument that we must believe in them nevertheless. Now if there are laws of nature we must approach science in one way, but if there are no laws, we are freed to leave behind a whole range of traditional problems. I make a proposal for how to do philosophy of science, and devote the third and fourth part to contributions to the semantic approach (as Frederick Suppe baptized it) which I advocate. The emphasis will be on *symmetry*, as a key to theory, though not in the sense that laws were.

This book was originally twice as long. When a general approach is announced and advocated, it remains hand-waving except to the extent that it is implemented. Accordingly, the now missing part was devoted to a detailed study of the structure and interpretation of quantum mechanics. It will appear separately, as *Quantum Mechanics: An Empiricist View*.

My colleagues in philosophy of science may be a little surprised by Part I, for the accounts of law taken up belong to the area of metaphysics which they generally ignore. They are of course free to ignore that part and turn to the chapters on inference to the best explanation, theories, symmetry, classical physics, and probability. I too lack sympathy for metaphysics, though not in general: only for pre-Kantian metaphysics—and then only if practised after Kant. I have tried to ensure nevertheless that none of my arguments rests on this rejection; otherwise my critique would not be very effective. Those who don't care to engage the metaphysicians on their own ground, I would like to urge at least this: do not rely on such concept as law without inquiring whether there is anything that could play the required role. For that is similar to other philosophers relying unquestioningly on fifty-year-old philosophy of science.

There are many people and institutions I want to thank. The original manuscript was written during a sabbatical leave (1986/7) supported by Princeton University and the National Science Foundation. Specific debts are acknowledged in many sections, but I wish to thank Ernan McMullen, David Lewis, Storrs McCall, Wesley Salmon, Michael Tooley, and Richard Foley for special help with Part I. For Part II I want to acknowledge the help and insights received from Gilbert Harman, Richard Jeffrey, and Brian Skyrms. Elisabeth Lloyd helped me to see new possibilities in the semantic approach to theories, and R. I. G. Hughes in the many uses of symmetry. Discussions with Brian Skyrms, Maria-Carla Galavotti, and Nancy Cartwright helped with causality, while Roger Cooke helped with the intricacies of both classical mechanics and subjective probability. Joint work with R. I. G. Hughes appears in the last chapter. Perhaps needless to say, it was my valued teacher Wilfred Sellars and my eventual colleague Gilbert Harman who started me on the subject of inference to the best explanation, a number of years ago. Margot Livesey suggested a good deal of stylistic improvement in the first three chapters and I hope this had its effect on the later ones as well.

Anne Marie DeMeo typed endless drafts of this material over a period of about three years, and kept my practical life functioning even while I travelled; she deserves special thanks.

Finally, I want to acknowledge with some nostalgia the places where I wrote the draft during that initial year: Vancouver, Victoria (BC), Montreal, Notre Dame, Bologna, Assisi, Rome, Jerusalem, Tiberias, Cambridge (Mass.), Cleveland, and London.

SUMMARY TABLE OF CONTENTS

CONTENTS

1
Introduction

WHEN philosophers discuss laws of nature, they speak in terms of universality and necessity. Science too knows the terminology of laws, both in title ('Ohm's law', 'the law of conservation of energy'), and in generic classifications ('laws of motion', 'conservation laws'). Scientists, however, do not speak of law in terms of universality and necessity, but in terms of symmetry, transformations, and invariance. You may open a scientific journal and read that some result was reached on the basis of considerations of symmetry—never that it was found through considerations of universality and necessity. Is the common terminology of laws still apt, or do we have here two discussions of entirely different subjects?

In the seventeenth century Descartes, Newton, and Leibniz founded modern physics as well as modern philosophy. They spoke freely in terms of laws of nature. There is now an informative historical study of the use of this word 'law' in science, due to Jane Ruby, which shows its manifold roots.[1] Its central use is of course to denote the codes of conduct and arbitration imposed by society or its rulers. But the cluster of its common uses has from antiquity included references to laws of logic, of geometry and number, of poetic form, of optics, and so forth. The elements of lawgiver and conscious subject are obviously not present in these peripheral uses. That should not surprise us. Our own daily conversation gains its verve and vividness from transposition and generalization, as well as from analogy and metaphor, all engaged in with little thought or scruple. Thus when Aquinas insisted that the participation of inanimate nature in eternal law was just metaphor, he was perhaps rightly guarding against philosophical confusion but overly strict about language.

What does beg for explanation however, is that for Descartes, Newton, and Leibniz, and many subsequent writers, this term had come to stand for the central object of scientific inquiry, and for a pre-eminent candidate for explanation of the charted phenomena. Why did the term not merely remain a manner of speaking—as

Boyle explicitly said he used it, 'for brevity's sake, or out of custom'[2]—but come to stand for the very key to scientific understanding of the world? I am not quite certain how such questions are to be answered; for it is not entirely clear what contrast they pose, or what relevant scope they allow for answers. But in response I will offer a little speculation. It will be brief and somewhat sketchy, but may serve as an introduction to more serious analysis.[3]

1. THE NEED FOR GLOBAL CONSTRAINTS

The great scandal of pre-modern natural philosophy was the distance between physics and astronomy. The continued inability to integrate them during our science's first two millennia, was chronicled in Pierre Duhem's *To Save the Phenomena* and is now a familiar story.

The success of astronomy put into doubt the most basic principles of Aristotelian physics. Those principles, we can say in retrospect, entail that a physical account must be entirely in terms of local conditions and local interaction. That nothing moves, unless it be moved by something else, is Aristotle's first principle of physics. This applies not only to local motion, but to all change. The local conditions which provide for the possibility of change, and of action of one substance on another, are in turn characterized entirely in terms of the properties of the individual entities involved, taken singly. These are the *complexio* (composition in proportions of the elements) and the occult properties (properties not derivable from the *complexio*), some essential and some accidental. What happens admits of scientific explanation only if it is a phenomenon which proceeds with necessity, from the natures of the individual substances involved, as they act upon each other.[4]

The successful astronomical theory of Ptolemy did not look like this, and physicists did not succeed in recasting it in this form. In the Renaissance, the situation did not at once appear to be much better for the New Sciences. For Copernicus' theory is still a global description of elegantly choreographed movement—while the mechanical philosophers insisted as much as the Aristotelians ever had, on an account purely in terms of local action and interaction. This new insistence was in effect on a much more narrowly construed locality, because final causes were banished, and only action by contact alone fits the mechanical mold. Robert Boyle later summed up the prescribed pattern of scientific account concisely as follows: 'I mean, by such corporeal agents as do not appear either to work otherwise than by virtue of the motion, size, figure, and contrivance of their own parts (which attributes I call the mechanical affections of matter).'[5] There did not appear to be any action-by-contact mechanism to drive the clockwork of the heavens. Sir Richard Blackmore, writing only shortly before all of England fell to Newton's sway, set the scandal to verse:

> You, who so much are vers'd in causes, tell
> What from the tropicks can the sun repel?
>
> If to the old you the new schools prefer,
> And to the famed Copernicus adhere;
> If you esteem that supposition best,
> Which moves the earth, and leaves the sun at rest;
> With a new veil your ignorance you hide,
> Still is the knot as hard to be unty'd;
>
> This problem, as philosopher, resolve:
> What makes the globe from West to East revolve?
> What is the strong impulsive cause declare,
> Which rolls the ponderous orb so swift in air?
> To your vain answer will you have recourse,
> And tell us 'tis ingenite, active force,
> Mobility, or native power to move,
> Words which mean nothing, and can nothing prove?
> That moving power, that force innate explain
> Or your grave answers are absurd and vain. . . .[6]

Alternatively (as Blackmore did not see or grant) physics had to be allowed to extrapolate from necessities inherent in local interaction, to global constraints.

By a global constraint I mean a principle that applies to a system as a whole, and is not equivalent to any principle that applies distributively to the localized particulars or point locations in that system. As extreme illustration, imagine a world in which the total mass is conserved, but by the happenstance that some bits of matter spontaneously appear in random locations, to balance the mass that disappears elsewhere. Here the global principle of conservation of total mass of the system is not derivable from principles that govern any proper part. The alternative which Blackmore gave no sign of perceiving, is that physics could introduce principles which are global in this sense, though perhaps not quite so extreme.[7]

Both the Aristotelian tradition in physics, and the mechanical philosophy, lacked justification for such a step. But there was another tradition, not so inhibited.

> The time was earliest morning and the sun
> was climbing upwards with those very stars
> that were its company when holy law
> gave primal motion to their loveliness.[8]

Here in Canto I of the *Inferno*, Dante echoes a long tradition in which the world is indeed subject to global constraints: God's decrees. The Scriptures and the Church fathers use the metaphor, or analogy, of law and lawgiving to describe God's relation to the world.[9] In the great synthesis of the Middle Ages, Aquinas had wedded this Judaeo-Christian world picture to Aristotelian philosophy.

But Aquinas had done this without giving any new licence to physics. On the contrary: Aquinas was at pains to contest a preceding scholastic view that everything which happens, does so because it is directly and individually willed by God.[10] This would seem to make science a pointless enterprise; according to Aquinas it also denigrates creation. Yet theology points to God as ultimate cause. The reconciliation Aquinas offered was this: to explain why phenomena happen as they do, requires showing why they must; this necessity however derives from the natures of the individual substances involved—which themselves are as they are because of God's original design.[11] Thus the necessity does derive *ultimately* from God's decrees for the world as a whole, made at the point of creation—but derives *proximately* from the local conditions and characters in the Aristotelian pattern:

> the death of this animal is an absolutely necessary consequence of its being composed of contraries, although it was not absolutely necessary for it to be composed of contraries. Similarly, the production of such and such natures by God was voluntary; but, having been so constituted, something having absolute necessity comes forth from them . . .[12]

So Aquinas rejects the liberty to construe divine decrees in any form beyond the Aristotelian. But the liberty was there, even if it was not fully exploited until Descartes insisted that God keeps the quantity of motion in the universe constant, and Leibniz postulated a pre-established harmony, that keeps all substances in synchronic, correlated evolution:

> in whatever manner God might have created the world, it would always have been regular and in a certain order. God, however, has chosen the most perfect, that is to say the one which is at the same time the simplest in hypotheses and the richest in phenomena, as might be the case with a geometric line, whose construction was easy, but whose properties and effects were extremely remarkable . . . (*Discourse on Metaphysics*, vi).

We may be tempted at this point to say that the seventeenth-century concept of a law of nature is simply Aquinas's concept of God's design, with the Aristotelian qualifications removed.[13]

If that were all, however, it would be altogether mysterious why this concept should have remained and been cherished in the Enlightenment and thereafter. Indeed, if we look more closely at the seventeenth century we see an insistence, even more adamant than Aquinas's, upon the autonomy of physics from theology. Descartes insists on it most stringently (*Principles*, pt. 1, xxvii) and so does Leibniz (*Discourse on Metaphysics*, x): 'the physicist can explain his experiments . . . without any need of the general considerations which belong to another sphere, and if he employs the cooperation of God . . . he goes out of his path . . .'

This insistence nevertheless does not keep Leibniz from reiterating the gloss of Divine decree when he uses the term 'law of nature' (*Discourse*, xvii), nor from discussing his own contribution of minimal principles in optics, or his rudimentary law of least action, in that terminology (ibid. xix, xxi, xxii). What are we to make of this? Only, I think, that the terminology of law was still soothing to the religious ear, while already taken to denote aspects of structure entirely present and immanent in the world.

2. THE SECULAR, GLOBAL, AND AXIOMATIC METHOD

The *Drang nach Autonomie* of physics, even as developed by such theological thinkers as Descartes, Newton, and Leibniz, needed an intermediate link between God's decree and nature. Aquinas had needed such a link to explain proximate causation, and found it in the Aristotelian substantial forms (individual natures). For the seventeenth century another link was needed, one that could impose a global constraint on the world process. In general terms this link was provided by the idea that nature has its inner necessities, which are not mere facts, but constrain all mere facts into a unified whole. The theological analogy and dying metaphor of law provided the language in which the idea could be couched.

This distinction between laws and mere facts suited philosophical reflection on science especially well, because science had rediscovered the axiomatic ideal of theoretical form. All of science was to be developed *more geometrico*, with each proposition about fleeting or replicable circumstances to be deduced as an instance of basic principles. Could this format not mirror the exact way in which the small but chaotic realm of the senses instantiates the necessary and universal law of nature?

Certainly Descartes's and Newton's great treatises on natural philosophy equate the laws with what the axioms are meant to capture. Thus Part 2 of Descartes's *Principles of Philosophy*, develops his physics from the principles:

xxxvii. The first law of nature: that each thing as far as in it lies, continues always in the same state; and that which is once moved always continues so to move.

xxxix. The second law of nature: that all motion is of itself in a straight line. . . .

xl. The third law: that a body that comes in contact with another stronger than itself, loses nothing of its movement; if it meets one less strong, it lapses as much as it passes over to that body.

which are preceded by a much stronger principle, at once theological and global:

xxxvi. That God is the First Cause of movement and that He always preserves an equal amount of movement in the universe.

Details in his deductions show that this is what we would now call a law of conservation of momentum, because movement is measured by the product of velocity and quantity of matter. When Leibniz disputes this (*Discourse*, xvii) it is only to replace it by another equally global principle: that the quantity not of movement, but of force, in the universe is conserved. Given how he measures force, this is in effect what we would now call conservation of energy.

It may be objected that I have told this story very prettily, but just in terms of exactly those physicists who lost the battle for the foundations of modern science. Did not Newton, the victor, restore locality by means of his forces, exerted by bodies on each other in equal action and reaction? And could not the 'laws' of conservation of momentum and energy be deduced from his laws of motion, which include no specifically global postulates?

Newtonians may well have said or believed this, as part of their constant claims of superiority, but it is not correct. I would first of all underscore the Cartesians' complaint that Newton's mathematics did not bear out his verbal gloss. This language of bodies exerting forces on each other is nullified by the fact that the 'effect' is instantaneous over any distance. Is there anything more than a verbal difference between instantaneous action at a distance and pre-established harmony, if there is no question of conscious agency? The mathematical analysis into equal but opposite vectors, whose magnitudes and directions change continuously with time, does not turn this verbal distinction into a real one. If a comet hits the earth, the moon wobbles at the very same moment as the earth does—to say that the moon too is made to wobble by the collision looked to the Cartesians, and I think rightly so, as merely a baseless gloss on the mathematical description of a perfect correlation.[14] And second of all I would emphasize the awareness—imperfect until the nineteenth century—that Newton's laws of motion and force do not tell the whole story.[15] Perhaps it was never fully realized before Helmholtz, but the global law of conservation of energy is an integral part of classical physics, not entailed by Newton's axioms but still always implicitly assumed. The deduction of the conservation laws is for conservative systems—and the universe, though by definition isolated, is not by definition conservative. Whatever had been Newton's original hope or intent, modern physics gave us a world globally constrained into harmony.

3. THE END OF METAPHYSICS?

The somewhat speculative story I have now told means to show how the concept of law could have come to be taken as the key to the structure of science, and how it could have continued to be so regarded in the secular atmosphere of the eighteenth century. It does not explain why it should have continued in this fashion until today, and indeed my story suggests that it may be something of an anachronism. For the end of the eighteenth century marks a great turning-point in philosophy, and the philosophical understanding of science could not remain unaffected.

If everyone agreed on the project of separating science from theology, there was also a strong sentiment in empiricist quarters to render it independent of metaphysics. By empiricism I mean the philosophical position that experience is our source of information about the world, and our only source. The metaphysics attacked by Kant's *Critique* was characterized by its conviction that reason can bring us to logical—or as Descartes sometimes says, moral—certainty of truths that transcend experience.[16] The understanding simply does not reach that far. The most one could achieve by reason alone is a deduction, from the conditions required for experience to be possible at all, of general truths concerning the structure of experience.

There was thus a point at the end of the eighteenth century when philosophers by and large agreed that metaphysics was dead. Kant, who dominated all of Western philosophy for a century, had purportedly shown up this enterprise as inherently and essentially mistaken. It involved after all the extrapolation of our concepts, familiar from daily application in experience, to applications outside the reach of experience—and there cannot be for us even the glimmer of a hope of the possibility of a warrant for such extrapolation. We can spin and weave our words into a rich and colourful tapestry to depict ourselves weaving a likeness of ourselves in the world. But the result must inevitably depict us as hopelessly ignorant of even the conditions under which the woven picture would be true. In science, theorizing can always be harshly brought to a stop, through confrontations arranged within our experience; but purported applications of our concepts outside experience can never be put to the question within experience.

Kant's *Critique* ended most decisively the relative placing of metaphysics and science which Descartes had described so strikingly: 'Thus philosophy as a whole is like a tree whose roots are metaphysics, whose trunk is physics, and whose branches, which issue from this trunk, are all the other sciences.'[17] In practice, as Kant must have perceived, the progressive separation of modern science from metaphysics was already clear: the trunk and branches grew without much attention to the shape of the roots, if any.

Metaphysics was not the discipline on which science drew for its first principles. Instead, quite to the contrary, metaphysical theorizing had implicitly begun to take as its touchstone that it should 'save' the sciences in much the way that the sciences must 'save' the phenomena. If Kant's *Critique* succeeded, the relation of philosophy to science was henceforth quite different also from this submission, and of metaphysics there survives only the critical archaeology of ideas to uncover the actual presuppositions in actual history of science, plus the analysis of possible presuppositions that could constitute a foundation for science.

But such foundations, as Wittgenstein said apropos of mathematics, support science only in the sense in which the painted rock supports the painted tower. The enterprise of philosophy of science so conceived is not essentially different from the philosophy of logic, of mathematics, of morals, of law, of religion, and of art. It allows no hierarchical relation, neither that of trunk to roots, nor that of roots to trunk.

The very name of metaphysics disappeared for a while from university curricula—history of metaphysics was the history of an illusion. But of course the illusion did not die so easily. Post-Kantian philosophy, beginning as a critique in imitation of Kant, soon thought it had found presuppositions which we must believe to be true, if reason—or at least philosophy—was not to crumble into dust, and which relate experience to what is not experience nor shown in experience. Under many guises, pre-Kantian metaphysics returned. Since internal consistency and human interest are the only criteria that really operate there, metaphysics can go on and on, forever amending its story, venturing a little here, withdrawing a little there—sometimes also producing a great artist, with a vision that unifies large vistas of human thought, as if seen from a great mountain. And who would deny that art brings insight and understanding, as well as beauty? But when it does we should credit the author, the muse, or Providence, and also ourselves; for insight is recognition of the truth as true. Metaphysics, however, purports not to be 'mere' art.

What I have now written is of course only one view of our history, and by no means agreed on every hand. Sympathies aside, it is clear that on such a view, no sense is made of science by depicting it as a part of metaphysics. And sympathies aside, we all want to come together in the joint enterprises of philosophy of art, of law, of religion, and of science. We can do this if we look for clues to the structure of science inside science, inside the human activities of theory-making, model-construction, experiment-design, hypothesis-testing, test-evaluation. In the preceding sections we have seen how the idea of laws, and their distinction from mere fact, could act as a clue to the structure of science. It may or may not continue to play this role effectively in an environment less hospitable to (pre-Kantian) metaphysics. We should in any case look for other clues; and who knows which will give us the better key to the aim and structure of science? The other great clue, which began to be discerned already in that same seventeenth century, and which has steadily grown in visibility since, is symmetry.

4. THE BIRTH OF SYMMETRY

> God, Thou great symmetry,
> Who put a biting lust in me
> From whence my sorrows spring,
> For all the frittered days
> That I have spent in shapeless ways
> Give me one perfect thing.[18]

Symmetry, like laws, is not an idea to be explained in one sentence. You can begin by thinking of a concrete example—Roman law for the one, and mirror symmetry for the other; or the Napoleonic code, and the five perfect solids. But then, with quickening interest, you will be struck by suggestive analogies—between law and necessity; between rotation, which allows you to see the solid from all different angles, and intellectual abstraction. And soon you may turn reflective, espying similar structure in your own thought—the necessity of logical consequence in an argument, the symmetry of parallel solutions to essentially similar problems.

These remarks are at best tantalizing, I know. In a later part of this book, the idea of symmetry is to be explored properly. I shall end my introductory speculations here with a brief look at how symmetry also appeared, though as yet unnamed and altogether unsung, in the context of seventeenth-century metaphysics.

In homage to the axiomatic ideal, Leibniz sketched a system of metaphysics, which begins with the proper concept of God, deduces first of all His existence, and then exactly how he has created the world. It follows from this concept of God, that God does not act except with sufficient reason. This entails that the general Principle of Sufficient Reason must also be obeyed in any true description of the world, such as the special sciences aim to give us. Indeed, Leibniz proposed, every last fact about the world must follow logically, though for many facts thé chain of demonstration will be infinitely long, and hence not to be arrived at a priori by a finite mind.

But we can read Leibniz subversively, and speculate how he has arrived, not at principles of physics by deduction from the Principle of Sufficient Reason, but at that Principle by analogy and generalization from physics. In the *Discourse on Metaphysics* he discusses a scientific development of great philosophical significance in his eyes. The ancients had already arrived at the correct law of reflection for light: angle of incidence = angle of reflection. For example, if a light travels from lamp *A* via a flat mirror to point *B* on the wall, then it was reflected at the point *M* of the mirror such that *AM* and *MB* cut the mirror at equal angles. The ancient demonstration arrived at this conclusion by deduction, from the prior principle that light will follow the shortest path between emission and arrival. (For an example of this sort of demonstration, see the beginning of Chapter 10.)[19]

In the seventeenth century, Fermat had used similar reasoning to deduce Snell's law of refraction, in a way that Leibniz regarded as a proper generalization: 'it appears [the rays] follow the easiest way . . . for passing from a given point in one medium to a given point in another medium' (*Discourse*, xxiii). The generalization was as follows: if in different media, the ray travels at different velocities, then the time of travelling from emission to arrival is minimized. Because distance travelled equals velocity multiplied by time, it follows that when the velocities are equal (e.g. travel through a single medium) this new law also minimizes the distance travelled, thus yielding the old law of reflection as a special case.

Leibniz was quite correct to point out the value for physical theory of such a principle of 'least effort'. It was Leibniz's peculiar contribution to see Fermat's demonstrations as of the same form as Hero's and Heliodorus' about reflection—and allied to Aristotelian and medieval uses of final cause explanations—and finally to place them correctly as *symmetry* arguments, though Leibniz does not use that term.

Leibniz's reconstruction of these arguments goes roughly like this. Let it be given that the light travels from point *A* to point *B*; demonstrate that its path will be the straight line *AB*, if these points lie within an entirely homogeneous medium. This is the problem; how does one approach its solution? The problem itself introduces a geometric figure: the pair of points *A*, *B*, and the direction from *A* to *B*. To define any other direction in space, not along the line *AB*, one would need to refer to some other point, line, plane or figure, which has not been introduced in the given. Any rule governing the motion of light in this case must therefore either (*a*) imply that light follows the line *AB*, or (*b*) draw upon some feature *X* of the situation which could single out a direction not along that line. But the assumption that the medium is homogeneous, rules out the presence of such a feature *X* to be drawn upon. Therefore. . . .

We cannot quite yet conclude that therefore light travels along the straight line. As Leibniz clearly perceived, we need some bridge to get us from the fact that we could not possibly formulate any other rule here to the conclusion that light—a real phenomenon in nature—must follow that rule. The bridge, for Leibniz, is that the world was created by Someone who was in the position of having to solve these problems in the course of creation, and who would not choose without a reason for choice. If there was no such feature *X* to be preferred, obviously none of the choices of type (*b*) could then have been made. That God does not act without sufficient reason, implies that any correct science of nature, satisfies the constraint of Sufficient Reason. In the above problem, the conclusion that we cannot formulate any rule for the motion of light under these conditions, except that of rectilinear motion, yields then the corollary that light can follow no other path. The Principle of Sufficient Reason is introduced to fill the gap in the sort of arguments (the above, and also Hero's and Fermat's) here represented, and is in turn grounded in a certain conception of creation.

Leibniz put this Principle of Sufficient Reason to great tactical and polemical use, especially in his controversies with Newton, carried on via Samuel Clarke. To give but one example, there can be no Absolute Space: if there were, God would have had to choose to locate the world here rather than a few yards further on, and could have had no sufficient reason for such a choice.[20] It will not be apparent from all this, why I have started using the term 'symmetry' here. In advance of our exploration of symmetry in Part III, I can only give this hint: in the above example, the problem situation described was symmetric about the directed line *AB*, and the process described in the solution did not introduce any new asymmetry into the picture. This is a little tantalizing, but the promise will be made good.

Leibniz did not create the symmetry argument, he celebrated it and raised it to the status of foremost clue to the structure of the universe. He was right to discern it in ancient physics, and in his contemporaries (we shall see many more examples during our inquiry). He did not, however, complete his insight. The Principle of Sufficient Reason harbours ambiguities, and is at best a crude formulation of the form of argument Leibniz had definitely recognized and utilized. And it is highly dubitable, in that it extrapolates from the activity of the intellect, to the structure of the world. For Leibniz, the pursuit of symmetry in theorizing mirrored God's method of design for creation. For Descartes and Newton, the division of propositions imposed by the axiomatic method mirrored the distinction of law and mere fact in the world. When the scientific wish for autonomy was fulfilled, and the theological underpinning had been discarded, the metaphors were dead—or rather petrified, and honoured as if carved in stone—but still accepted as showing how the structure of science mirrors the structure of the world.

There is an alternative approach to the understanding of science, as I emphasized before: to study its structure in and by itself, as a product of the intellect that strives to order and unify the deliverances of experience. Both the notion of law and that of sufficient reason served as 'transcendental clues'. Departing from structural features of theory, they delineate the structure of any possible world allowed by physical theory—that is, the structure of its *models*. Both also were honoured or distorted—one's philosophical sympathies are crucial here—by a reification which accepts them as clues to the structure of the *world being modelled*. The alternative is to reject that reification.

However that may be, we must now enter upon a critical inquiry, first into laws and then into symmetry. Whether the roles that laws were meant to play in philosophical thought can really all be assigned to a single concept, and to what extent some (or indeed, any) of these roles can be illuminated by symmetry, we need to examine in detail.

PART I
Are There Laws of Nature?

Introduction

In Parts I and II I shall be concerned with the philosophical approach to science which has laws of nature as its central concern. After an initial discussion of criteria, Part I will focus on specific theories of laws of nature, recently proposed and defended. I shall argue that these theories face an insuperable dilemma—of two problems whose solutions must interfere with each other—which will occur for all theories of similar stripe. Epistemological arguments to the effect that we must believe in the reality of laws regardless of such difficulties will be broached in Part II. Throughout I shall keep in view the main question: can this approach to science, which looks to such deep foundations, possibly be adequate to its subject?

2.
What Are Laws of Nature?

THIS question has a presupposition, namely that there are laws of nature. But such a presupposition can be cancelled or suspended or, to use Husserl's apt phrase, 'bracketed'. Let us set aside this question of reality, to begin, and ask what it means for there to be a law of nature. There are a good half-dozen theories that answer this question today, but, to proceed cautiously, I propose to examine briefly the apparent motives for writing such theories, and two recent examples (Peirce, Davidson) of how philosophers write about laws of nature. Then I shall collect from the literature a number of criteria of adequacy that an account of such laws is meant to satisfy. These criteria point to two major problems to be faced by any account of laws.

1. THE IMPORTANCE OF LAWS

What motives could lead a philosopher today to construct a theory about laws of nature? We can find three. The first comes from certain traditional arguments, which go back at least to the realist–nominalist controversy of the fourteenth century. The second concerns science. And the last comes from a reflection on philosophical practice itself; for while in the seventeenth century it was scientific treatises that relied on the notion of law, today it is philosophical writings that do so.

The motive provided by the traditional arguments I shall spell out in the next section, drawing on the lectures of Charles Sanders Peirce.

The second and much more fashionable motive lies in the assertion that laws of nature are what science aims to discover. If that is so, philosophers must clearly occupy themselves with this subject. Thus Armstrong's *What Is a Law of Nature?* indicates in its first section, 'the nature of a law of nature must be a central ontological concern for the philosophy of science'.

This does indeed follow from the conception of science found among seventeenth-century thinkers, notably Descartes. Armstrong elaborates it as follows. Natural science traditionally has three tasks: *first*, to discover the topography and history of the actual universe; *second*, to discover what sorts of thing and sorts of property there are in the universe; and *third*, to state the laws which the things in the universe obey. The three tasks are interconnected in various ways. David Lewis expresses his own view of science in such similar comments as these:

> Physics is relevant because it aspires to give an inventory of natural properties. . . . Thus the business of physics is not just to discover laws and causal explanations. In putting forward as comprehensive theories that recognize only a limited range of natural properties, physics proposes inventories of the natural properties instantiated in our world. . . . Of course, the discovery of natural properties is inseparable from the discovery of laws.[1]

But what status shall we grant this view of science? Must an account of what the laws of nature are vindicate this view—or conversely, is our view of what science is to be bound to this conception? We know whence it derives: the ideal of a metaphysics in which the sciences are unified, as parts of an explanatory, all-embracing, and coherent world-picture (recall Descartes's 'philosophy as a whole is like a tree whose roots are metaphysics, whose trunk is physics, and whose branches, . . . are all the other sciences'). But this ideal is not shared throughout Western philosophy, nor ever was.

By its fruits, of course, shall we know this tree. If, starting with this conception, philosophers succeed in illuminating the structure of science and its activities, we shall have much reason to respect it. I do not share this conception of science, and do not see prima facie reason to hold it.

On the other hand, if metaphysics ought to be developed in such a way that the sciences can be among its parts, that does indeed place a constraint on metaphysics. It will require at least a constant series of plausibility arguments—to assure us that the introduction of universals, natural properties, laws, and physical necessities do not preclude such development. But this observation yields, in itself, only a motive for metaphysicians to study science, and not a motive for philosophers of science to study metaphysics.

The third and final motive, I said, lies in our reflection on philosophical practice itself. Even in areas far removed from philosophy of science, we find arguments and positions which rely for their very intelligibility on there being a significant distinction between laws and mere facts of nature. I can do no better than to give an example, in section 3 below, of one such philosophical discussion, by Donald Davidson, about whose influence and importance everyone is agreed.

2. PEIRCE ON SCHOLASTIC REALISM

The traditional arguments are two-fold: to the conclusion that there must be laws of nature, and quite independently, to the conclusion that we must believe that there are such laws. The first argues from the premiss that there are pervasive, stable regularities in nature [illegible]

to the audience, Peirce displayed a stone (piece of writing chalk?):

Suppose we attack the question experimentally. Here is a stone. Now I place that stone where there will be no obstacle between it and the floor, and I will predict with confidence that as soon as I let go my hold upon the stone it will fall to the floor. I will prove that I can make a correct prediction by actual trial if you like. But I see by your faces that you all think it will be a very silly experiment.

Why silly? Because we all know what will happen.

But how can we know that? In words to be echoed later by Einstein, Podolsky, and Rosen, he answers 'If I *truly know* anything, that which I know must be real.' The fact that we know that this stone will fall if released, 'is the proof that the formula, or uniformity [which] furnish[es] a safe basis for prediction, is, or if you like it better, *corresponds to*, a reality'. A few sentences later he names that reality as a law of nature (though for him that is not the end of the story).

Do we have here the first or the second argument, or both? We very definitely have the second, for Peirce clearly implies you have no right to believe that the phenomena will continue the same in the future, unless you believe in the reality in question. But the reality cannot be a mere regularity, a fact about the future 'ungrounded' in the present and past, for that could not be known. Peirce did recognize chance, and agreed that anything at all could come about spontaneously, by chance, without such underlying reasons. Therefore he does not subscribe to the validity of 'There is a regularity, therefore there must be a reason for it, since no regularity could come about without a reason.' However he does not allow that we can know the premiss of that argument to be true, unless we also know the conclusion—nor to believe the premiss unless we believe the conclusion. This is a subtle point but important.

He gives the example of a man observed to wind his watch daily over a period of months, and says we have a choice: (*a*) 'suppose that some *principle* or *cause* is *really* operative to *make* him wind his watch daily' and predict that he will continue to do so; or else (*b*) 'suppose it is mere chance that his actions have hitherto been regular; and in that case regularity in the past affords you not the slightest reason for expecting its continuance in the future'. It is the same with the operations of nature, Peirce goes on to say, and the observed regularity of falling stones leaves us only two choices. We can suppose the regularity to be a matter of chance only, and declare ourselves in no position to predict future cases—or else insist that we can predict because we regard the uniformity with which stones have been falling as 'due to some *active general principle*'.

There is a glaring equivocation in this reasoning, obscured by a judicious choice of examples. Sometimes 'by chance' is made to mean 'due to no reason', and sometimes 'no more likely to happen than its contraries'. Of course, I cannot logically say that certain events were a matter of chance in the second sense, and predict their continuation with any degree of certainty. That would be a logical mistake. Nor do I think that a person winds his watch for no reason at all, unless he does it absent-mindedly; and absent-mindedness is full of chance fluctuations. But I can quite consistently say that all bodies maintain their velocities unless acted upon, and add that this is just the way things are. That is consistent; it asserts a regularity and denies that there is some deeper reason to be found. It would be strange and misleading to express this opinion by saying that this is the way things are by chance. But that just shows that the phrase 'by chance' is tortured if we equate it to 'for no reason'.

Perhaps we should not accuse Peirce of this equivocation, but attribute to him instead the tacit premiss that whatever happens either does so for a reason or else happens no more often than its contraries. But that would mean that a universe without laws—if those are the reasons for regularities—would be totally irregular, chaotic. That assertion was exactly the conclusion of the first argument. Hence if this is how we reconstruct Peirce's reasoning, we have him subscribing to the first argument as well. His indeterminism would then consist in the view that individual events may indeed come about for no reason, but not regularities.[4]

Peirce knew well the contrary tradition variously labelled 'nominalist' and 'empiricist', which allows as rational also simple extrapolation from regularities in past experience to the future. He saw this represented most eminently by John Stuart Mill, and attacked it vigorously. The following argument appears in Peirce's entry 'Uniformity' in Baldwin's Dictionary (1902).[5] Of Mill, Peirce says that he 'was apt to be greatly influenced by Ockham's razor in forming theories which he defended with great logical acumen; but he differed from other men of that way of thinking in that his natural candour led to his making many admissions without perceiving how fatal they were to his negative theories' (ibid. 76).

Mill had indeed mentioned the characterization of the general uniformity of nature as the 'fact' that 'the universe is governed by general laws'.[6] (He did not necessarily endorse that form of language as the most apt, though he does again use it in the next paragraph.) Any particular uniformity may be arrived at by induction from observations. The peculiar difficulty of this view lies in the impression that the rule of induction gives, of presupposing some prior belief in the uniformity of nature itself. Mill offered a heroic solution:

the proposition that the course of nature is uniform is the fundamental principle, or general axiom, of Induction. It would yet be a great error to offer this large generalization as any explanation of the inductive process. On the contrary I hold it to be itself an instance of induction, and induction by no means of the most obvious kind. (*Collected Works*, 392)

According to Peirce, Mill used the term 'uniformity' in his discussions of induction, to avoid the use of 'law', because that signifies an element of reality no nominalist can admit. But if his 'uniformity' meant merely regularity, and implied no real connection between the events covered, it would destroy his argument. Thus Peirce writes:

It is, surely, not difficult to see that this theory of uniformities, far from helping to establish the validity of induction, would be, if consistently admitted, an insuperable objection to such validity. For if two facts, *A* and *B*, are entirely independent in their real nature, then the truth of *B* cannot follow, either necessarily or probably, from the truth of *A*. (*Collected Papers*, 77)

But this statement asserts exactly the point at issue: why should *A*, though bearing in itself no special relation to *B*, not be invariably or for the most part be followed by *B*? It is true that there would be no logical necessity about it, nor any probability logically derivable from descriptions of *A* and *B* in and by themselves. But why should all that is true, or even all that is true and important to us, be logically derivable from some internal connection or prior circumstance?

The convictions expressed by Peirce are strong, and have pervaded a good half of all Western philosophy. Obviously we shall be returning to these convictions, in their many guises, in subsequent chapters. A law must be conceived as *the reason which accounts for* uniformity in nature, not the mere uniformity or regularity itself. And the law must be conceived as something real, some element or aspect of reality quite independent of our thinking or theorizing—

not merely a principle in our preferred science or humanly imposed taxonomy.

3. A TWENTIETH-CENTURY EXAMPLE: DAVIDSON

Concepts developed or analysed in one part of philosophy tend to migrate to others, where they are then mobilized in arguments supporting one position or another. From the roles they are expected to play in such auxiliary deployment, we should be able to cull some criteria for their explication. A good example is found in recent philosophy of mind.

Is there mind distinct from matter? Peter felt a sudden fear for his safety, and said 'I know him not'. The first was a mental event, the second at least in part a physical one. But materialists say that the mental event too consisted solely in Peter's having a certain neurological and physiological state—so that it too was (really) physical. Donald Davidson brought a new classification to this subject, by focusing on the question whether there are psychophysical laws. Such a law, if there is one, might go like this: every human being in a certain initial physiological state, if placed in certain circumstances, will feel a sudden fear for his or her safety. Davidson denies that there are such laws, yet asserts that all mental events are physical.

> It may make the situation clearer to give a fourfold classification of theories of the relation between mental and physical events that emphasizes the independence of claims about laws and claims of identity. On the one hand there are those who assert, and those who deny, the existence of psychophysical laws; on the other hand there are those who say mental events are identical with physical and those who deny this. Theories are thus divided into four sorts: *nomological monism*, which affirms that there are correlating laws and that the events correlated are one (materialists belong in this category); *nomological dualism*, which comprises various forms of parallelism, interactionism, and epiphenomenalism; *anomalous dualism*, which combines ontological dualism with the general failure of laws correlating the mental and the physical (Cartesianism). And finally there is *anomalous monism*, which classifies the position I wish to occupy.[7]

This last position is that every strict law is a physical law, and most if not all events fall under some such law—which they can obviously do only if they admit of some true physical description.

Therefore most if not all events are physical. This is consistent provided that, although every individual mental event has some physical description, we do not assert that a class of events picked out by some mental description—such as 'a sudden feeling of fear'—must admit some physical description which appears in some strict law.

This point of consistency is easy enough to see once made. It does not at all depend on what laws are. But whether the position described even could be, at once, non-trivial and true—that does depend on the notion of law. If, for example, there were no distinction between laws and true statements in general, then there obviously are psychophysical laws, even if no interesting ones. Imagine an omniscient being, such as Laplace envisaged in his discussion of determinism, but capable also of using mental descriptions. Whatever class of events we describe to It, this being can list all the actual members of this class, and hence all the states of the universe in which these members appear. It can pick out precisely, for example, the set of conditions of the universe under which at least one of these states is realized within the next four years. Davidson must object that what It arrives at in such a case is in general not a law, although it is a true general statement.[8]

The form of objection could be anthropomorphic: although It could know that, we humans could not. Then the cogency of the objection would hinge on the notion of law involving somehow this distinction between what is and is not accessible (knowable, confirmable, . . .) to humans. The position of anomalous monism would no longer have the corollary 'Therefore most if not all events are physical', but rather something like: every event which we humans could cover in some description that occurs in a humanly accessible (knowable, or confirmable, or . . .) general regularity, is physical. In that case the position would seem to have no bearing at all on the usual mind-body problems, such as whether the mental 'supervenes' on the physical (which means, whether our mental life being otherwise would have required the physical facts to be otherwise).

Davidson's objection to the story about this omniscient genie would therefore need to be non-anthropomorphic. It would have to insist on a distinction between what the laws are and truths in general, independent of human limitations. The reason this being would not automatically arrive at a law, by reflection on just any class of events we mentioned to It, would have to be due to a law being a special sort of fact about the universe.

Davidson himself notes this presupposition of his argument, and places the burden of significance squarely on the notion of law. What he then goes on to say about laws is unfortunately in part predicated on the logical positivists' very unsuccessful approach to the subject, and in part deliberately non-committal: 'There is (in my view) no non-question begging criterion of the lawlike, which is not to say that there are no reasons in particular cases for a judgment' (*Essays*, 217). This statement, which begins his discussion of laws, itself presupposes the positivists' idea that laws are simply the truths among a class of statements (the 'lawlike' ones) singled out by some common element of form or meaning, rather than by what the world is like. (Davidson comments 'nomologicality is much like analyticity, as one might expect since both are linked to meaning' (p. 218). This presumption was later strongly criticized, for example by Dretske; at this point we should note only that it is dubitable, and not innocuous. I do not mean to go further into how Davidson discusses laws here; the point I wanted to make should now be clear.

The assumptions involved are that there is a significant concept of natural law, that the distinction between laws and truths in general is non-anthropomorphic and concerns what the world is like, and that the correct account of laws must do justice to all this. These are indispensable to Davidson's classification of philosophical positions on mind and matter, to the arguments for his position, and for the significance of that position.[9] This is a striking illustration of how general philosophy had, by our century, learned to rely on this notion of law.

4. CRITERIA OF ADEQUACY FOR ACCOUNTS OF LAWS

If we do have the concept of a law of nature, this must mean at least that we have some clear intuitions about putative examples and counterexamples. These would be intuitions, for example, about what is and what is not, or what could be and what could not be, a law of nature, if some sufficiently detailed description of the world is supposed true. It does not follow that we have intuitions of a more general sort about what laws are like. But when we are offered ideas of this more general sort, we can test them against our intuitions about specific examples.

The use of such examples and our intuitive reactions to them serves at least to rule out overly simplistic or naïve accounts of laws of nature. Their use has also led to a number of points on which, according to the literature, all accounts of laws must agree. None of these points is entirely undisputed, but all are generally respected.

Disagreements about the criteria should not dismay us at the outset. As Wittgenstein taught, many of our concepts are 'cluster concepts'—they have an associated cluster of criteria, of which only *most* need be satisfied by any instance. The more of the criteria are met, the more nearly we have a 'clear case'. This vagueness does not render our concepts useless or empty—our happiness here as elsewhere depends on a properly healthy tolerance of ambiguity.

In what follows I shall discuss about a dozen criteria found in the literature. Some are less important, or more controversial than others. We can use them to dismiss some naïve ideas, especially cherished by empiricists—and in subsequent chapters bring them to bear on the main remaining accounts of law. Nowhere should we require that all the criteria be met; but any account should respect this cluster as a whole.

Universality

The laws of nature are universal laws, and universality is a mark of lawhood. This criterion has been a great favourite, especially with empiricists, who tend to be wary of nearly all the criteria we shall discuss subsequently. There is indeed nothing in the idea of universality that should make philosophical hackles rise, nor would there be in the idea of law if a law stated merely what happens always and everywhere. The hope that this may be so must surely account for the curiously uncritical attitude toward this notion to be found in even the most acute sceptics:

Whitehead has described the eighteenth century as an age of reason based upon faith—the faith in question being a confidence in the stability and regularity of the universal frame of Nature. Nothing can better illustrate Hume's adherence to this faith, and its separation in his mind from his philosophical scepticism, than his celebrated Essay *Of Miracles*. The very man who proves that, for all we can tell, anything may be the 'cause' of anything, was also the man who disproved the possibility of miracles because they violated the invariable laws of Nature.[10]

That does not make Hume inconsistent. If what a law is concerns only what is universal and invariable, the faith in question could hardly impugn Hume's scepticism about mysterious connections in nature beyond or behind the phenomena. For in that case it would merely be a faith in matters of fact, which anyone might have, and which would not—unlike the 'monkish virtues'—bar one from polite society (the standard Hume himself so steadfastly holds out to us).

Unfortunately this mark of universality has lately fallen on hard times, and that for many reasons. Let us begin with the point that universality is not enough to make a truth or law of nature. No rivers past, present, or future, are rivers of Coca-Cola, or of milk. I think that this is true; and it is about the whole world and its history. But we have no inclination to call it a law.[11] Of course we can cavil at the terms 'river', 'Coca-Cola', or 'milk'. Perhaps they are of earthly particularity. But we have no inclination to call this general fact a law because we regard it as a merely incidental or accidental truth. Therefore we will have the same intuition, regardless of the terms employed. This is brought out most strikingly by parallel examples, which employ exactly the same categories of terms, and share exactly the same logical form, yet evoke different responses when we think about what could be a law. The following have been discussed in various forms by Reichenbach and Hempel:[12]

1. All solid spheres of enriched uranium (U235) have a diameter of less than one mile.
2. All solid spheres of gold (Au) have a diameter of less than one mile.

I guess that both are true. The first I'm willing to accept as putatively a matter of law, for the critical mass of uranium will prevent the existence of such a sphere. The second is an accidental or incidental fact—the earth does not have that much gold, and perhaps no planet does, but the science I accept does not rule out such golden spheres. Let us leave the reasons for our agreement to one side—the point is that, if 1 could be law, if only a little law, and 2 definitely could not, it cannot be due to a difference in universality.

Another moral that is very clear now is that laws cannot be simply the true statements in a certain class characterized in terms of syntax and semantics. There is no general syntactic or semantic feature in which the two parallel examples differ. So we would go wrong from the start to follow such writers as Goodman, Hempel, and Davidson in thinking of the laws as the true 'lawlike' statements.

We can agree in the intuitions invoked above, before any detailed analysis of universality. But we have also already discerned some reason to think that the analysis would not be easy. In fact, it is extremely difficult to make the notion precise without trivializing it. The mere linguistic form 'All . . . are . . .' is not a good guide, because it does not remain invariant under logical transformations. For example, 'Peter is honest' is in standard logic equivalent to the universal statement 'Everyone who is identical with Peter, is honest.' To define generality of content turns out to be surprisingly difficult. In semantics, and philosophy of science, these difficulties have appeared quite poignantly.[13] Opinions in the literature are now divided on whether laws must indeed be universal to be laws. Michael Tooley has constructed putative counterexamples.[14] David Armstrong's account requires universality, but he confesses himself willing to contemplate amendment.[15] David Lewis's account does not require it.[16] In Part III we shall find an explication of generality allied to concepts of symmetry and invariance. While I regard this as important to the understanding of science, the generality we shall find there is theory-relative.

The criterion of universality, while still present in discussion of laws, is thus no longer paramount.

Relations to necessity

In our society, one must do what the laws demand, and may do only what they do not forbid. This is an important part of the positive analogy in the term 'laws of nature'.

Wood burns when heated, because wood must burn when heated. And it must burn because of the laws which govern the behaviour of the chemical elements of which wood and the surrounding air are composed. Bodies do not fall by chance; they must fall because of the law of gravity. In such examples as these we see a close connection between 'law' and 'must,' which we should stop to analyse.

Inference. The most innocuous link between law and necessity lies in two points that are merely logical or linguistic. The first is that if we say that something is a law, we endorse it as being true. The inference

(1) It is a law of nature that *A*
Therefore, A

is warranted by the meaning of the words. This point may seem too banal to mention—but it turns out, surprisingly, to be a criterion which some accounts of law have difficulty meeting. One observes of course that the inference is not valid if 'of nature' is left out, since society's laws are not always obeyed. Nor does it remain valid if we replace 'law of nature' by 'conjecture' or even 'well-confirmed and universally accepted theory'. Hence the validity must come from the special character of laws of nature. In Chapter 5, the problem of meeting this criterion will be called the problem of inference.

Intensionality. The second merely logical point is that the locution 'It is a law that' is *intensional.* Notice that the above inference pattern (1) does remain valid if we replace 'a law of nature' by 'true'. But something important has changed when we do, for consider the following argument:

(2) It is true that all mammals have hair.
All rational animals are mammals.
Therefore, it is true that all rational animals have hair.

This is certainly correct, but loses its validity if we now replace 'true' again by 'a law of nature'. Another example would be this: suppose that it is a law that diamonds have a refraction index >2, and that as a matter of fact all mankind's most precious stones are diamonds. It still does not follow that it is a law that all mankind's most precious stones have a refraction index >2. Here we see the distinction between law and mere truth or matter of fact at work.

Our first two criteria are therefore merely points of logic, and I take them to be entirely uncontroversial.

Necessity bestowed. The moon orbits the earth and must continue to do so, because of the law of gravity. This illustrates the inference from *It is a law that A* to *It is necessary that A*; but this must be properly understood.

The medievals distinguished the *necessity of the consequence* from the *necessity of the consequent*. In the former sense it is quite proper to say 'If all mammals have hair then whales must have hair, because whales are mammals.' The 'must' indicates only that a certain consequence follows from the supposition. For law this point was therefore already covered above. The criterion of necessity bestowed is that there is more to it: if *It is a law that A* is true then also, rightly understood, *It is necessary that A* is true. This necessity is then called physical necessity or nomological necessity (and is now often generalized to physical probability).

Empiricists and nominalists have always either rejected this criterion or tried to finesse it. For they believe that necessity lies in connections of words or ideas only, so ultimately the only necessity there can be lies in the necessity of the consequence. This is not altogether easy to maintain, while acknowledging the preceding points of logic. Yet their persistent attempts to reconstrue the criterion of necessity bestowed, so that it is fulfilled if 'properly' understood, show the strength of the intuition behind it.[17]

Necessity inherited. There is a minority opinion that what the laws are is itself necessary.[18] This point definitely goes beyond the preceding, for logic does not require what is necessary to be necessarily necessary. More familiar is the idea that there are many different ways the world could have been, including differences in its laws governing nature. If gravity had obeyed an inverse cube law, we say, there would have been no stable solar system—and we don't think we are contemplating an absolute impossibility. But we could be wrong in this.

Of course, if laws are themselves necessary truths, their consequences would inherit this necessity. Therefore the strong criterion of *necessity inherited* entails that of *necessity bestowed.* And since what is necessary must be actual, the criterion of *necessity bestowed* entails that of *inference*. The entailments do not go in the opposite direction. So three of the criteria we have formulated here form a logical chain of increasing strength.

Explanation

Such writers as Armstrong insist that laws are needed to explain the phenomena, and indeed, that there are no explanations without laws. This is not in accord with all philosophical theories of explanation.[19] A more moderate requirement would be that laws must be conceived as playing an indispensable role in some important or even pre-eminent pattern of explanation.

There does indeed appear to be such a pattern, if there is an intimate connection between laws and necessity (and objective probability). It may even be the pre-eminent pattern involved in all our spontaneous confrontations with the world. Witness that Aristotle made it the key to narrative and dramatic structure in tragedy:

> And these developments must grow out of the very structure of the plot itself, in such a way that on the basis of what has happened previously this particular outcome follows either by necessity or in accordance with probability; for there is a great difference in whether these events happen because of those or merely after them. (*Poetics*, 52ᵃ17–22)

This account of tragedy bears a striking resemblance to Aristotle's account of how science must depict the world, in his *Physics*.[20] The parallel is no accident, though one must admit that Aristotle's demands upon our understanding of nature persisted longer than those he made upon our appreciation of literature.

What exactly is this criterion, that laws must explain the phenomena? When a philosopher—as so many do—raises explanation to pre-eminence among the virtues, the good pursued in science and all natural inquiry, he or she really owes us an account of why this should be so. What is this pearl of great price, and why is it so worth having? What makes laws so well suited to secure us this good? When laws give us 'satisfying' explanations, in what does this warm feeling of satisfaction consist? There are indeed philosophical accounts of explanation, and some mention laws very prominently; but they disagree with each other, and in any case I have not found that they go very far toward answering *these* questions.[21]

Hence we should not get very far with this criterion for accounts of laws, if its uses depended greatly on the philosophical opinions of what explanation is. Fortunately there are two factors which keep us from being incapacitated here. The first factor is the very large measure of agreement on what counts as explanation when we are confronted with specific, concrete examples. The other factor is the great degree of abstraction which characterizes many discussions of law. In Chapter 6, for example, we shall be able to take up Dretske's and Armstrong's arguments concerning what is for them the crucial argument form of Inference to the Best Explanation—and its relation to laws—without ever having to reproach them for the fact that they nowhere tell us what an explanation is.

We shall encounter a certain tension between the criteria regarding the connections of law with necessity on the one hand, and with science on the other. Here the concept of explanation could perhaps play an important mediating role: If explanation is what we look for in science, while necessity is crucial to explanation and law crucial to necessity, then that tension may perhaps be '*aufgehoben*' in a higher unity. We shall have to see.

Prediction and confirmation

That there is a law of gravity is the reason why the moon continues to circle the earth. The premiss that there is such a law is therefore a good basis for prediction. The second traditional argument which I briefly sketched above—and illustrated from Peirce's lecture—goes on: and if we deny there is such a reason, then we can also have no reason for making that prediction. We shall have no reason to expect the phenomenon to continue, and so be in no position to predict.

If there is a problem with this argument today, it is surely that we cannot be so ready to equate *having reason to believe that A* with *believing that there is a reason for A*. Linguistic analysis in philosophy makes us very wary of such pretty rhetoric. But the equation might perhaps hold for the special case of empirical regularities and laws. Certainly, a form of this second traditional argument is found very prominently in Armstrong's book. After canvassing some views on what laws are, he notes a possibility which he says was brought to his attention by Peter Forrest:

> There is one truly eccentric view. . . . This is the view that, although there are regularities in the world, there are no laws of nature. . . . This Disappearance view of law can nevertheless maintain that inferences to the unobserved are reliable, because although the world is not law-governed, it is, by luck or for some other reason, regular.[22]

Armstrong replies immediately that such a view cannot account for the fact that we can have *good reasons to think* that the world is regular. He gives an argument for this, which I shall discuss in Chapter 6.

A little of the recent history of confirmation appeared along the way in the article by Davidson which we discussed above. While Davidson attempts no definition or theory of what laws are, he says among other things that laws are general statements which are confirmed by their instances—while this is not always so for general statements. This makes sense if laws are the truths among lawlike statements, and if in addition we (who assess the evidence) can distinguish lawlike statements from other generalities. For else, how can instances count for greater confirmation?

But this idea receives rather a blow from the parallel gold and uranium examples we discussed above. These parallel examples are so parallel in syntactic form and semantic character that the independent prior ability to distinguish lawlike from other general statements is cast into serious doubt.

We should also observe that for writers on laws there is—and perhaps must be—a crucial connection between confirmation and explanation. For consider the following argument: that it is a law that *P* could be supported by claims either of successful explanation or of successful prediction (or at least, successful fitting of the data). But prediction cannot be enough, for the second sort of claim works equally well for the bare statement that *A*: *It is a law that A* entails or fits factual data only in so far as, and because, *A* does. Hence confirmation for the discriminating claim *It is not only true but a law that A* can only be on the basis of something in addition to conforming evidence. One traditional candidate for this something extra is successful explanation.

This observation gives us, I think, the best explanation of why advocates of laws of nature typically make Inference to the Best Explanation the cornerstone of their epistemology.

Counterfactuals and objectivity

Philosophy, being a little other-worldly, has always been fascinated with the conditional form *If* (*antecedent*) *then* (*consequent*). When the antecedent is false ('the conditional is contrary to fact' or 'counterfactual') what speculative leaps and fancies are not open to us? If wishes were horses then beggars would ride; if gravity had been governed by an inverse cube law there would have been no stable solar system; if Caesar had not crossed the Rubicon, ... Being also a little prosaic, the philosopher sets out to find the bounds of fancy: when must such a conditional be true, when false?

There is one potentially large class of cases where the answer is clear. If *B* follows from *A* with necessity, then *If A then B* is true and *If A then not B* is false. Thus if iron must melt at 2000°C, it follows that this iron horse-shoe would melt if today it were heated to that temperature . . . this is clear even if the horseshoe remains at room temperature all day. At midnight we will be able to say, with exactly the same warrant, that it would have melted if it had been heated to 2000°C. Many other such conditionals command our intuitive assent: Icarus' father too would have fallen if his wings had come loose, and so would I if I had stepped off the little platform when I went up the cathedral tower in Vienna. We observe that in all such cases we intuitively agree also to describe our warrant in terms of laws. These facts about iron and gravity are matters of law; if it is a law then it must be so; and if it must be so then it will or would be so if put to the test.

This large class of cases falls therefore very nicely under the previous criterion of necessity bestowed. But the requirement that laws be the sort of thing that warrant counterfactuals, has a much greater prominence in the literature. Is there more to it?

In the mid-1940s, Nelson Goodman and Roderick Chisholm made it clear that there are mysteries to the counterfactual conditional, which had escaped their logical treatment so far. This treatment did indeed fit necessary implications. Typical sanctioned argument patterns include

> Whatever is *A* must be *B*.
> Therefore, if this thing is (were) both *A* and *C*, then it is (would be) *B*.

But can all conditionals derive from necessities in this way? Consider: if I had struck this match, then it would have lit. It does not follow that if I had struck this match, and it had been wet at the time, then it would have lit. Nor, if I agree that the latter is false, do I need to retract the former. I can just say: well, it wasn't wet. We see therefore that counterfactual conditionals violate the principles of reasoning which govern 'strict' or necessary conditionals.

How are we to explain this mystery? Goodman did not explain it, but related it to laws.[23] We can, he said, support a counterfactual claim by citing a law. We cannot similarly support it by merely factual considerations, however general. For example it is a fact (but not a matter of law) that all coins in Goodman's pocket were silver. We cannot infer from this that if this nickel had been in his pocket then, it would have been silver. On the other hand it is also a fact and a matter of law that silver melts at 960.5°C. Therefore if this silver had been heated to that degree, it would have melted. This observation, Goodman thought, went some way toward clearing up the mystery of counterfactual conditionals. The mystery was not thereby solved, so the connection was inverted: giving warrant for counterfactual conditionals became the single most cited criterion for lawhood in the post-war literature.

But the mystery was solved in the mid-1960s by the semantic analysis due to Robert Stalnaker and extended by David Lewis. Unfortunately for laws, this analysis entails that the violations of those principles of inference that work perfectly well for strict conditionals are due to context-dependence. The interesting counterfactuals which do not behave logically like the strict ones do not derive from necessities alone, but also from some contextually fixed factual considerations. Hence (I have argued elsewhere) science by itself does not imply these more interesting counterfactuals; and if laws did then they would have to be context-dependent.[24] Robert Stalnaker has recently replied to this that science does imply counterfactuals, in the same sense that it implies indexical statements.[25] An example would be:

> Science implies that your materialist philosophy is due to a dietary deficiency.

This is a context-dependent sense of 'implies' (not of course the sense which I had in mind), because the referent of 'you' depends on context. Stalnaker's point is quite correct. But it leads us to conclude at best that the speaker may believe that some law is the case, and holds its truth-value fixed in a tacit *ceteris paribus* clause (which gives the counterfactual sentence its semantic content in this context). This is certainly correct, but is equally correct for any other sort of statement, and cannot serve to distinguish laws from mere truths or regularities. I suspect that the real use of Goodman's requirement concerned counterfactuals considered true in cases where the corresponding physical necessity statement is also implied. If so, the requirement coincided in philosophical practice with the requirement of bestowed necessity.

Context-independence. In view of the above, however, it is important to isolate the sense in which law statements cannot be context-dependent. Stalnaker's sort of example leaves us with the requirement:

> If the truth value of statement *A* is context-independent, then so is that of *It is a law that A*.

Related to the context-independence of the locution 'It is a law that', but not at all the same, is the point that laws are to be conceived of as objective.

Objectivity. Whether or not something is a law is entirely independent of our knowledge, belief, state of opinion, interests, or any other sort of epistemological or pragmatic factor. There have definitely been accounts of law that deny this. But they have great difficulty with such intuitively acceptable statements as that there may well be laws of nature which not only have not been discovered and perhaps never will be, but of which we have not even yet conceived.[26]

Relation to science

We come now to a final criterion which is of special importance. Laws of nature must, on any account, be the sort of thing that science discovers. This criterion is crucial, given the history of the concept and the professed motives of its exponents.

This criterion too is subject to a number of difficulties. First of all, there is no philosophically neutral account of what science discovers, or even what it aims to discover. Secondly, although the term 'law' has its use in the scientific literature, that use is not without its idiosyncracies. We say: Newton's laws of motion, Kepler's laws of planetary motion, Boyle's law, Ohm's law, the law of gravity. But Schroedinger's equation, or Pauli's exclusion principle, which are immensely more important than, for example,

Ohm's law, are never called laws. The epithet appears to be an honour, and often persists for obscure historical reasons.

Attempts to regiment scientific usage here have not been very successful. Margenau and Lindsay note disapprovingly that other writers speak of such propositions as *copper conducts electricity* as laws.[27] They propose that the term should be used to denote any precise numerical equation describing phenomena of a certain kind.[28] That would make Schroedinger's equation, but not Pauli's exclusion principle, a law. Even worse: it would be quite easy to make up a quantitative variant of the rivers of Coca-Cola example which would meet their criterion trivially.

Faced with this situation, some writers have reserved 'law' for low-level, empirical regularities, thus classifying the law of conservation of energy rather than Boyle's law as terminological idiosyncrasy. To distinguish, these low-level laws are also called phenomenological laws, and contrasted with basic principles which are usually more theoretical. Science typically presents the phenomenological laws as only approximate, strictly speaking false, but useful. According to Nancy Cartwright's stimulating account of science, the phenomenological laws are applicable but always false; the basic principles accurate but never applicable.[29] It is therefore not so easy to reconcile science as it is with the high ideals of those who see it as a search for the true and universal laws of nature.

The criterion of adequacy, that an account of laws must entail that laws are (among) what science aims to discover, is therefore not easy to apply. Certainly it cannot be met by reliance on a distinction embodied in what scientists do and do not call a law. Nor, because of serious philosophical differences, can it rely on an uncontroversial notion of what the sciences (aim to) discover. The criterion of objectivity we listed earlier, moreover, forbids identification of the notion of law with that of a basic principle or any other part of science, so identified. For if there are laws of nature, they would have been real, and just the same even if there had been no scientists and no sciences.

It appears therefore that in accounts of law, we must try to discern simultaneously a view of what science is and of what a law is, as well as of how the two are related. These views must then be evaluated both independently and in terms of this final criterion, that they should stand in a significant relationship.

Earlier in this century, the logical positivists and their heirs discussed laws of nature, and utilized that concept in their own explications and polemics. I shall not examine those discussions in any detail. If we look at their own efforts to analyse the notion of law, we find ourselves thoroughly frustrated. On the one hand we find their own variant of the sin of psychologism. For example, there is a good deal of mention of natural laws in Carnap's *The Logical Syntax of Language*. But no sooner has he started on the question of what it means to say that it is a law that all *A* are *B*, than he gets involved in the discussion about how we could possibly verify any universal statement. On the other hand there is the cavalier euphoria of being involved in a philosophical programme all of whose problems are conceived of as certain to be solved some time later on. Thus in Carnap's much later book *Philosophical Foundation of Physics*[30], we find him hardly nearer to an adequate analysis of laws or even of the involved notions of universality or necessity—but confident that the necessary and sufficient conditions for lawhood are sure to be formulated soon. The culmination of Carnap's, Reichenbach's, and Hempel's attempts, which is found in Ernest Nagel's *The Structure of Science*, was still strangely inconclusive. In retrospect it is clear that they were struggling with modalities which they could not reduce, saw no way to finesse, could not accept unreduced, and could not banish.

Having perceived these failures of logical empiricism, some philosophers have in recent years taken a more metaphysical turn in their accounts of laws. I shall focus my critique on those more recent theories.

5. PHILOSOPHICAL ACCOUNTS: THE TWO MAIN PROBLEMS

Of the above criteria, never uniformly accepted in the literature, five seem to me pre-eminent. They are those relating to necessity, universality, and objectivity and those requiring significant links to explanation and science.

But apart from the more or less piecemeal evaluation these allow, of all proffered philosophical accounts of laws, there will emerge two major problems. I shall call these the *problem of inference* and the *problem of identification*. As we shall see, an easy solution to either spells serious trouble from the other.

The problem of inference is simply this: that it is a law that *A*, should *imply* that *A*, on any acceptable account of laws. We noted this under the heading of necessity. One simple solution to this is to equate *It is a law that A* with *It is necessary that A*, and then appeal to the logical dictum that necessity implies actuality. But is 'necessary' univocal? And what is the ground of the intended necessity, what is it that makes the proposition a necessary one? To answer these queries one must identify the relevant sort of fact about the world that gives 'law' its sense; that is the problem of identification. If one refuses to answer these queries—by consistent insistence that necessity is itself a primitive fact—the problem of identification is evaded. But then one cannot rest irenically on the dictum that necessity implies actuality. For 'necessity', now primitive and unexplained, is then a mere label given to certain facts, hence without logical force—Bernice's Hair does not grow on anyone's head, whatever be the logic of 'hair'.

The little dialectic just sketched is of course too elementary and naïve to trip up any philosopher. But it illustrates in rudimentary fashion how the two problems can operate as dilemma. We shall encounter this dangerous duet in its most serious form with respect to objective chance (irreducibly probabilistic laws), but it will be found somehow in many places. In the end, almost every account of laws founders on it.

Besides this dialectic, the most serious recurring problem concerns the relation between laws and science. The writers on laws of nature by and large do not so much develop as presuppose a philosophy of science. Its mainstay is the tenet that laws of nature are the sciences' main topic of concern. Even if we do not require justification for that presupposition, it leaves them no rest. For they are still required to show that science aims to find out laws *as construed on their account*. This does not follow from the presupposition, even if it be sacrosanct.

While I cannot possibly examine all extant accounts of laws, and while new ones could spring up like toadstools and mushrooms every damp and gloomy night, these problems form the generic challenge to *all* philosophical accounts of laws of nature. In the succeeding three chapters, we shall see the three main extant sorts of account founder on them.

3.
Ideal Science: David Lewis's Account of Laws[1]

THE account of laws David Lewis offers us is the least metaphysical of all those we shall examine. It is nearest to the straight empiricist 'regularity' view, and attempts from the outset to put laws in touch with science. Lewis is well known to be a realist with respect to alternative possible worlds, but we shall see that this realism is not crucial here. The only metaphysics crucially involved is anti-nominalism: that is, a realist construal of the difference between 'natural' and 'merely arbitrary' classifications.[2] In addition, we shall see that Lewis's account has *prima facie* considerable success in meeting the criteria listed in the preceding chapter. But I shall argue that the successes are, in the end, only apparent.[3]

I. THE DEFINITION OF LAW

Lewis first presented his account in *Counterfactuals*. There he refers to F. P. Ramsey's 1928 account of laws as 'consequences of those propositions which we should take as axioms if we knew everything and organized it as simply as possible in a deductive system'.[4] John Earman points out that Ramsey was perhaps following John Stuart Mill's *System of Logic* which says about the expression 'Laws of Nature':

> Scientifically speaking, [it] is employed . . . to designate the uniformities when reduced to their most simple expression. . . . According to one mode of expression, the question, What are laws of nature? may be stated thus: What are the fewest and simplest assumptions, which being granted, the whole existing order of nature would result? Another mode of stating the question would be thus: What are the fewest general propositions from which all uniformities which exist in the universe might be deductively inferred?[5]

We shall look at Lewis's refinements in a moment, but note first that this sort of approach gives no indication of dependence on realism about possible worlds, or modal realism. While Lewis may present certain aspects of the account in terms of possible worlds, he can be followed there very well by someone who regards those as the theoretical fictions of semantics. Mill offers us here an account which seems really a sort of pattern for accounts of law, that could be elaborated by almost anyone, regardless of his views concerning the nature of necessity or of science.

The first difficulty to be faced is that there are innumerable true theories about the world, all of which entail the uniformities there actually are in nature. Mill suggests that we must pick out the theory which can be axiomatized by 'the fewest and simplest' or (equivalently?) 'the fewest general' propositions. This rather vague response to the difficulty leaves a number of open questions. Will there be a unique such theory? And is simplicity the only thing that matters? And is entailing the uniformities the single factual desideratum besides truth? And what is a uniformity anyway, or simplicity?

Lewis's refinement meets this difficulty as follows: There are innumerably many true theories (in the sense of: deductively closed sets of true sentences). Some of these are simpler than others, some are stronger (i.e. more informative) than others. What we value in science is both simplicity and strength, so we wish for a properly balanced combination. The *laws* are those sentences describing regularities which are common to all those true theories that achieve a best combination of simplicity and strength. (If there can be better and better combinations *ad infinitum*, this definition needs a technical adjustment, which I leave aside.)

I have written here as if simplicity, strength, and balance are as straightforward as a person's weight or height. Of course they are not, and the literature contains no account of them which it would be fruitful to discuss here. Strength must have something to do with information; perhaps they are the same. Simplicity must be a quite different notion. Note also that we have here *three* standards of comparison: simplicity, strength, and balance. The third is needed because there is some tension between the first and second, which cannot be jointly maximized. Sometimes a simple theory is also more informative. But if we have a simple theory, and just add a bit of information to it, so as to make it stronger, we will almost always reduce its simplicity. These are intuitive considerations that surely everyone shares. As soon as we reflect on balance, however, the certainty of our intuitions dwindles fast. A person who weighs 170 pounds is overweight if he is five foot and underweight if he is six foot three—there we have a notion of balance. But how shall we gauge when a gain in simplicity is well bought for a loss of information? When does gain equal loss for two such disparate virtues?

So simplicity, strength, and balance are not straightforward. To utilize these motions uncritically, as if they dealt with such well-understood triads as 'under five foot five, over 200 pounds, overweight' may be unwarranted. Still I shall leave the legitimacy of these notions unchallenged, in as far and for as long as I can.[6]

As Lewis himself notes, there is a much more serious difficulty.[7] This difficulty appears exactly after we have agreed to the use of simplicity as a criterion. In what language(s) are these theories formulated? Suppose that we form a new language, in which simple predicates correspond to long and cumbersome constructions in our own. If we then translate two of our theories into this new language, the verdict of simplicity between them may be reversed. This is not merely an academic problem. When Poincaré asserted that, despite the logical possibility of doing otherwise, physics would always remain wedded to Euclidean geometry, he based this assertion on a verdict of simplicity. But he spoke at a time before the exploitation of differential geometry. Thereafter, these very considerations of simplicity told against the retention of Euclidean geometry, just because geometric language had become so much more general and rich in its descriptive powers. Philosophers of course have discussed more academic examples, with novel predicates like: 'grue', meaning 'green if examined before the year 2000 and blue if not'.

Should we count as best theories those whose translations win the competition in every language we could construct? Then we cannot expect to find any best theories at all. Even if one true theory in our language has, among theories in that language, the optimum combination of simplicity and strength, its translation into other language will not, in general, preserve this virtue. If on the other hand we ask for those theories which each have that pre-eminent status in *some* language, we must expect to obtain such a large class that they may have little more than tautologies in common.

The only remedy open to Lewis, as he himself explains, is to restrict the language(s) in which the candidate theories are allowed to be formulated. On what basis could this restriction be made? Lewis assumes that the 'correct' language has the simple extensional structure studied in standard logic, with predicates as only non-logical terms. Each predicate has an *extension*: the class of things to which this predicate applies. Assert now that some classes—for example, the class of stars—are marked by a real distinction, and other classes—for example, that of people whose names begin with M—are merely an arbitrary grouping. Then a good way to select a 'correct' language appears: the basic predicates must each have as extension such a 'natural' class. Other predicates may be introduced by definition only. Simplicity must be judged before any definitions are introduced.

By adopting this remedy, Lewis makes it once more plausible—at least prima facie—that the class of laws, as defined, will be an appreciably informative set of sentences. The remedy certainly has historical precedent in metaphysics. Indeed, this insistence on the distinction between real and verbal or arbitrary classifications has often enough been taken as a defining difference between late medieval realists and nominalists.[8] So I shall call the position here adopted *anti-nominalism*.

The account so far already has notable virtues. A law is a statement; but since theories are logically closed, any logical equivalent of a law is also a law. A law is true; but true statements, however general or otherwise syntactically or semantically privileged, are not always laws. What is a law is an objective question, whose answer is independent of what science is actually developed in history, and indeed, independent of any other merely historical, psychological or otherwise anthropocentric fact.[9] And surely science attempts to find for us true, strong, simple theories; so *a fortiori*, by definition even, it seems that science is in pursuit of laws as defined by Lewis. Whether that is really so, we must now investigate.

2. THE DEFINITION OF NECESSITY

So far Lewis's account answers the question of what the laws are—in our world or, *mutatis mutandis*, in any world. But if something is a law, then it is not only true, but necessary. How does Lewis honour this criterion? Since he is well known for his realism about possible worlds, we expect to see this come into play now. And so it does; but it does not play a crucial role for the account of laws as such. For the upshot is simply—as we shall see—that *It is necessary that A* is said to be true if *A* is implied by the laws of nature. This is, in effect, the definition of necessity. Since the notion of law has been defined previously, without reference to what is true in any other world, this way of honouring the criterion is open to even the most anti-metaphysical empiricist as well.

I shall be brief here about the connection between necessity and possible worlds, because I shall be discussing it at greater length in the next chapter (where it will play a crucial role). The logical warrant for the ideas which I shall now briefly describe will also be discussed at greater length there.

The logic of the word 'necessary' is such that any definition of the form '*It is necessary that A* is true in a world *x* if and only if *A* is true in every world which is possible relative to *x*', meets the logical criteria. There are distinct senses of 'necessary' which in this way correspond to different relations of relative possibility. It suffices therefore for Lewis to define a particular such relation, *relative physical possibility*. For this he offers:

> world *y* is *physically possible relative to* world *x* exactly if the laws of *x* are all true in *y*.

Note that he does not require that the laws of *x* also be laws of *y*; they need merely be true statements in that other world. The relevant sense of 'necessary' is introduced by

> *It is physically necessary that A* is true in world *x* if and only if *A* is true in every world which is physically possible relative to *x*.

And now we can *deduce* from these definitions:

> *It is physically necessary that A* is true in world *x* if and only if *A* is implied by the laws of *x*.

Here 'implies' means 'semantically implies'; that is, certain premisses imply *A* exactly if *A* is true in every possible world in which those premisses are true.

This is a wonderful result. We may harbour a little doubt, due to its strength. Could not some individual matter of fact be physically necessary without being entailed by the laws? But that question aside, a major desideratum has apparently been met.

Is this part of the account metaphysical, in the sense abhorrent to empiricists? I think not. We have here only a semantic analysis, a proposal for truth conditions which, in effect, define the clause 'It is necessary that'. If one is not a realist about possible worlds, one may tacitly read 'world' as 'model of our language'. The definitional approach Lewis uses here is available to all. Anyone who feels he has an adequate notion of law, can go on to equate physical necessity with the status of being entailed by the laws. The use of this move by someone who believes all possible worlds to be real, neither weakens nor strengthens its benefits. Whether the benefits of this move are real, and not merely apparent, for the general account of laws, is an independent question.

This completes Lewis's account of laws and necessity. Its assumptions seem almost entirely acceptable even to empiricists (perhaps entirely, to many) and its prima-facie successes are remarkable. So why not be content to accept this reconstruction as conclusive?

3. LAWS RELATED TO NECESSITY

As we have found, on Lewis's account, the assertion that it is a law that *A* entails that it is physically necessary that *A*. This meets one of our main criteria. As we also saw, Lewis shows us here by example how the criterion can be met, through a stipulative definition, by anyone who feels he has already an adequate notion of law.

It is hard, however, to escape the feeling that if the criterion can be met satisfactorily in this way, then it must be devoid of all probative force. Doesn't Lewis meet the criterion by robbing it of significance?

The intuition behind the criterion is that the existence of a law that *A* makes it physically necessary that *A* (and, *a fortiori*, makes it true that *A*). I do not quite know what to make of this notion of making something necessary or true. Of course I know well enough the traditional terminology of the 'ground of necessity', and the recently sprung-up terminology of 'truth-makers'. I also know the Aristotelian tradition of real necessities (as opposed to verbal necessities) grounded in real, substantial natures. So I, and you, gentle reader, have much circumstantial evidence to suspect that if laws are merely definitionally connected with necessity, then they won't do the right job. But none of this needs bother Lewis, if it only points to possible unhappiness with his account among eventual advocates of archaic or anachronistic ideas.

To some extent, the complaint may be formulated in terms of explanation: that something is a law should *explain* why something is physically necessary—and no fact can explain anything to which it is definitionally equivalent. I would like to postpone this form of the objection to another section below.

What if Lewis replied that we should not reproach him for honouring by means of a definition any equivalence which we independently accept? Then we may still fault him for not having tried to account for this equivalence, if there is one. Why should anyone accept it, let alone take it to be so basic that it might as well be built into our very language? This is not an idle question, I think. Let me give an example.

Consider the view that spatio-temporal relations among events are not *sui generis* but derive from physical relations such as connection by signals, physical contact, and identity through time. (We need not assume that the reduction is definitional, nor any *particular* version of the relational theory of space–time.) On such a view, one might wish to assert that it is physically impossible for any signal to connect events *E* and *F*. One would wish to assert this exactly if one held that *E* and *F* are simultaneous in some frame of reference. But this assertion cannot follow from any independent facts about *E* and *F* via general laws. For the only relevant facts concern their space-like separation, which derives from facts about signal connectability—the very subject of our statement.

Perhaps some will find this example fanciful, because they consider relational theories of space–time absurd. So let us delve into our uneasiness about the definitional link between law and necessity in yet another way.

To say that we have the concept of a law of nature must imply at least that we can mobilize intuitions to decide on proffered individual examples. Let us then consider a possible world in which all the best true theories, written in an appropriate sort of language, include the statement that all and only spheres are gold. To be concrete, let it be a world whose regularities are correctly described by Newton's mechanics plus law of gravitation, in which there are golden spheres moving in stable orbits around one another, and much smaller iron cubes lying on their surface, and nothing else. If I am now asked whether in that world, all golden objects are spherical because they must be spherical, I answer *No*. First of all it seems to me that there could have been little gold cubes among the iron ones, and secondly, that several of the golden spheres could (given slightly different initial conditions) have collided with each other and thus altered each other's shapes. If my intuitions are a bit strong for your taste, perhaps you will at least grant that you feel no intuitive inclination to say *Yes* or to assert that the generalization is a law. But on Lewis's view it is a law in this world that all golden objects are spherical, and also physically necessary. I say this on the basis of intuitive judgements about simplicity and strength and their balance; but for such a simple world it does not seem difficult to find the best true theory.

Could it be argued that some presumption of laws was involved in my use of the terms 'gold' and 'iron' (along the lines perhaps of Wilfrid Sellars's 'Concepts as Involving Laws and Inconceivable Without Them')? I think the response is not open to Lewis, because his account of laws requires that the truth-values of all non-modal sentences be settled beforehand, before the laws can be identified. Could it be argued instead that the world I have described is indeed possible, but that I am wrong to balk at Lewis's conclusions about what its laws are? This would mean presumably that my intuitions are warped by my knowledge of gold and iron in our own world. But suppose I said the large spheres of that world were made not of gold, but of some substance I do not know, call it *S*. I am then willing to say that, as far as I can see, the simplest true descriptions of this world all contain the statements 'All spheres are *S*' and 'All *S*-things are spheres'. But do I feel I have the warrant to say that, in such a world, *S*-things *must be* spherical? I do not. Truth and simplicity just do not add up to necessity, as far as my intuitive reactions are concerned. No, I think that the consequences of Lewis's view for this sort of example can be swallowed only if one downgrades radically the connection between law and necessity.

4. DO LEWIS'S LAWS EXPLAIN?

At first sight, it seems obviously true that laws of nature, in Lewis's sense, explain the phenomena. But after a second and third look, the mystery is rather that this should have seemed so at all. What could have led us to think so?

We may imagine the following train of thought (though Lewis definitely does not present it to us). Science explains; more generally, scientific theories explain. The best theories give the best explanations. But laws are the common part of all best theories. Therefore they are the ingredients present in any best, overall explanation of what the world is like. Surely that earns them the right to be called explanatory?

There are several assumptions here whose examination I wish to leave aside for now.[10] Even granting the assumptions, the 'therefore' and 'surely' hide a great deal. Consider: tautologies are a common part of all theories; but they are not explanatory. Also some very uninformative, near-tautologies are common to all the best theories, but would not be called explanatory, even if they mark the beginning of any explanation.

This is to the point for we do not know about the set of laws in Lewis's sense—i.e. the common part of all the best theories—how informative it is. That is because we do not know the diversity of best theories. So what could make us say at once: the common part of all the best theories must have the pre-eminent explanatoriness, which has always been claimed for laws of nature?

To give an analogy, suppose that the three material goods are money, houses, and land. I have as much of these as all rich men have—does it follow that I am rich? No, for one has little money but much land, one has only a small garden but many houses, etc. What they *all* have is at least a little money, at least a little garden, and at least a little house. So do I, and am not rich at all.

This analogy is most troubling if one thinks of explanation as requiring, as a minimum, the provision of sufficiently much relevant information (*sub specie* whatever criteria of sufficiency and relevance you like). The difficulty remains, I think, even if we connect explanation essentially with unification rather than information—the unification of loads of little theories and bits of factual information, which are subsumed and no longer isolated and separate. Thus to show why Newton's theory was such an achievement, we explain that Galileo's law for falling bodies and Kepler's laws follow from Newton's theory together with only a few simple factual assumptions. The strength of Newton's theory (in the logical sense, informativeness) is crucial to this point. If the laws form by themselves a very weak theory, there will be no parallel. The trade-off between simplicity and strength, carried out in different ways to produce different 'best' theories, might well result in exactly this situation for what Lewis calls the laws.

Let us turn to a second problem. We must still raise the independent question whether the best theories themselves really are the most explanatory. In a first, perhaps trivial form, this is the question whether simplicity and strength, properly balanced, make for explanatory power. (If so, the above point that the common element—i.e. the set of laws—cannot be expected to have those virtues, poses a real problem.) To this, Lewis could respond that if he could be convinced they did not, he could revise the standard of comparison. Then he would say that the laws are what is common to the most explanatory, true theories, whatever 'explanatory' means. But that is not such an easy way out as it looks. For there was a reason why Lewis chose simplicity and strength to begin: the evidence from reports about science, that something like those virtues are actually pursued there. Similar reports of course reach us from philosophers of science about the pursuit of explanation. But if this were indeed a third pursuit, might there not be a further trade-off, with scientists sometimes forgoing explanation for the sake of the other two desiderata? And if so, might it not be that what is common to the best theories is not only weaker than they, but drastically less explanatory, because the trade-off between explanation and the other virtues differs from best theory to best theory? That is exactly what is to be expected in such a case—and so the apparent way out is not a good way at all.

Would it matter if the laws by themselves do not form a very explanatory theory? Well, it matters if you take the conceptual link with explanation to be crucial to the idea of law. Here I have my third problem. If I write down the law of radioactive decay, it is simply a sentence that might, as far as looks and content are concerned, be a mere truth (so Lewis would say). Could the fact that this sentence is a theorem of all the best theories be cited as the explanation of the present behaviour of the Geiger counter in the presence of radium?

Let us turn the question around.[11] Why would it not be regarded as the right sort of fact about the world? Well, if *this* fact explains why radium behaves in that fashion, should we not be able to say that in the absence of this fact, *ceteris paribus*, radium need not display this regularity? Suppose the contrary, therefore; suppose there is a best theory in which this sentence is not a theorem. That means presumably that some ultimate science treats the equation describing radioactive decay as an ancillary fact, theoretically isolated, which may be used in conjunction with deep principles of a totally different sort to explain the behaviour of Geiger counters, in a footnote. Suppose also that only one, or at least a small minority, of the best theories are like this; so if we could see them, we would regard them as admirable logical trickery, achieving the aims of science by far-fetched logical devices. How would this look to someone who takes the idea of law seriously, someone who is strongly inclined to insist that not mere facts but only laws can explain the phenomena? Could he see the existence of such a theory as showing that the putative law of radioactivity decay is not a law? He would instead say, I think, that the *appearance* of explanation can be produced by the logically *outré*, but not *real* explanation.

To sum up: I have four serious doubts about whether the laws of this world, in the sense of Lewis, explain what happens. The first is that explanation is crucially dependent on information, and that what is common to all the best explanations may not be informative enough to be explanatory itself. The second is that the best theories, by the criteria of simplicity and strength, may not be the best explanations. (They might not be, namely if explanation is crucially dependent on some other feature, which requires sacrifice of simplicity, strength, or the balance between them.) The third doubt is that the fact that something is a law—in the sense of Lewis—is perhaps not the sort of fact that gives us an explanation at all, at least of the type laws were meant to give. Finally, I noted one additional doubt in passing: the initial impression here that *of course* Lewis's laws explain must largely come from their location 'at the ideal end of science', so to say; and I distrust that. This last doubt is the most weighty to me; but before turning to its examination, we must still stop to look at the properly metaphysical ingredient of Lewis's view.

5. LEWIS'S ANTI-NOMINALISM[12]

Anti-nominalism is the view that some classes correspond to real distinctions, and others do not. This view has appeared in many varieties, since Plato introduced it in his theory of Forms.

Thus some philosophers say that a class corresponds to a real distinction if its members have some *property* or *universal* in common, which nothing else shares. For example, they might say, all green objects have a property in common, namely the colour green. But the objects which are grue—which means, examined before 2000 AD and green, or else not examined before 2000 AD and blue—do not have a special property in common. The predicate 'green' stands for a real property and the predicate 'grue' does not. This is one possible account of the idea, and it involves, besides anti-nominalism, a definite further idea about the the existence of a certain kind of abstract entities and their relation to ordinary objects. Other philosophers speak instead of natural kinds and say that mice do, and humans do, constitute natural kinds (mouse-kind and humankind) but their sum does not (there is no mouse-or-humankind).

Lewis pointed out quite correctly in his paper 'New Work for Universals' that his account of laws could be saved from a serious problem by the addition of some such anti-nominalism. (See section 1 above.) The laws are to be taken as the theorems of all the best theories formulated in a *correct* language. A correct language is one whose predicates all correspond to real distinctions. If we call a class *natural* exactly if it marks such a natural classification, corresponding to a real distinction, then the requirement is: each predicate of a correct language applies to all and only the members of a certain natural class.

As Lewis also saw, it does not matter in the present context what form of anti-nominalism is embraced. Any of them will save him. So I shall also limit myself in this discussion to the minimum tenets of anti-nominalism. The division between natural classes and merely arbitrary or artificial classifications is just assumed to be drawn somehow.

Now if laws are to be what science hopes to provide in the end, then science had better hope to formulate its theories in a correct language. And the guardians of this correctness can only be the scientists themselves. What basis could there be for this hope? Is there anything in the process of scientific theorizing, theory choice, or theory evaluation which would tend to lead it to correct language?

A priori only two types of affirmative answers could be given here. We could suggest that humans have a special insight into the difference between natural and unnatural classes, and that this insight is one of the guiding factors in science. On the other hand, we could suggest that, without any such insight, scientists will tend to end up with natural predicates due to the ruthless weeding out of theories by empirical and/or theoretical success and failure. Let us look at each alternative in turn.

Is it plausible to think that we humans are naturally fit to distinguish real distinctions among all the ones we can describe? Recall that whatever we say must be combined with the following assertion: the most basic predicates of science will in the long run tend to correspond to real classes. But the distinctions which we use so easily—green vs. blue, hard vs. soft, mouse vs. cat—do not at all belong to the basic categories of physical science. Nor are they likely to do so in the future. Indeed, science has progressively undermined the primacy of those categories which have priority for us. Colours have had second-class citizenship for centuries, and the biological species—paradigm for Aristotle's forms—have lost their theoretical status with the advent of evolution.

Indeed, evolution suggests a status for the distinctions we naturally make, that removes them far from the role of fundamental categories in scientific description. Classification by colour, or currently stable animal-mating groups is crucial to our survival amidst the dangers of poison and fang. This story suggests that the ability to track directly certain classes and divisions in the world is not a factor that guides scientists in theory choice. For there is no such close connection between the jungle and the blackboard. The evolutionary story clearly entails that such abilities of discrimination were 'selected for', by a filtering process that has nothing to do with successful theory choice in general. Indeed, no faculty of spontaneous discrimination can plausibly be attributed a different status within the scientific account of our evolution. Even if successful theory choice will in the future aid survival of the human race, it cannot be a trait 'selected for' already in our biological history.[13]

Perhaps the distinctions we are able to track directly, and those which science honours in its basic terminology, are both natural. Perhaps so; but the question was whether theory choice could tend to favour correct language because the scientists can tell directly what is a natural predicate and what is not. We can't say *Yes* to this on the basis that scientists have a good eye for natural classes of a sort which do *not* correspond to the basic scientific predicates. To be fit for one task (avoidance of the common poisons, snakes, etc.) does not make one automatically fit for another.

So let us look at the second alternative. Is it possible that the selection of the more successful theories—*vis-à-vis* experimental data and theoretical criteria—will tend to favour formulation in a correct language? We must do a thought experiment here first. Suppose that at a certain point in history, all the primitive scientific predicates are natural ones. Now suppose that one scientist devises a theory which is simpler and more informative than any to be had so far—but only by the use of new theoretical terms which do not stand for natural classes. Why should we think that his theory should be judged inferior? New theoretical terms are typically not definable in terms of the old, and on the other hand, are typically required for radical theoretical innovation. No, I expect that this would be the end of the natural classes' winning streak—the incorrect language would take over.

How could we designate this as an evil day for science? Should we predict that scientific progress will be held up? But it is quite conceivable that this errant new theory is part of one of the best theories, formulated in *its* language. And it is conceivable that this new theory is simpler when formulated in its language, than any of its translations into correct languages. After all, that was the reflection that set David Lewis on this round to begin!

The suspicion I have at this point is this: if there really is an objective distinction between natural classes and others, and if laws in the sense of Lewis are what science hopes to formulate in the long run, then the only possible evidence for a predicate being natural is that it appears in a successful theory. If that is so, then science can never be guided even in part by a selection of natural over unnatural predicates. For the judgement of inferiority of any terminology on such a basis can then be made only in retrospect, on the basis of some other lack of success. But in the absence of any selection for natural predicates, in independent fashion, at the time of theory choice or evaluation, we can have *no* reason to expect that science will tend to develop such a 'correct' language.

But even worse follows. To be precise: if the *only* link we have is that a predicate is more likely to be natural if it occurs in a successful theory, then we shall never have warrant to think that any predicate is natural. This sounds paradoxical, but consider the following example. Suppose

(a) 1 per cent of all available predicates have feature F
(b) 2 per cent of all predicates which appear in successful theories have feature F
(c) feature F is not correlated with any independently checkable characteristic.

Then it is clearly true that a predicate is more likely to be natural if it occurs in a successful theory—indeed, *twice* as likely. Yet we shall never have reason to have any but an extremely low opinion of any predicate's claim to naturalness.

I submit that there is no plausible way to improve on this dismal picture. To think that our opinion of such a claim could cumulatively improve would require something like this: every time a predicate survives theory change, we must raise our opinion of its claim to naturalness. But that is exactly what would be plausible if independent selection in favour of naturalness were going on in theory change—the opposite of our present hypothesis.

Could we stand the problem on its head and identify the natural predicates, in Peircean fashion, as just those which will in the long run remain part of the evolving scientific account of the world? In that case, what the natural predicates are comes to depend on the actual history of our science, and perhaps on counterfactual judgements about how it would continue to evolve in the absence of nuclear holocausts, Armageddon, and the like. A major desideratum for the account of laws—that it makes them independent of any historical, psychological, or other anthropocentric factors—appears to be abandoned. In any case, we cannot make any such suggestion to Lewis, who, quite properly, (a) wants an account of what the laws are of any world, inhabited by scientists or not, and (b) would evaluate counterfactuals in terms of what the laws are and not define laws by means of counterfactuals.

We see therefore that the anti-nominalist manœuvre, by saving Lewis's account from one peril, has precipitated it into another. For it has produced unchartable distances between Lewis's best theories—and hence laws—and the theories we could reasonably hope for at the ideal end of science. The two ideals have been radically separated. We turn now to a very different line of thought that will point to the same separation.

6. LAWS RELATED TO THE PURSUIT OF SCIENCE

> One of the features of [Lewis's] account is that, on the assumption that scientific theorizing is an attempt to achieve the best overall deductive system, it explains why we are normally justified in believing that the axioms and theorems of the best available scientific theories are (or approximate) laws.
>
> John Earman, 'The Universality of Law'[14]

The last, but one of the most important criteria for an account of laws is this: the account should make it plausible that laws of nature are the truths which science aims to discover. My phrasing should not be too strictly or prejudicially construed. If the account makes it plausible that the laws, as defined, are part of the theoretical description of the world provided by science in the long run, if all goes ideally well—that is enough. At first sight, Lewis's account has this very important virtue. For prima facie, our hope for science, and our expectation of its achievement if all should go ideally well, is that science reach one of the best true theories in our world. And by definition, all the ways in which this hope could be realized will lead to the laws of nature, in Lewis's sense.

That this prima-facie virtue should be a real one, we found to be a crucial concern for the other desideratum of explanation. Perhaps laws in the sense of Lewis can be expected to explain only in the sense that anyone tends to grant at once that scientific theories explain. (We contrast this with the idea that laws should explain why the phenomena are necessarily what they are, in some more substantial sense, which certain authors refuse to grant for laws in the sense of Lewis.) But that expectation too requires a previous conviction, that laws are identifiable as just the sort of truths we may ideally expect to find in our scientific theories.

Does Lewis's account fare well with this desideratum, upon due reflection? To reach the set of laws, on Lewis's account, we successively narrow down the sets of true statements by means of criteria of selection conceived of as purely syntactic and semantic (as opposed to pragmatic, which would admit as relevant also historical, psychological, or other contextual factors). Everything will be well, therefore, only if we can maintain *either* that actual theory choice in the history of science is by such criteria, *or* that it should be, *or* that it would be under ideal conditions. If not, we cannot plausibly expect science to reach one of Lewis's best theories, even if all goes ideally well. But can we maintain some such thesis about the history of science?

I have three reasons for saying *No*. The *first* is that the criteria for theory choice in science are not Lewis's criteria of selection, and do not have the same general character. The *second* is that even if Lewis's selection criteria were among those guiding scientific theory choice, his purpose would be defeated by the presence of additional criteria. The *third* is that even if Lewis's selection criteria were the actual and sole actual criteria utilized in theory choice, reflection on our starting-point will make it impossible to conclude that science tends toward one of Lewis's best theories as end point, even *ceteris paribus*.

Lewis's selection criteria, to separate out the best theories, are four: truth, simplicity, strength, balance. There is a fifth, or perhaps I should say zeroth, criterion: the selection is made from theories formulated in a correct language (languages with natural predicates). Now actual science begins with theories not known to be true, but in any case, not very simple, not very strong, with regrettable sacrifices of simplicity for strength or vice versa, and formulated with predicates for which we claim no virtue beyond familiarity. The progress of science will not choose among these; it will modify them. We envisage therefore two processes: one a logical subdivision of the whole class of theories, and the other a trajectory through that class, starting from a specific point. Question: should we expect the trajectory to land in the target area which the selection marks out?

First of all we suspect that when theories are in competition, and one has the advantage of simplicity, that this advantageous simplicity is a human, historically conditioned one. At this point in time, the language used is not entirely correct, but it is familiar to the contesting scientists. David Lewis's criterion of simplicity would be applied (by someone outside history) as follows: first translate both theories into the correct language, then compare those two new formulations as to simplicity. Even if the *general* criterion of simplicity is the same, the verdict may well be reversed by the translation. But that real simplicity, which becomes apparent only upon formulation in a language which the contesting scientists neither have nor know, cannot affect the outcome of the contest! That outcome will only be affected by the simplicity felt and appreciated by the contestants.

This problem arises even if we think that the *general* notion of simplicity is the same for the actual scientists as for someone not historically conditioned in the same way. Of course, the problem is much worse if scientists' peculiar education or aesthetic sensibilities enter their judgements of simplicity. In that case—and I fear it may be so—the pious sound of the word 'simplicity' may be the only link between the two sorts of evaluation.

We also suspect that if two theories are in competition, and one has the advantage of strength (that is, informativeness), the strength is peculiarly historical. For—information about what? Any sort of information? Information has a generally agreed upon measure in the simple context of communication engineering. But if we laud a theory for informativeness, that measure is not intended, I am sure. For in practice we call a theory more informative if it answers more of *our* questions—and we are highly selective in what questions we pose. I think all scientists agree on the value of accurate prediction of empirical phenomena. But even there Thomas Kuhn has charted historical variations in what empirical information scientific theories have been required to give.[15] In addition, theories may be more or less informative about what goes on behind the phenomena. The putative information it gives there is evaluated quite differently by different scientists, at least until it issues in new empirical predictions. This is amply illustrated by the differing nineteenth-century views on the value of atomic and molecular underpinnings for thermodynamics.

The general point is this: even if the measure of information is objective, and is just what Lewis thinks it is, the operative principle of theory-evaluation will be in terms of *valuable* information. When the scientific community apparently judges that one theory is more informative than its competitors, and should therefore be favoured, that favoured theory may well be one that is really less informative all told. The reason is that historically conditioned values are modifying the judgement tacitly. The perceived superiority with respect to desired and valued information, will issue in that apparent judgement of greater informativeness pure and simple.

So far my first reason. Now I turn to the second reason: even if the actual evaluation of simplicity and strength in the history of science coincide with the historically unconditioned evaluation, things may still not work out. The reason is that there may be additional criteria operative in the historical evaluation—and I think there are. I would especially mention the advantage a theory may have if it is more easily combined with theories outside the context. For example, Lord Kelvin objected to Darwin's theory that natural selection needed more time than physics could allocate to the age of the earth. Darwin's theory was not competing with physics; but incompatibility with the historically given physics could have disqualified it from competition in biology, if Kelvin had turned out to be right.

Perhaps there are still further criteria at work in the history of science, beyond those considered by Lewis. All we need is some suspicion of this sort. For then we have immediate reason to suppose that the process of theory choice will go awry, from Lewis's point of view. All it needs is *some* extra criteria. Consider this parallel. One child says: the best objects in this room are the largest. A second child says nothing but begins to compare the objects it finds two by two. If it always discards the smaller, we may reasonably expect that—if it is not interrupted or deceived or whatever—it will end up holding one of those items which the first child considered best. But if the second child has an additional criterion—if, for example, it regretfully puts aside any object, however large, if it is red—we no longer have that expectation. The largest objects may be a very different class from the set of largest non-red objects. We can no longer expect their selections to be the same, if the second child displays *any* decided preference guided by colour—for example, if it always puts aside the red object unless it is at least twice as large as the other. As long as any other proclivity is at work, the outcome will depend a great deal on the actual composition of the room's contents, and we have no logical way to speculate about that.

Finally I turn to my third reason. Even if the criteria of historical theory choice were all and only those described by Lewis in his theory of laws, all would not be well. For the evolution of science as a whole is historically conditioned by its starting-point, and by the schooled imaginations of its practitioners. Again, an analogy. Let one extra-terrestrial visitor to earth judge that the most beautiful animals here are the largest, most active ones. He, she, or it must have some criterion of balance in mind; but obviously the class he thus selects contains elephants and perhaps a few others. Now let a second such alien begin to breed mice, always selecting from each generation the largest, most active ones. He will eventually have large active mice, but not large, active animals. For a large mouse is still a small animal.

This ends my critique of a law-oriented eschatology of science. Should we now add that if I am right, so much the worse for science? Are all reasons to think that science would not even ideally arrive in Lewis's target area *ipso facto* reasons to expect that science will fail in its proper task?

By no means. As I see it, science aims to give us theories which are empirically adequate. The practitioners commit themselves to one theoretical framework rather than another, if they judge, to the best of their cognitive ability, that this is more likely to serve the aim of empirical success. They are acting in good faith if their selection criteria do indeed, by their own lights, help rather than hinder, and at least do not sabotage, the pursuit of this aim. That they should always act as best as they can, by their own lights, is their ethic and their conscience. They have also been very successful in this pursuit, and have as much reason as anyone to believe in their enterprise. What I have been arguing is that this positive trust in the actual process of science, establishes no link between its eschatology and Lewis's laws. That the process of science leads to greater empirical success always gained through more beautiful intellectual constructions, if all goes well, is implied (*modulo* the meaning of 'all goes well'). But that it leads to laws in Lewis's sense, is not implied. Thus there is no reason to equate Lewis's laws with what science pursues.

7. A PARABLE

High in the mountains by the eastern sea, the magicians have their own kingdom. It is small, compared to ours, not much larger than our largest city, but rich with the gifts of magic and nature. High up it lies, on a still plateau where the rising sun brings warmth early every morning before it turns to us. The magicians who live there seek to draw us with their subtle powers, but are hindered by the frailties of our own intellect and flesh.

In our kingdom, all manner of weakness of the eyes is hereditary. Our kings, whom few have seen, were always the most far- and clear-sighted creatures on earth. In the sky they saw—so it is told—stars of fire to which they gave many wonderful names, likening them to warriors, beasts, and jewel-studded girdles. Our soldiers too were always far-sighted, and—then as now—strike fear in the heart of all that lives and moves beneath the sun. They detect enemies before they come within stone-throwing distance, and signal each other with mirrors glinting in the sun. We ordinary people of lesser stock, the craftsmen, fishers, and scholars, see as much as we need; though compared to them we live as if in mist and haze.

This story is told of long ago. The magicians sent a dream to three kings, three soldiers, and three scholars. The dream revealed the magical kingdom in all its glory, with such felt hope and grace as to be at once infinitely desirable. Each dreamer resolved to seek the kingdom. But our minds are clouded in proportion to our eyesight, so the kings, soldiers, and scholars did not learn equally much. The kings saw clearly the magicians' houses and castles, the high mountains, and a brilliant star which they recognized, at its zenith. The soldiers saw only a mountainside, and green meadows in the dawn; by the shadows they judged that the place must lie due east. They could not discern houses from rocks, nor see any star. But such was the longing this dream inspired, that they knew it held a prize beyond what any campaign could bring. Lastly, the scholars, as captivated as the others, received no inkling of whether the place was high or low, though they too saw how the rising sun cast the shadows. Each group began its journey east, quite unbeknownst to the others.

Many obstacles lay in the kings' path: rivers and ravines, hunger-maddened goblins and wolves, cliffs too steep to climb and lakes too wide to swim. Almost every day they were diverted, now left then right, out of their way. But each night the kings saw their guiding star and each dawn set out towards it. After three years and a day, the kings ascended the eastern mountains, and were welcomed into the magicians' city.

The soldiers, trained to find their way across difficult terrain, and to judge direction accurately from shadows cast by sun and moon, struck east. Coming upon the hills, they ascended. But the hills proved low, judging by their memories of old campaigns, and they knew they had not come to the place they sought. Eventually, climbing almost unscalable cliffs, they came to the top of a mountain. As far as they could see, there were no heights comparable to this. The high meadows were green and berries abounded, a lake held trout. In the earth they found silver and gold, the bees gave up their honey, the trees gave them wood for building. In their dream they had not seen the great magical castles, nor did they have the kings' grasp of how high the eastern mountains are. So there they stayed, still a year's journey from the east, in bounty undreamed of in their old soldiers' life—but still in poverty and want compared to the kings.

The journeying scholars did not have the kings' eyesight, nor the soldiers' fieldcraft. They did not know the place they sought was high in the mountains. The magicians' kingdom could after all have been as glorious and rich if it had been in a valley, and the east would still have been east if the land had run everywhere level to the sea. So they sought only the east and indeed, if they had journeyed due east they would have arrived. To guide them they had a lodestone compass, fashioned by our finest craftsmen. They attached a small light to the lodestone, which they sighted through narrow slits in a screen, so as to draw a line with true direction. Thus their determination of the compass points was exceedingly fine by night and day. Always after an obstacle they used a small sand-clock to gauge the time they had needed, departing from true; set up their compass again, and adjusted their path. Yet at every turn, some minute angle was lost, whether to south or north. The proportion of deflections favoured, ever so slightly overall, the south. After five years of travel they came upon the sea, where they found a land of milk and honey, warmth and welcome among a friendly people. There were green fields and the sweet taste of dates ripened in the sun. To the north, across an arid desert, there lay soaring mountains, they were told. But they had come to the easternmost shore, and there they stayed. A half year's journey to the north, lay the incomparable intellectual splendours of the magicians' land, where scholarship had already bloomed for ten thousand years.

Many generations have repeated this tale, which could only have come to us from a returning king, still shining with the magicians' knowledge. The soldiers remained, happy, in the lower mountains, and the scholars, also happy, by the eastern shore. Are we right to describe our fellow scholars of so long ago, as in error? They truly travelled east, by the finest determination human hands and sight allowed them. Of course they realized that their instrumentation was not infinitely fine, and that such a journey could not have a single, pre-ordained end. But what they found, at the easternmost point by their reckoning, was paradise by their lights—they would not have been content with less. Yet we sigh; their light seems dim and poor to us who, though of the same benighted kin, have pictured to ourselves magicians, kings, and stars. Some say the tale is not a history of long ago, but a vision of our far future. In these republican days, some even say that our kings never had their fabled power of sight, and no one ever will. Whatever be true, we pity those scholars, our brothers, who only found happiness, but never that true home with its true riches.

8. CONCLUSION: DECEPTIVE SUCCESS

The reason why I liked Lewis's theory of laws must have been clear from the beginning. First of all, the account involves very little that could be associated specifically with (pre-Kantian) metaphysics. Lewis himself is a realist about possible worlds, but his account of laws could be accepted word for word by someone who regards possible worlds as (semantic) theoretical fictions. The second reason is that the account makes a real effort to establish a link with science. The laws, as defined, should be good candidates for what science will ideally arrive at, and the fundamental principles of science should be good candidates for laws. By defining laws in terms of good, better, and best theories about our world, Lewis makes a sincere effort to honour this desire.

So what are the difficulties that render the account inadequate? They are of two sorts—the ones I have taken up, and the ones that Lewis himself points out in later writings.

In his *Philosophical Papers*, vol. ii, David Lewis proposes an amendment to his account of laws. He introduces the notion of objective chance, in the first instance to broaden his account to cover the probabilistic theories of contemporary physics. This generalization of the notion of law he concludes, ruefully, not to admit of the sort of treatment given to non-probabilistic laws. So he admits chance as a separate category. Then, in his definition of law, he replaces the set of true theories, by the set of those theories which never had any chance of being false. He writes 'The field of eligible competitors is thus cut down. But then the competition works as before. The best system is the one that achieves as much simplicity as is possible without excessive loss of strength, and as much strength as is possible without excessive loss of simplicity. A law is a regularity that is included . . . in the best system' (p. 126).

I have chosen to concentrate on Lewis's original theory for three reasons. The first is that the difficulties I see for the original, more limited account seem to me to persist almost entirely intact for the recent, amended account. It is true that after cutting down the field of competitors, the criteria of simplicity, strength, and balance have less work to do. But we can't really tell how much less; so we cannot evaluate the import of this remark at all. Secondly, I wanted to make clear that difficulties faced by an account of laws are not brought on by its recourse to metaphysics. In Lewis's original account, there is an absolute minimum of metaphysics, and I did not need to raise an objection to this minimal presence as such, to find what I regard as debilitating difficulties. This will make clear to ametaphysical philosophers, I hope, that accounts of law turn to more metaphysics out of need, not idiosyncratic preference. With Lewis's amended account, this would not be nearly so clear, because with chance as a separate and irreducible notion, the reality of possible worlds does become crucial. Finally, I conjecture that new difficulties introduced by the ontological reification of chance will affect Lewis's new account as much as some others to be discussed. (These last two reasons will be clearer, I think, after the discussions of chance and its relation to opinion in the next three chapters.) A conjecture is not a firm reason, but it may incline.

To complete our overview let me summarize the problems discussed in this chapter which already affect the earlier version of the account.

It is true that the account does not presuppose modal realism. Unfortunately, the moderation with respect to metaphysics made the account vulnerable to charges that it does not respect real necessity, in several ways.[16] Secondly, the laws of this world, in Lewis's sense, are not at all guaranteed to be explanatory. If the best theories are the best explanations, then those laws are part of every best explanation of the world as a whole. But the laws themselves may well lack those very features that make the best theories explanatory. And thirdly, the attempt to link up with science founders, in my opinion, inevitably. For the criteria for better and best theories utilized, must be such as to leave it an objective matter, independent of history and psychology, what truths are laws. That means that the equation we are tempted to trust—the best theories are those theories which science could or might reach, should all go ideally well—is simply divorced entirely from the equation that defines best theories for Lewis. I see no remedy for this.

Most of all, we see here the dilemma posed by the problems of inference and of identification, which I discussed at the end of the preceding chapter. Lewis formulated his definitions in such a way that there can be no question about the validity of 'It is a law that *A*; therefore, *A*.' The inference problem is thus successfully handled. And to begin, it seemed that identification too was unproblematic. But that turned out not to be so, because the criteria for better and best theories—crucial to the definition of law—were not translation invariant. The consequent introduction of the notion of natural classes and predicates, led to an identification problem which I believe to be unsolvable. The attempts at identification examined put laws out of touch with science even if otherwise granted to be workable.

As we go on now to other accounts of laws, we shall find more and more pre-Kantian metaphysics, and at the same time, less and less contact with science. For Lewis's account there was still a point to serious discussion of the eschatology of science—there will be little point to it later in this part. The notion of necessity, and the idea of very strict criteria for explanation of *what is* as *what has to be*—these will be honoured all the more. I cannot hide my conviction that if Lewis's account had been more successful, it would have been foolish to look further—but there it is. The last hope for an empiricist account of law, that a little sacrifice to anti-nominalism would ward off peril, is gone.

4.
Necessity, Worlds, and Chance

> The conceptual framework in terms of which we have been operating points to the following definition of natural law: *A natural law is a universal proposition, which holds in all histories of a family of possible histories. . . .*
>
> 'Concepts as Involving Laws and Inconceivable without them'.
> Wilfred Sellars, p. 309.

THE accounts of laws of nature to which we shall now turn, I call *necessitarian*. For in these accounts, unlike in Lewis's, necessity comes before law in the order of definition. They are also less ambitious, for they do not attempt to characterize necessity and laws by means of commonly understood relations between theory and fact. Instead they begin with a substantial assumption of reality: the reality of other possible worlds besides our own. I shall begin with a critical assessment of that assumption; thereafter I will mostly grant it for the sake of argument.

This chapter will focus also on physical probability, which gives a new shape to the idea of law. This will introduce questions which, I maintain, necessitarian accounts cannot answer—questions which we shall confront again later, however, for they arise today for any philosophy of science.

1. ARE THERE OTHER POSSIBLE WORLDS?

Since realism about possible worlds will now play such an important part, I propose a suspension of our disbelief for most of the discussion. But here, and finally again at the end, I shall briefly examine this realism and its support, and outline (even more briefly) the corresponding anti-realist stance.

The recent respect for possible worlds derives from their use in semantics. Let us first see how they are used there, and then consider an argument for their real existence.

Modal terms like 'necessarily', 'possibly', 'actually' obey certain logical rules. The venerable modal square of opposition (see Fig. 4.1) summarizes the main ones. The vertical arrows signify implication or valid inference. For example, as the medievals codified it: necessity implies actuality and actuality implies possibility. The diagonal lines marked 'cont' link mutually *contradictory* propositions: each is exactly the other's denial. 'Cntry' means that the linked statements are *contraries* (could not both be true), and 'subcntry' that they are *subcontraries* (could not both be false). To know this diagram is to have a very good initial grasp of the valid inference patterns in modal discourse.

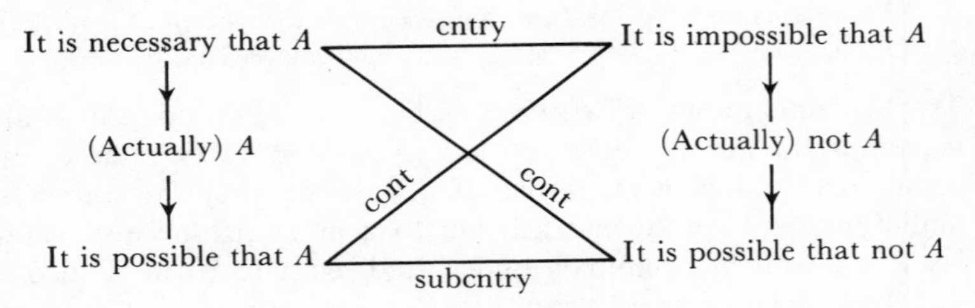

FIG. 4.1. The modal square of opposition

This understanding about arguments is thus available before we have asked, what the conditions are under which modal statements are true. A semantic theory will answer that question, at least at some level of generality. The answer it gives must bear out the correctness of the above square of opposition—that is, a semantic theory must save the phenomena of inference! And a clue to how to do this has also been available for many centuries: it is the similarity of the above diagram to the *quantifier square of opposition* (see Fig. 4.2). Here an assumption is clearly present: something, correctly designated as *This B*, exists.

The following theory now suggests itself. There are other ways the world could have been—briefly, there are possible worlds, the actual one and some others. 'Actual' is like 'this', 'necessary' like 'all', and 'possible' like some. To be precise, call a world an *A-world* exactly if proposition *A* is true of it. Then Fig. 4.3, which is a specific instance of the quantifier square of opposition, indicates clearly how to translate modal discourse into discourse about possible worlds. This is a graphical summary of the truth-conditions

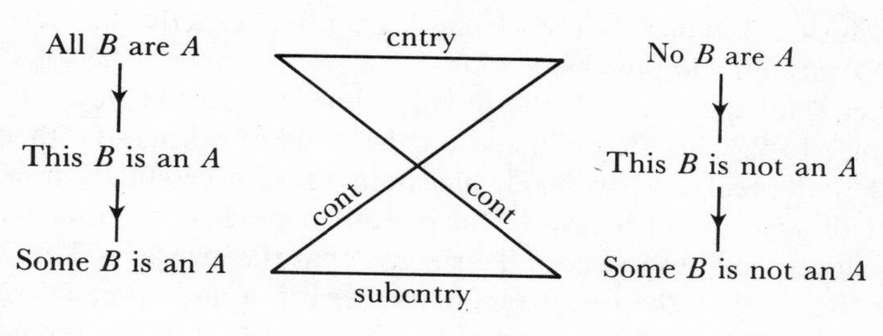

FIG. 4.2

for modal language which the semantic theory presents. Thus *It is possible that there are chimaeras* is true if and only if *there are possible worlds of which 'There are chimaeras' is true*, or more perspicuously, *There are possible worlds in which there are chimaeras*, is true.

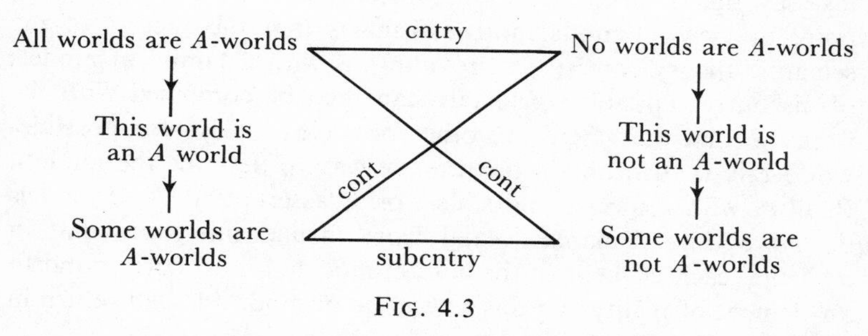

FIG. 4.3

This semantic theory—standard reference, truth, and possible world semantics—was greatly elaborated beyond the initial stage. In the 1960s and early 1970s especially, it went from success to success in philosophical logic, and theoretical linguistics. Consider now the argument:

1. There is a proposition *A* such that both *It is possible that A* and *It is possible that not A* are true.
2. There are at least two possible worlds.
3. There is at least one other possible world, besides the actual one.

The first is a premiss offered for your acceptance. The second follows then by the above semantic theory (given 1, there must be an *A*-world and also a world which is not an *A*-world). The conclusion 3, which follows from 2, expresses exactly the view we call *realism about possible worlds*.

Let's not quarrel with the premiss, but ask instead how we get from 1 to 2. This step is made on the basis of a semantic theory, and that theory, as we said, has been very successful. But what sort of success did it have? Did it lead to predictions, that could be checked by observation? To suggest that the answer is *Yes* (also for their other theories) adherents began to use 'predict' as a synonym for 'imply' or even for 'allows for'. But the predicted phenomena are, in any case, all about how people speak, and about what they regard as grammatical, correct, tautological, valid, or alternatively, invalid or absurd. The phenomena saved, if any, are the accepted patterns of inference in a certain area of discourse.

Is this sort of success sufficient to force us from an acceptance of 1 to acceptance of 2? I will not argue the point here, but will just say that it does not seem sufficient to me.[1] The alternative point of view—a modal anti-realism—is that the success of the semantic theory consists in providing us with a family of models of discourse. Possible-world talk can then be combined with the robust denial that there are other possible worlds—for possible-world talk is then only a picturesque way to describe the models. Realism with respect to possible worlds asserts that these models do more than demarcate valid from invalid inference—that in addition, each element of the correct model(s) must correspond to an element of reality. To this *reification* of models I shall return in the last section.

2. LAWS RELATED TO WORLDS

Law implies necessity: if it is a law that wood burns when heated, then wood *must* burn when heated. This traditional connection was also elaborated so as to make it stronger: the law is the reason for the necessity, the necessity is there because of the law, and not conversely. The terms 'reason' and 'because' are mysterious, however. It is not clear whether something could be necessary without following from a law—the directedness of 'because' may or may not be reflected in that of 'implies'.

David Lewis, as we saw, simply equates being (physically) necessary with being (implied by) a law. That equates law with necessity, since on his account whatever is implied by laws is also a law. In this course he was not egregious. Reichenbach defined a fact *P* to be physically necessary exactly if 'the sentence describing *P* is a nomological sentence in the wider sense', the indicated class of statements being intended to consist of laws of logic, laws of nature, and their consequences. A few years later Fitch defined the corresponding modal connective *It is naturally necessary that* to mean *It is (logically) necessary that if L then* where *L* stands for, he says, the conjunction of all laws of nature. Montague's treatment in 1960 presented a corresponding semantic characterization. The form of presentation in all three cases clearly shows that the authors take themselves to be merely making precise a common notion.[2]

The important innovation that gave flexibility to possible world semantics was Saul Kripke's insight that the same square of opposition which fit 'bare necessity':

1. *Necessarily A* is true in world *x* if and only if *A* is true in all worlds

would also fit any 'restricted necessity':

2. *Necessarily A* is true in world *x* if and only if *A* is true in all those worlds which are possible relative to *x*.

Here 'possible relative to' is a relation, of whatever character. We require only that it be reflexive—any world *x* must be possible relative to itself. For then it will follow that if *Necessarily A* is true, so is *A* itself, regardless of which world we focus our attention on.

The relation of relative possibility is also called the 'access relation', in which case 2 takes the form

2′. *Necessarily A* is true in *x* exactly if *A* is true in each world to which there is access from *x*.

But what is this relation, and what is it like? It is instructive to look at this a little more, from the point of view which defines necessity in terms of law. It appears that we have three candidates for the relation of (physical or nomological) relative possibility. Each will give rise, via equation 2, to a distinct sense of (physical or nomological) necessity:

World *y* is *possible* relative to world *x* exactly if:

R_1: no law of *x* is violated in *y*

R_2: every law of x is also a law of y
R_3: y has exactly the same laws as x.

These three relations have different characteristics. It must follow from what is meant by 'law' that a law of x is not violated in x, so all three relations are reflexive (x bears R_i to x). The second and third are also transitive, and finally, the third is in addition symmetric. Among the earliest results of Kripke's pioneering work we find that the logic of 'Necessarily' obeys three distinct, previously studied logical systems in these three cases. The differences between them come out only when we place modal phrases within each other's scope, as in *It is necessarily the case that it is not necessary that* . . . Philosophers, unlike logicians have largely ignored the complexities of such 'nesting' at least in discussions of physical necessity.

So far I have followed the pattern, also used by Lewis, of characterizing necessity in terms of law. But now we must ask the question: can we invert this order, and define the notion of law in terms of such notions as those just introduced: worlds, and relations among worlds?

It will take the remainder of this chapter to arrive at an answer; but here I shall chart the main alternatives.

In 1948, Wilfrid Sellars introduced the use of Leibniz's metaphysical story of possible worlds into the twentieth-century discussion of laws and necessity.[3] His paper raises the most important issues involved and gives a precise formal treatment foreshadowing the now familiar possible-world semantics. Of course, Sellars was acutely aware of Kant's impact on the metaphysical *problématique*, and characterizes what he does as a philosophically useful 'exploration of naïve realism' (p. 302). The writers who followed him—though perhaps unaware of his work—mostly did not share this attitude.

The simplest sort of analysis of law, of this sort, merely postulates that (*a*) there are possible worlds, (*b*) there is a relation of relative possibility between them (the *access* relation). The definition then follows: it is a law of world x that A if and only if it is true that A in every world which is possible relative to x—i.e. exactly if *It is necessary that A* is true in that world. Special postulates may then be added, to describe the access relation, and say that it is reflexive and so forth. This analysis was proposed by Robert Pargetter.[4]

A somewhat more complex sort of analysis would require some special features for any proposition *A* to be a law, in addition to its necessity. Such a special feature could be universality, explained in one way or another. An alternative sort of sophistication could lie in attempts to relate the access relation to features of the worlds themselves, such as their history. An analysis with both sorts of elaboration has been developed in a number of publications by Storrs McCall.[5]

I did not raise the issue in the preceding chapter, but a special effort is needed to accommodate probabilistic laws. This has become crucial, since physics now includes irreducibly statistical theories. The assertion that a radium atom has a 50 per cent probability of decaying within 1600 years, does not have the form that something specific *must* happen, or even will. Thus at least the simplest type of possible worlds account will not work for them. Peter Vallentyne (and also McCall) draws on theories of objective chance—notably, David Lewis's—to generalize the necessitarian notion of law.[6] He does so in the framework of possible-worlds models incorporating time, along McCall's lines.

There are important differences between the accounts of law and necessity given by Pargetter, McCall, Vallentyne, and other authors who have explored this avenue of approach. Some of these differences are of merely technical interest, others are more philosophical. I must emphasize, with apologies, that I shall not try to replicate their individual accounts. I shall only try to explore general alternatives, and to show that there are two problems that beset all of them. Both concern the relation of model to reality. In later chapters I will detail how I recognize a legitimate place for possible-worlds models, and how their use can be meaningfully integrated into the epistemic process. But what I say there will not require that we have a way of referring to real, unactualized possibilities, or to one (purportedly denoted, but undescribed) relation among them, or that every element in the model used have a counterpart in reality. However that may be, the metaphysical accounts of laws of nature, based on realism about possible worlds, have their own peculiar problems, to which we now turn.

3. THE IDENTIFICATION PROBLEM

Here is the simplest necessitarian account. There really are possible worlds; which is not to say that all logically and consistently describable worlds are among them. Perhaps there are worlds in which pigs fly, or in which Bertrand Russell—like Jourdain's R*ss*ll—was torn apart by suffragettes; but perhaps there are not. Secondly, there is a relation of relative possibility between worlds—the *access* relation. To say that a proposition is a law in a given world is the same as saying that it is necessary in that world. It means that this proposition is true in all worlds which are possible relative to that given world.

As Pargetter, who proposed such an account, explicitly recognized, we are immediately faced with a problem. *Which* relation among worlds is that relation of relative possibility? Since law is now a derivative notion, it cannot be used to identify the access relation. Calling it the relation of relative possibility also does not help: this is to baptize it *in absentia*, so to speak.

Certain characteristics of that relationship may be postulated, for example that it is reflexive. If we make the list of postulates long enough, will that single out a unique relation? No, it won't, unless it is one of those trivial relations which either hold between all worlds or between none. Otherwise we can always find a distinct, isomorphic relation, which satisfies the same postulates. (Indeed, metalogical results teach us that if there are infinitely many worlds, there will also be non-isomorphic relations that fit.) We can't single out the relation by description; and obviously we also can't by pointing to it. This is the *identification problem*.

Suppose that when I say that there is a law of gravity, I *mean* that something or other is true about the access relation. If I don't know what relation that is, then I don't know what I'm saying.

Pargetter proposes a solution. The solution is again anti-nominalism—the realist position concerning classes to which Lewis also turned—but with respect to relations among worlds. To the mathematical mind, a relation is simply a class of pairs—e.g. kinship is the class of pairs x, y such that x is a kin of y. But not all classes correspond to *real* relations. A natural class of pairs does, and an unnatural class does not. There is, says Pargetter, only one *real* relation among worlds. So there is no problem about identifying the access relation. It is the only one there is.

He need not have been quite so radical. After all he had postulated some characteristics for the access relation. So to solve the identification problem he need only add to his postulates: and there is only one real relation among worlds which has these characteristics. That does not mean that he has given us a complete or even very informative list; the extra postulate would only be that the list contains enough information to identify the access relation—as fingerprints might identify a murderer.

Before turning to less simple necessitarian accounts, let us see where we now stand. First consider the person who decides to believe the above account, including some such postulates as Pargetter's. Suppose in addition that he believes that what science says is our best guess at what the laws of nature are. This person has of course no difficulty (in principle) in finding out what the access relation is like. Every bit of science he studies will tell him a little more. There is a problem of under-determination: he will never learn enough to identify the access relation by description. But that's why he has his postulate: if we ask him what relation he is talking about, he can say *But there's only one.*

For those of us who do not believe this postulate—either because we have some little doubt about the reality of other worlds, or because we believe in worlds, but are not anti-nominalists of the same sort—the situation is a little hopeless. We don't know what he is talking about.

Could we finesse our communication difficulties in some way? Suppose we translate his utterance *It is a law that A* as *There is a relation R among worlds, which has all the properties he has explicitly laid down, and it is the case that A in every world which bears R to ours.* Then, if we take all classes of pairs to be real relations, we have trivialized his assertion—it has lost all informational value for us. If on the contrary we say that only some classes of pairs are real relations, but we don't know which, we are back in our original quandary. We have to ask him: if there are several such relations, what distinguishes the one you are talking about from the others? And he has no reply except the, to us so unsatisfactory, *But there's only one.*

So far the simplest necessitarian account. Let us see if more sophisticated variants fare better.

4. TIME AND THE BRANCHING UNIVERSE

The identification problem would be solved entirely if the access relation could be defined, in terms of characteristics of the worlds themselves. Since the access relation determines all modal facts, the definition would have to be in terms of what actually happens and actually is the case in each world—in other worlds, in terms of their histories.

There are many possible worlds whose history is the same up until now. They too started with a Big Bang, or were created in 4004 BC (as Newton thought), contained dinosaurs but no Piltdown men, and issued the race of Attila, Rembrandt, and Mother Teresa. But from now on they diverge: in some I continue to write this with a smile, in others with a frown, and in still others I stop. If we draw a picture of these diverging histories, it looks like a tree, with its trunk the settled past, and its branches the many possible futures open before us.

This picture of a settled past and open future perennially contends, in our ontology, with the view of history *sub specie aeternitatis*, from beyond the end of time so to say. In the former, the past has gained its own unalterable necessity—*Wesen ist was gewesen ist*—but determines the future only within certain limits. Prevision is but guessing; what we see has already turned to stone behind us.

> A Klee painting named 'Angelus Novus' shows an angel looking as though he is about to move away from something he is fixedly contemplating. His eyes are staring, his mouth is open, his wings are spread. This is how one pictures the angel of history. His face is turned toward the past. Where we perceive a chain of events, he sees one single catastrophe which keeps piling wreckage upon wreckage and hurls it in front of his feet.[7]

Besides this angel there must be another, facing forward, his face shining with hope as he beholds the *embarras de richesse* of possibilities in our open future.[8] But let us see how this view of time and history could aid the necessitarian programme for explicating law of nature.

The identification problem may be solved if we define relative possibility in terms of shared history.[9] Call two worlds x and y *t-equivalent* exactly if they have the same history through time t. Thus, at time t, they share their entire past and present. This is a defined relation among worlds—though one that is time-indexed—and we can take it to be the access relation. The corresponding sense of 'necessary' is customarily expressed by another word, 'settled'. The worlds which are t-equivalent to world x together form the *t-equivalent* of x, or as we can also say, the *future cone* of x at t. We can illustrate this as in Fig. 4.4. The worlds y, x, w belong to the t-cone of x; and all those, plus z and v, belong to the $t°$-cone. The other worlds w' and v', which share no history with x, do not belong to either cone.

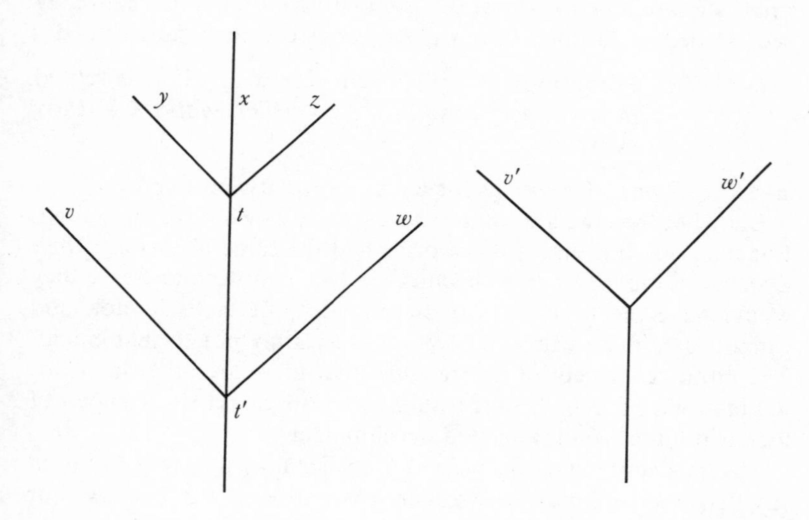

FIG. 4.4. Possible-world histories

Definition. Proposition A is *settled* in x at t if A is the case in every world which is t-equivalent to x (i.e. every world in the t-cone of x).

This being settled is a kind of necessity, for if it is settled that A, that means that A is true regardless of how our future develops, no matter how history goes from now on.

What we have now introduced is certainly well enough defined, but does it give us the access relation needed for an account of laws? There are two reasons why it cannot. First of all, it is clear that if it is settled that A, it is still not generally true that it is a law that A. Many facts have been settled by what we call the accidents of history. This can be so even though A concerns the future—it may have been settled by 1900 that the twentieth century would see the devastating effects of population explosion, world wars, and social disease. To say that this was settled means that certain alternative possible futures were ruled out, could no longer come into being, after 1900. Even if that was due in part to laws of nature, it was partially due to what had happened already.

To eliminate the element of historical accident, we could require what is a law now to be not only settled now, but to have already been settled at all times in our past. Let us call that 'fully settled':

Definition. Proposition A is *fully settled* in x at t if A is settled in x at every time $t' \leqslant t$ (at which world x already existed).

(See *Proofs and illustrations* for a further discussion.)

But what the laws are, cannot be the same as what is fully settled. For that would mean: if two worlds had the same history for their first two minutes, or two seconds, or two nanoseconds, . . . they would have the same laws of nature. But the facts do not, and cannot, determine what the laws are—certainly not to that extent. The intuitive concept of law requires that there are other laws this world could have had, that would allow at least a short epoch of identical initial conditions and development.

This is a very telling problem. We realize now that we can't even say that if something is a law then it is settled that it is a law. For if it is a law, then it won't be contravened in any possible future *that is allowed by our laws*. It may be violated only in some possible future, belonging to a world that is just like ours till now, if that world is subject to different laws. This realization puts a complete stop to the idea that we could solve the identification problem by defining the 'correct' access relation in terms of shared historical structure.

Indeed we have now ruled out another option along the way. We also cannot hope to characterize the laws as a special class of settled propositions. But as we have seen, something can be a law, and not settled. That is shown by the example of two worlds x and y which have, say, the first two minutes of history in common, but have different laws of nature. For example, in x but not y it is a law that A; no part of x's future violates A but some part of y's future does. Then it is not settled that A in x or y, during those first two minutes.

We have thus seen that the sense of necessity appropriate to laws does not even imply that of being settled (let alone fully settled) and hence also cannot be defined as a special case thereof. I wanted to explore this very fully, so that the hope of seeing 'nomic necessity' explicated in terms of the structure of branching histories would not remain hovering in the background.[10]

McCall and Vallentyne both speak of a special nomic necessity and indicate that this does not supervene on the historical structure which they describe; so they cannot be accused of a lack of insight here. But that does mean that they are in no better position with respect to the identification problem. Which necessity is nomic necessity? Each necessity can be defined in terms of an access relation among worlds—but which is the nomic access relation?

Proofs and illustrations[11]

The model of possible worlds developing in time has three basic ingredients: Time—which I will here take as represented by the real number continuum $\mathbb{R}$; the set of worlds W; and the set of states that worlds can have at a given moment—I'll call that K. Each world x has a history: a function s which gives world x an instantaneous state $s(x, t)$ at time t, during a certain interval T. That state is the location of x at time t in the 'state-space' K, and $s(x, \text{-})$ has as graph the 'trajectory' of this world in that space. Fig. 4.4, which looks like a tree, shows a number of such trajectories, plotted against time; the limitations of the printed page require K to be represented there by the straight horizontal line which is not drawn at the bottom.

A proposition is identified by the set of worlds in which it is true. We may call a proposition *historical* if its truth-value is determined by a world's history—i.e. if two worlds have the same history, then such a proposition is either true in both or true in neither. The model allows for non-historical propositions as well, but they can't be represented in such diagrams.

Strictly speaking, a proposition is simply true or false in a world, and it does not make sense to say that something is true at one time and not at another. But the less strict way of speaking is easily explained. It is true of world x that it has state $s(x, t)$ at t. If $s(x, t)$ is inside region K' inside k, we can similarly say that x is inside K'

at t. The proposition which is true in a world y exactly if $s(y, t)$ is inside K', can be very perspicuously be written $K'(t)$, and the propositional function $K'(\,-\,)$ may be called a (temporal) proposition which can be true in world x at one time t—when $s(x, t)$ is in region K'—and false at another. And finally, any proposition which is not historical at all—for example, that the world was created or $2 + 2 = 4$—may be said to be true at all times if it is true at all, and false at all times otherwise.

Suppose now that A is fully settled in x at t. That means that A is settled in x at all $t' \leqslant t$. Now take an arbitrary time t* at which x exists. If $t^* \leqslant t$, then it follows automatically that A is settled in x at t^*. If $t^* \geqslant t$, could A be *not* settled at t^*? That would require a world y, whose history is the same as x through t^*, in which A is not true. But that world y has a history which is *a fortiori* the same as x's history through t, so that would imply that A was not even settled at t. Therefore we see that if A is fully settled at t, then it is settled at all times. But that entails in turn that it is fully settled at all times. (Henceforth we can just say 'fully settled', and leave out 'at time . . . '.)

It follows now that if x and y share *any* initial segment of history, then what is fully settled in the one is so in the other. For let A be fully settled in x, and let y be in x's t-cone, for any time t, even at the very beginning of their history. This means that A is settled in both x and y at all times $\geqslant t$, since—having the same history so far—they have the same future cones, at least through time t. But then A is fully settled in y at t, and hence, as we have seen, at all times.

5. PROBABILITY: LAWS AND OBJECTIVE CHANCE[12]

Radium has a half-life of approximately 1600 years. That means that half of an initial amount of radium decays into radon, within 1600 years.[13] For an individual radium atom, the decay is a matter of probability. It has a 50 per cent probability of remaining stable for 1600 years, 25 per cent for 3200 years, 12.5 per cent for 4800 years, and so on. The general law has the form

$$P(\text{atom stable for an interval of length } t) = e^{-At}$$

where A is the relevant decay constant. This shows how the progression works: e^{-A2t} is the square of e^{-At}, so probability $\frac{1}{2}$ over 1600 years becomes $\frac{1}{4}$ over 3200 and so forth.

Physics gives us such statistical information, and in contemporary physics the probabilities are essentially irreducible. That is, they cannot be regarded as merely measuring our ignorance of factors which really determine the exact time of decay beforehand. With this probabilistic turn in science, the discussion of laws suited in form to deterministic worlds only, becomes inadequate.

What if, as in the preceding section, we attempt to see this in terms of what our possible futures may bring? Intuitively, the atom decays within 1600 years in half of these possible futures. Hence we can suggest that the real probability—*objective chance* as opposed to our subjective likelihood—is a measure of proportion among the possible worlds which share our history so far. Greater objective chance of happening actually, *means* really happening in more possible futures.

Unfortunately the identification problem now reappears in an especially striking and difficult form. If our possible futures are only finitely many, perhaps it could be insisted that each is equally likely (though, why?) and that calculation of chances should proceed on that basis. If they are infinitely many, but only countable, they *cannot* be equally likely. For their probabilities must add up to 1 (i.e. 100 per cent), and any number, greater than zero, becomes greater than 1 if added to itself sufficiently often. And finally, if they are infinitely many and form a continuum (surely the most plausible idea) then it literally makes no sense to say: the objective chance is *the* measure that treats them all as equally likely.

Let me explain that. I do not mean simply that each individual future, in such a continuum, must receive probability *zero*. That is so, but that does not rule out a positive probability for, for example, the class of futures in which this atom decays within 1600 years. However, to give each individual point in the continuum the probability *zero*, treats them equally but leaves open entirely what the probability of such a class shall be. Fig. 4.5 represents various probability distributions over a continuum of possible worlds. We see curves C_1, C_2, C_3 and I define probability measures P_1, P_2, P_3 to assign a number to any class of worlds represented on the horizontal line, by their means. Measure P_i gives to class A_j a number proportional to the area above A_j but below curve C_i. Now each such measure gives zero to every individual point (world), but

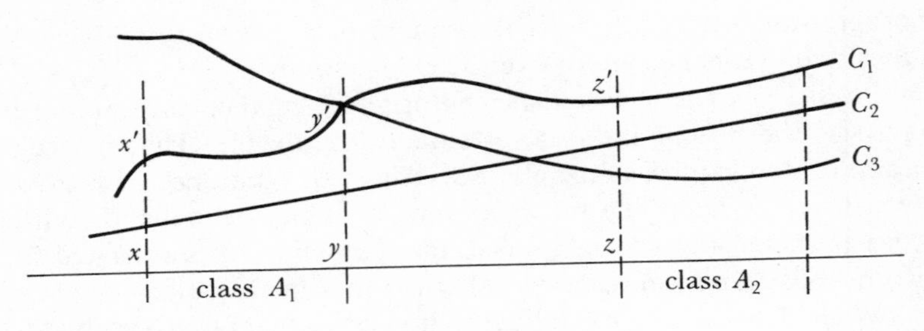

FIG. 4.5

they disagree very much otherwise. According to one, class A_1 of worlds is 'larger' (has greater probability) than class A_2, and according to another, it is 'smaller'.

In the diagram, classes A_1 and A_2 are represented by equal intervals on the horizontal line that represents the worlds. Should they therefore receive equal probability? But that makes sense only if we have a measure of distance between worlds, which is of course no easier to come by. I could have just as well described the diagram differently, and said that the points on curve C_3 represent the worlds. Why should the distances between x, y, z be a more faithful representation than those between x', y', z' above them?

To have a way of representing chance, as a measure on the possible futures, is nothing. That we can so represent it is a trivial logical fact, and helps not at all to tell us what chance is, or what the real chances are. *Which* measure is chance?

Two courses seem open here. We can say that it is something quite different, and independent of the world histories. In that case, we must add some move like Pargetter's to close the identification gap. Needed is an anti-nominalism, in the very specific form of a postulate which says: there is only one *real* measure function on the possible futures, and the remainder are mathematical fictions.

The second course is to look into past history, and identify factors in terms of which chance can be identified. This would be like the attempt to define the relative possibility relation in terms of shared history. We would have at once what I noted as a problem before. If world history is only two minutes old, say, does what has happened already determine uniquely all the objective chances of everything that could happen later? Are there really no two worlds, agreeing on the first two minutes—two seconds, two nanoseconds—of history, but disagreeing in objective chances for their future?

We can certainly *postulate* that chance supervenes on actual history—including of course the structure of the substances involved therein. But this supervenience will be an illusory gain if it is via a law—such as the law that if an atom has a certain structure then its chances of decay are thus and so. For as we saw before, the identification problem for non-probabilistic laws could only be solved—if 'solved' is apt at all!—by sheer postulation. Merely to postulate that chance supervenes on history or structure, without saying how, would of course also be no help at all in identifying chance. Nevertheless, there are two things we cannot forbid. One is to postulate that chance, though not identifiable, is real. The other is to postulate, in anti-nominalist fashion, that only one probability measure on the possible worlds or histories is real, so that identification is easy by fiat. But then we must ask whether this newly introduced notion, about to bear the entire weight of the concept of probabilistic law, can do its job. That will turn out to be surprisingly difficult.

6. THE FUNDAMENTAL QUESTION ABOUT CHANCE

How and why should beliefs about objective chance help to shape our expectations of what will happen?

This is the fundamental question about the concept of chance.[14] I am going to argue that within the metaphysical point of view, the question cannot be answered at all.

Indeed, in this context, that question about chance appears as a generalization of the *inference problem*: show that on the advocated account of laws, the assertion *It is a law that A* entails *A*. That this must be shown follows from the most minimal criterion concerning how law relates to necessity. Neither David Lewis's nor the necessitarian accounts of law have any difficulty with it in this form. (Matters will be otherwise for universals accounts, examined in the next chapter.) However, once we begin to look at probabilistic laws, and necessity is generalized to chance, we must pose the problem in a new and more general form.

The assertion that there is an objective chance of 50 per cent that a radium atom now stable will decay within 1600 years, does not imply that it will actually do so. What information does it give us then about what will happen? Or, if that question cannot be answered without repeating the assertion itself, let me rephrase what we need to know: how and why should beliefs about objective chance help to shape our expectations of what will happen?

Probability has two faces.[15] On its subjective side, probability is the structure of opinion. But when physics today tells us the probability of decay of a radium atom—for example—it does not in the first instance purport to say something about opinion, or to give advice, but to describe a fact of nature. This fact being a probability, we are looking upon probability's objective side—physical probability, or objective chance.

There must be a connection between the two. Given that the objective chance is thus and so, my opinion must follow suit, and I must align my expectations accordingly. This summary of the connection between the two is generally called Miller's Principle:

(*Miller*) My subjective probability that A is the case, on the supposition that the objective chance of A equals x, equals x.
Symbolically: $P(A|\mathrm{ch}(A) = x) = x$

This is meant to hold for me, or anyone whose opinion is rational, and who grasps the concept of chance.[16]

Principles of rationality can at times be warranted by coherence arguments. Their general form is: if someone does not form his opinion in this way, then he is sabotaging himself, even by his own lights. But how could we have a coherence argument for *Miller*? The coherence would have to be between opinions about what will happen, and opinions about chance—but then the latter would have to include opinions about the connection between chance and what happens.

It is not easy to see what those opinions could be. If a coin is fair, then the objective chance that it comes up *heads* if tossed, equals 50 per cent. We should like to infer something about how often it will come up *heads* if tossed repeatedly. But of course, it may come up *tails* every time. And if it does come up *tails*, say ten times, the objective chance of its coming up *heads* on the eleventh toss is no greater—to think otherwise is the gambler's fallacy.

We appear to have two alternatives here. The *first* course is to deny my cavalier 'of course, it [the fair coin] may come up *tails* every time'. That is, we could propose an account of chance that identifies it, or links it very intimately, with actual relative frequency. The *second* course is to accept the possibility of radical divergency between chance and actual frequency, but show that it is not likely.

The first course I shall not discuss here.[17] It does not fit well with the concept of chance as generalized from that of law. Laws are meant to explain regularities, hence they can't be (mere) regularities—on that the tradition insists. Similarly then, chance is meant to explain frequency (and statistical correlation in general) so it cannot be (mere) actual relative frequency.

The second course faces the immediate question: which face of probability does 'likely' signify here? If it means subjectively likely, we have simply restated Miller's Principle. But if it means objectively likely, how would that get us any farther? There is in fact a beautiful general theorem concerning probability in any of its senses: *The Strong Law of Large Numbers*. Let me explain its implication for the example of coin tossing, and then see if it helps us.

Suppose a coin is fair, and each toss is independent of all other tosses. I take 'fair' to mean that the objective chance of *heads* in any one toss equals $\frac{1}{2}$.

Now this entails at once an objective chance for getting two *heads* in two tosses, namely $\frac{1}{4}$. In just the same way it entails an objective chance for getting exactly two *heads* in three tosses namely ($\frac{3}{8}$) and for getting no *heads* at all in ten tosses namely $(\frac{1}{2})^{10}$. What is the objective chance, so calculated, of getting tosses half the time, in the long run, if you keep tossing forever? The answer, according to our theorem, is 1, that is, 100 per cent; total objective certainty that objective chance and long run frequency are the same.

Could this result warrant Miller's Principle? Not at all, unless we assume that our expectations should be based on this 100 per cent objective chance. But the question at issue was exactly: *what makes it rational to base our subjective expectation on our beliefs about our objective chance*? A theorem, which assumes only that objective chance is a probability measure (and extends the measure from individual events to sequences) simply cannot answer such a question. After all there are many probability functions; suppose that the probability$_1$ that the coin comes up *heads* on any one toss equals ($\frac{1}{2}$), and that the probability$_2$ of its doing so equals ($\frac{1}{3}$). Then the theorem tells us that there is 100 per cent probability$_1$ that it will come up *heads* half the time in the long run, and also 100 per cent probability$_2$ that it will come up *heads* one-third the time in the long run. These corollaries to the theorem give no guidance as to whether I should base my expectation on my beliefs about probability$_1$ or on those about probability$_2$.

Returning now to the McCall–Vallentyne programme for explicating laws of nature, in which objective chance is conceived as a kind of graded necessity, we can pose the inference problem very concretely. Suppose chance is a sort of proportion among our possible futures—what does it have to do with frequency of occurrence in our actual future? In this form I call it the *horizontal–vertical* problem because of its pictorial illustration (see Fig. 4.6).

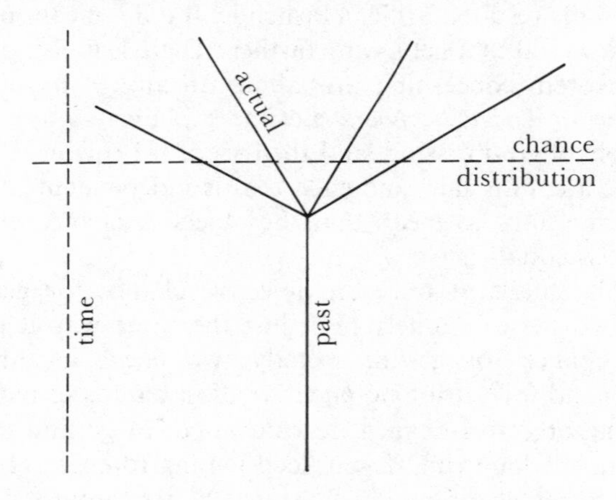

FIG. 4.6. The horizontal–vertical problem

Relative frequency is measured by counting along the vertical line which represents our actual history. But chance is a distribution over the horizontal dotted line, which represents the spread of possible worlds. So how could one possibly show that the one should help to determine the other?

This sort of problem has two interesting historical aspects in physics. In nineteenth-century statistical mechanics it appears as the ergodic problem, which is to relate the 'equiprobablility' metric of Maxwell–Boltzmann statistics to the proportion of sojourn times in a system's trajectory. The problem was solved by the development of ergodic theory in the twentieth century. But the solution takes roughly the form: under certain special conditions, which may or may not be realized in nature, the ergodic hypothesis is correct.[18] In addition, those conditions concern the probability measure in question. Hence we would have the same problem as with the Law of Large Numbers if we tried to appeal to the ergodic theorem to establish a relation between the two faces of probability.

The second place we see such a problem is more obviously connected with the present context: it is in the Everett–DeWitt 'many-worlds' interpretation of quantum mechanics. Quantum mechanics is an indeterministic theory in some sense: it gives us irreducible probabilities for measurement outcomes. On the many-world interpretation, whenever a measurement occurs, each of the possible (mutually incompatible) outcomes actually occurs, but in a different possible world. The world we are in 'splits'. This is exactly the sort of story we have just been hearing from McCall and Vallentyne. But for Everett and DeWitt this takes a very concrete form, which derives from the Hilbert space models of quantum mechanics.[19]

One of the advantages that was claimed for the many-worlds interpretation was that the rule (Born's rule) for calculating probabilities of measurement outcomes would be derivable. The derivation was supposed to show that the numbers assigned by this rule were exactly the relative frequencies to be expected (for the measurements in question) in the actual world. Since it is easy enough to see a probability function for this sort of situation, as it were 'encapsulated' in the quantum mechanical state, we can see what this claim amounts to: it is exactly the claim that the horizontal–vertical problem has *here* a demonstrable conclusion.

In actual fact, however, this topic constitutes one of the failures of the many-worlds interpretation. The purported derivation of Born's rule was from premisses, in addition to the mathematical description of the quantum mechanical state, which smuggled in the conclusion itself. The horizontal–vertical problem was solved by assumptive fiat.[20]

These sidelights on recent history of physics thus only underline the seriousness of this *horizontal–vertical problem*. It appears that if the metaphysical point of view is maintained, in which the problem arises in this form, *then no solution is possible*.

I do not want to leave this as an open question. There is no way at all to answer the fundamental question about chance within the present confines. For consider the options. First of all as I noted, no coherent argument could justify the principle, since it would have to rest on premisses about rational opinion of chance—but Miller's Principle is the basic such premiss. Secondly, as David Lewis noted at the end of his 'Subjectivist Guide', the Principle would be derivable if both rational opinion and chance derived from a single objective probability measure. That measure could only be what Carnap called logical probability, the unique probability that respects all principles of logical symmetry ('indifference'). But there is no such unique measure (as we shall see in Part IV), therefore logical probability does not exist. Finally Miller's Principle is derivable if both opinion and chance derive from a single subjective probability.[21] But that would make chance itself subjective, which is also not a feasible option. Yet as long as chance is postulated as a feature of the universe, which is not identifiable in terms of what happens, the fundamental question cannot be circumvented, dismissed, or finessed.

We will however return to this question later, with a different approach.

7. POSSIBLE WORLDS AND EXPLANATION

Laws of nature are what—according to the sympathetic—science aims to discover. If laws are what Pargetter, McCall, or Vallentyne says they are, is that contention plausible? Is the character of relations among possible worlds, or of functions on sets of worlds, what science aims to discover?

The answer, made perhaps most clearly by Pargetter, has two parts. The first part is that science aims to discover what is possible and impossible, probable and improbable. The second part is that what is possible and probable depends entirely on what is the case in this and related other possible worlds. Then it follows indeed that science aims to discover what not only our, but also those other related worlds are like. The plausibility of this two-part answer is certainly improved if we say not 'possible worlds' but 'possible futures open to us now (or in the past)'.

Plausibility diminishes, however, when we gather that the possible worlds have no interaction or influence on each other, and that the other possible futures that we might have had will make no difference at all to what will actually happen. If it is true that wood burns in all possible worlds, then it is true that wood burns here, because our actual world is possible. But why is that reflection called explanatory?[22] If I ask why this animal is aggressive, and you tell me 'because all wolves are, and this is a wolf', I may be satisfied. But my satisfaction surely derives in this case from my background opinion which now adds to this that aggression, if prevalent in wolves, has a common cause, either because it is an inherited trait in mammals or because it tends to be induced by learning in the pack. No such back-up is available in the 'actual because true in all possible' case. What if I had said 'This animal is aggressive because it is a wolf and all wolves are aggressive, although this is known not to be an inherited trait in mammals, nor ever induced by learning'? You would surely complain that I had raised a much greater mystery than you had already?

This objection must seem strong to anyone who takes causal accounts as needed for explanation. I do not have such stringent criteria.[23] But I do think that to be an explanation, the proffered information must provide the missing piece in the puzzle that preoccupies the questioner. This presupposes that he has already pieces in place, which the newly offered piece fits into. By my weaker criteria, the above reflection remains an objection, unless we can construct a question to which 'It is so in all accessible possible worlds' is a relevant and informative answer.

McCall and Vallentyne appear to be sensitive to this issue and their talk of the 'nomic structure of the world' may be an attempt to speak to it. Suppose all wood burns not only now, and always in the past, but also in every possible future. Is that not a good *meaning* for the assertion that it is causally necessary for wood to burn? Is it not intuitively plausible to think that what happens in all possible futures, *reveals* causal relations and causal necessity?

Yes, to the extent that these terms are clear, that is indeed plausible. But surely the plausibility derives from the fact that all our possible futures are 'rooted' in our settled past, and the past determines important features of the future to come? In other worlds, 'It is so in all possible futures' points to a causal explanation, which it does not itself provide. McCall and Vallentyne take away this very basis of the plausibility, it seems to me, if they deny that the possible futures' common character is there *because of* what the past is like. And that is what they do, if they explicate the past truth of modal statements, as amounting simply to the common character of the futures.

But perhaps objective chance is something that resides in the past, in addition to what actually happened? And it is *due to* this past objective chance that the futures are the way they are? Then it is not the way the alternative possible futures are, even considered collectively as a family, which accounts for the regularities in actual world history—it is rather that objective chance from which they too derive their character. In that case we need to be told what this objective chance is, and how it influences the events to come. The answer given, however, is that to say what the chance is, consists in delineating the proportions of possible futures in which various events come about. So we are led full circle, again, with the second part of the answer taking away the plausibility accrued to the first.

8. RELATION TO SCIENCE: PANDORA'S BOX

Whether possible-world models are truly explanatory has now been put in doubt. But suppose we agreed that they are. What riches and resources would this not open up for scientific explanation, in a new and higher key? I recall a news item in the *Fortean Times* a few years ago, of a physicist, intrigued with the many-worlds interpretation, who speculated that worlds might interact after all. Specifically, he reflected, this could explain why psychics tend to be wrong so often in their predictions (they 'see' into another possible world and their predictions are right about *that* world). But I think we need not go so far as to postulate interaction. It is a common view that entities and individuals existing in this world also exist in some others (and that we will continue to exist into most of our possible futures).[25] That is enough, I think, to create new and interesting sorts of explanations in physics, if we take it seriously.[26]

Imagine a world which is not entirely deterministic. Specifically, there is a type of crystal, let us call it the *Q*-crystal, which begins to glow *in vacuo* at temperature 100°C, with probability $\frac{1}{2}$. There are no conditions besides atmospheric pressure and temperature which have any effect at all relevant to this phenomenon. Yet some of these crystals always do so, some never, and some in apparently random fashion half the time. Let us add that these three types of behaviour appear among naturally occurring *Q*-crystals in proportions $\frac{1}{4}, \frac{1}{4}, \frac{1}{2}$. Thus if a randomly selected natural *Q*-crystal is subjected to the test, it has a probability $\frac{1}{2}$ of glowing. It is important that the three types of behaviour form a classification only applicable in retrospect and not predictable on the basis of any independent detectable characteristic.[27]

The example is meant to be schematic, and is about crystals and glowing only to aid our visual imagination. The important point is that the scientists in the imagined world regard this account as ultimate, with its probabilities irreducible.

In popular science articles, the writers there make such profound observations as that it is as if the crystals remember their past history. If they have glowed every time so far, they are more likely than not to glow again next time. But if they have ever glowed and also sometimes failed to glow under those conditions, they are exactly as if they flip a perfectly fair coin each time to determine their behaviour. All this despite the complete absence, according to the physicists there, of any mechanism or structure that could act like a memory bank. You may now have read some actual popular book or article about elementary particles in quantum mechanics, elaborated with references to telepathy, time-travel, Oriental mysticism, and the like—the imagined world is sure to have its counterpart in its bookstores.

But modal realism now opens the door for a truly explanatory account, going beyond what the reputable scientists have said so far. You see, this imagined world is physically possible relative to other worlds. Call our central imagined world *W*. It is physically possible relative to the completely deterministic *W*+, in which each *Q* crystal *must* glow under those conditions of pressure and temperature. World *W* is also physically possible relative to deterministic world *W*−, where each *Q* crystal *must fail* to glow under those conditions. If now a *Q* crystal in our world *W* exhibits the first (always glowing) behaviour pattern, *that is because* it is a crystal which also exists in *W*+. And if it never glows, that is because it is also one of the crystals of world *W*−. Crystals that exist only in *W* behave randomly.

The explanation is powerful, simple, and explains something which has no other sort of explanation—and which is just too regular to be plausible as a coincidence. Shall we not recommend this pattern of explanation to science? Indeed, in our very own world, in this century, physicists must deal conceptually with patterns of apparent pre-established harmony in indeterministic phenomena. Should they not be made to realize the extraordinary explanatory power of hypotheses about worlds from which there is access to ours? Is this the great gift analytic philosophy can finally bring to science?

9. THE PERILS OF A REIFIED MODEL

There is good to be found in, or behind, the possible-worlds story, and our critique should not end without some attempt to uncover it.

Under what conditions is it true, or false, to say that it is a law of nature that, for example, momentum is conserved in collisions? There are three possibilities. The *first* is that there is such a law (in our world), the *second* is that this law is violated—again, in this world—because in some collision the total momentum is not the same before and after. But if that were all, if actual violation was the only way the law claim could be false, then law would be mere regularity. That is not in accordance with the concept of a law of nature, so we must add the *third* possibility: the law is never violated in actuality, yet it is not a law of our actual world. The very idea that laws are more than mere regularities entails that a law claim could be false, although nature harbours no actual counterexample.

In the extreme, this reflection becomes: this world could proceed, in all its occurrences and throughout its history, in just the way it does although it has no laws at all. Pargetter calls a world without laws a *Hume world.* In the possible-worlds jargon this is equivalent to: a world relative to which all worlds are physically possible.

Imagine I have a trans-world travel machine, and take you on a tour. I can program it to take us to another world, provided our world bears Pargetter's access relation to it. I can also program it so as to take us to a world which has, or lacks, specific laws which I type into the console. To begin I type: *delete all laws*. I press the *Entry* button, and as far as you can see, there is no change in our situation at all. 'Don't be afraid,' I say. 'We are now in a Hume world, but it is one which has exactly the same world-history, past and future, as ours.' We quickly gather some experimental evidence for this claim—for example, I release a pencil and lo! it does fall to the floor. From a Hume world, we can go to any world at all, so I choose one where pigs fly. I press *Entry*, and we see a pig happily flying by outside. 'That is the only lawlike fact about this world,' I say, 'but happily it is very much like the Hume world we just came from, and you are not likely to have other surprises.' Just then you catch a glimpse of a white rabbit hurrying by with a pocket-watch, but I press *Entry* again and tell you we are arriving safely back in that Hume world. Now, after all these adventures, don't you wish I'd quickly get us back to our own safe home? I can do it, of course; but tell me, is there any reason for us to return?

That Hume world is just like ours, all the same things happen in it, but it has no laws. My world-travel machine has actually introduced one experimentally verifiable difference: from that Hume world it can take us in one step to where pigs fly, but from our original world it needs two steps. Unfortunately for the story, this machine is a fiction; it not only does not but cannot exist.

So, there is no *occurrent* difference between the two worlds at all. There is no observational or experimental evidence anyone could gather, that would have any bearing on whether we are in that Hume world, or in our supposed original. Equivalently: no such evidence could bear at all on the question whether we don't really live in a Hume world already.

This makes the possible-world story a fancy tale indeed. Its friends add the coda: but that there are laws which make the actual regularities necessary is *the best explanation* of why these regularities should be there at all. Hence we should believe that we are not in a Hume world, but in one with laws. No *experimentum crucis* is possible, that is true; but the conclusion that there are laws is deduced by inference to the best explanation.

We will have another chapter to discuss this point of epistemology. Now I want to present a different view: the idea, that science describes what is possible (and necessary, probable, and improbable), is correct but means something quite different. According to my view, this story of possible worlds reifies a model—that is, it attributes reality to what is only a means of representation.

In a discussion of science, possibility and necessity enter in two ways. Taking the point of view of an accepted theory, speaking *ex cathedra* so to say, we declare impossible whatever this theory denies. That is what we do if we say flat out that no perpetual-motion machine can exist, or that no body can be accelerated to the speed of light. Secondly, still in this same way, we may declare something possible if the theory allows it, under certain conditions, and we have no information to the contrary. This is the possibility of ignorance, *sub specie* the belief involved in theory acceptance. For our assertion signals that the facts in question are accommodated in some model our theory provides, and we have no evidence at odds with that model.

In addition, inside the theory's models, we may find alternative possibilities as well. This can happen only if the theory is not deterministic. For example: in its models, certain events happen with some probability, and other contrary events happen with some positive probability too. In that case, if someone who accepts the theory says that either type of event may happen, he is not merely signalling ignorance. For he admits that with only the belief involved in accepting this theory, plus maximal information about relevant conditions, he would still admit these alternatives as possible.

So far, our ontological baggage includes only persons, theories, models, and actual events. To say that of two events only one can occur actually, but both are possible, does *not* imply that both are real. The possible worlds accounts we have now inspected *add* that both are real. For they entail that on the correct understanding of 'possible' this is so.

We can see the main motive for this idea in our brief, intuitive discussion of how possibility talk arises. For the person speaking *sub specie* acceptance of a theory clearly believes that the way the world is, is correctly represented by some model of the theory. Now what is this 'correctly represented'? If we take it to mean not only that what is real is represented in the model, *but also that every element of the model represents some part or aspect of reality*, then we are at home with the realists. For then, if the model contains alternative possible courses of events, it follows that mutually incompatible events are all real. The more grandiloquent 'possible world' is not essentially less innocuous than 'possible event'.

This characterization of the relation of a good model to reality reifies the model. It equates model and reality, in effect. And suddenly, the goodness of good models goes way beyond any relation to what we can verify, even in principle, for it must include correctness about what did and also about what does not happen in actuality. No great esteem for the notion of verification is needed to see the problems this raises. Make your requirements of verifiability or testability as weak, as modest as you like, and they will still *never* allow you to say: we have checked that this model is correct about what did not happen, about the experiment we did not carry out, as well as about what did happen. There is undoubtedly more to be said about the relations of models to phenomena. Especially, the question of how we can become rationally confident that a model does fit the phenomena—which after all stretch beyond our ken in space and time—will need to be taken up. But with that postponed for now, the reification of alternative possibilities has not proved fruitful for us. It required the introduction of relations and functions defined on possible worlds, which (*a*) could not be identified, (*b*) did not appear to warrant prediction or rational expectation, (*c*) seemed not to explain, (*d*) gave authority to pseudo-science, and (*e*)—*c'est le bouquet*—brought us into an excess of metaphysics. However plausibly the story begins, the golden road to philosophy which possible-world ontologies promise, leads nowhere.

5.
Universals: Laws Grounded in Nature

ALL bodies near the earth fall if released because such bodies *must* fall—and this is because of the *law* of gravity. Naïve as this may sound, it presents the paradigm pattern of explanation by law. If we take it completely seriously, it signals that a law is not itself a necessity but accounts for necessity. The law itself, it seems, is a fact about our world, some feature of what *this* world is actually like, and what must happen is due to that fact or feature.

The accounts we shall now examine begin with a robust anti-nominalism: there are real properties and relations among things, to be distinguished from merely arbitrary classifications. There are also real properties of and relations among these properties. Such doctrines have been well known in Western philosophy since Plato and Aristotle, and were revived in the last hundred years by the Neo-Thomists and Bertrand Russell. Current *universals accounts of laws*, as I shall call them, are due to Dretske, Tooley, and Armstrong among others.[1] As we saw, the accounts we examined earlier also had to embrace some form of anti-nominalism, if not always so robust. Their adherents should therefore have little way to disagree with Dretske's conclusion that 'in such barren terrains [as the nominalists' ontology] there are not laws, nor is there anything that could be dressed up to look like a law'.

1. LAWS AS RELATIONS AMONG UNIVERSALS

Consider Boyle's law, expressed briefly in the equation $PV = kT$. We explain what it says as follows: for any ideal body of gas, the product of pressure and volume is proportional to its absolute temperature. But perhaps that is not what it says, but rather what it *implies* for the things (if any) to which it applies. Perhaps what it says is rather that a certain relation holds between the quantities P, V, T. These quantities are (determinable) properties, while the gases are their instances (a gas at temperature 200°K is an instance of the property *having a temperature of 200°K*). On the present suggestion, the law is not about the instances, at least not directly, but about the quantities themselves. It states a relation between them.

What is this relation? Dretske symbolizes it as $\rightarrow$, and reads it as 'yields'; Armstrong usually says 'necessitates', though sometimes he uses Plato's phrase 'brings along with it'. In our present example, we would then say: the joint property of having volume V and pressure $P \rightarrow$ the property of having temperature VP/k. To use another example: to say that it is a law of nature that diamonds have a refraction index of 2.419 *means* here that the property of being a diamond $\rightarrow$ (yields, necessitates) the property of having refraction index 2.419. 'Laws eschew reference to the things that have length, charge, capacity, internal energy, momentum, spin, and velocity in order to talk about these quantities themselves and to describe *their* relationship to each other.' (Dretske, 'Laws of Nature' 1977, p. 263.)

There is at first something very appealing about this view. Certainly we recognize the typical symbolic expression of science here, in equations relating quantities. But that symbolic expression comes from mathematical usage where it has explicit definition in the theory of functions:

$f = hg$ *means* that for any argument x, $f(x) = hg(x)$

which then allows us to abstract, and apply algebraic reasoning to functions directly.[2] And science appears to find use for this mathematical practice because it represents physical quantities by means of mathematical functions. Temperature in a room is not uniform; it may be represented by means of a function that maps the set of points (spatial locations) in the room, into the Kelvin scale, or into another temperature scale. The universals account of laws begins by insisting that the representation of a physical quantity by means of such a function, is really *only* the representation of an accidental feature of that quantity—namely, of how it is instantiated in the world of things. The appeal is therefore at once undermined by the story itself.

Nevertheless, the story is intelligible: it says that the assertion of general law concerning things is really a singular statement of relation between specific properties. We have learned a good deal in the two preceding chapters, so we can immediately raise two problems.[3]

identification problem: which relation between universals is the relation → (necessitation)?

inference problem: what information does the statement that one property necessitates another give us about what happens and what things are like?

The two problems are obviously related: the relation identified as necessitation must be such as to warrant whatever we need to be able to infer from laws. It is equally clear what the paradigm inference must be: if *A* and *B* are properties, then *A necessitates B* must entail that any instance of *A* is an instance of *B*—or perhaps even that any *A* must necessarily be a *B*.

We shall look carefully at various attempts to solve or circumvent these problems. But we can already see certain perils that must beset any such attempts.

The first peril concerns 'necessary'. If the account is to meet the necessity criterion, then it must entail that, if it is a law that all *A* are *B*, then it is necessary for any *A* to be a *B*. In order not to become merely a necessitarian account with universals added in, the necessity cannot have any very basic status in the account. Nor could it be merely verbal or logical necessity: no matter what else is true about, for example, diamond-hood and refraction index, it cannot be a *logical* truth that the Koh-i-noor diamond has refraction index 2·419. Since the law must be the ground of the necessity, *the general* line has to be something like: if an *A* is a *B*, there are two further possibilities, namely that *A* necessitates *B* or that it does not. If, and only if, the former is the case, then the *A* is necessarily a *B*—that is what 'necessarily' means here. But note well: once necessity is given this derivative status, it can be of no use in the development of the universals account itself. For example, if we want to ask whether *A* necessarily necessitates *B*, we had better realize that this question will now have to mean *either* whether '*A* necessitates *B*' is a logical truth, or *else* whether some higher property that *A* has necessitates the property of necessitating *B*. Uncritical use of modal language, and its convenience, must be strictly eschewed.

The second peril concerns the obvious idea to play the two problems off, one against another, by a sort of bootstrap operation. In the preceding paragraph we saw that we had better not ask of *A necessitates B* that it entails more than that any *A is* a *B*. But this role *necessitate* must be able to play, and if it can, then the inference problem is solved. So why not solve the identification problem by postulating that there exists a relation among universals with certain features which logically establish that it can play this crucial role? Call this relation 'necessitation', or if there may be more than one, call them 'necessitation relations' or 'nomological relations'. Then both problems will have been solved.

The difficulties which beset this ploy come as a dilemma. Define the relation of *extensional inclusion*: *A* is extensionally included in *B* exactly if all instances of *A* are instances of *B*. Then *A is extensionally included in B* entails that any *A* is a *B*. But if this qualifies as a necessitation relationship, then all ordinary universal regularities become matters of law. To avoid this trivialization, the envisaged postulate must include among the identifying features something more than is needed to solve the inference problem. Actually, 'something more' is not quite apt—for if any relation which consists of extensional inclusion *and* something else also qualifies as nomological, the account is still trivialized. So the identifying features must be, not something *more* but something *else*, other than extensional inclusion. Now the second horn of the dilemma looms: how could features so distinctly different from extensional inclusion still solve the inference problem?

This is not just an open question, for we have considerable reason to think that they cannot. The law as here conceived is a singular statement about universals *A* and *B*. The conclusion to be drawn from it is about another sort of things, the particulars which are instances of *A* and *B*. True, the instances are, by being instances, intimately related to the universals. This is not enough, however, to make the inference look valid to us. We are intimately related to our parents, but that does not make us regard the inference

(1) *X* knows *Y*
therefore
(2) All children of *X* are children of *Y*

as a valid inference. There are ways to turn this argument into a valid one, but they are either too trivial to be of comfort:

(1) *X* has the same children as *Y*
therefore
(2)

or else require a special extra premiss that makes the connection:

(1) *X* has carnal knowledge only of *Y*
(1.5) All a person's children are born of someone of whom he or she has carnal knowledge
therefore
(2)

What we need from a universals account at this point, is such an extra premiss to make the connection. It would be very disappointing if we find merely another postulate that asserts the connection to be there, and does not explain it.

There is a certain simplicity to the minimum postulate needed: if *A necessitates B then any A is* (*necessarily*) *a B*. But its intuitive flavour may just result from the pleasing choice of words. As David Lewis put it: to call the relation 'necessitation' no more guarantees the inference than being called 'Armstrong' guarantees having mighty biceps.

We shall see in a later section that Armstrong's book on laws of nature contains a sustained attempt to solve the inference problem. But earlier versions of the universals account of laws were sanguine. Tooley writes 'Given the relationship that exists between universals and particulars exemplifying them, any property of a universal, or relation among universals . . . will be reflected in corresponding facts involving the particulars exemplifying the universal.'[4] That this cannot be a matter of logic, is clear from the parallel example of parents and children born to them. Perhaps relations among parents are reflected in *corresponding* relations between their children, but it will take more than logic to find the correct correspondence function! With not much less sang-froid, but appreciating the problem, Armstrong himself writes: 'The inexplicability of necessitation just has to be accepted. Necessitation, the way that one Form (universal) brings another along with it as Plato puts it in the *Phaedo* (104D–105), is a primitive, or near primitive, which we are forced to postulate.'[5] But what exactly is the postulate? And how will it fare with our dilemma? Dretske writes:

I have no proof for the validity. . . . The best I can do is offer an analogy. Consider the complex set of legal relationships defining the authority, responsibilities, and powers of the three branches of government in the United States. . . . The legal code lays down a set of relationships between the various *offices* of government, and this set of relationships . . . impose legal constraints on the individuals who occupy these offices—constraints that we express with such modal terms as 'cannot' and 'must'. . . . Natural laws may be thought of as a set of relationships that exist between the various 'offices' that objects sometimes occupy.[6]

Yes, that is the analogy we know. But in the analogy we also know that what gives the legal code its force is the continued agreement to enforce it. This passage highlights the inference problem by reminding us of the similar problem of values in a world of facts.[7] Given that chastity is a value, how does it follow that we should value it? The noun and verb had better be more intimately connected than by spelling and sound, if the answer is to satisfy us. And the reminder is not felicitous. We know the sad fortunes of attempts to solve the problem of facts and values, by reifying values as abstract entities!

In this section I have presented the basic idea of the universals account of laws, and its two basic problems. It would be a great injustice not to examine the careful attempts by Tooley and Armstrong which implicitly react to what I call the identity and inference problems, and their attempts to extend the account into a satisfactory view of both deterministic and probabilistic laws. So we shall. Along the way, we shall encounter still further interesting problems.

2. THE LAWGIVERS' REGRESS

I painted a decidedly bleak picture, when I discussed the identification and inference problems for the universals account of laws. Here I shall describe a proposal which solves the identification problem—*sub specie* certain postulates of course—but in doing so, I shall argue, it renders the inference problem insoluble. The proposal is essentially due to Tooley (I shall ignore refinements of his theory, but, I believe, without loss in the present context). Nevertheless, I can best introduce it by continuing the quote from Dretske which elaborated the constitutional law analogy:

> Natural laws may be thought of as a set of relationships that exists between the various 'offices' that objects sometimes occupy. Once an object occupies such an office, its activities are constrained by the set of relations connecting that office to other offices and agencies; it *must* do some things, and it *cannot* do other things. In both the legal and the natural context the modality at level n is generated by the set of relationships, existing between the entities at level $n + 1$. Without this web of higher order relationships there is nothing to support the attribution of constraints to the entities at a lower level.[8]

What hierarchy of levels does he have in mind? Level 1, at the bottom, contains particulars (things and persons); level 2 contains the offices which they may occupy. That entities at level 1 must do such and such is determined by the fact that their offices are related so and so. But that their offices are so related is determined by something still higher. That is the legal code or Constitution, presumably, on the government side of the analogy; what is it on the nature side? And why have the levels stopped at 3—is it because on the government side we find a single higher entity? Why has Dretske fallen into terminology that suggests higher and higher levels, a useless generalization if the levels stop at the third storey?

Fruitless, surely, to push this analogy too hard; but the discussion which follows makes Dretske's surprising choice of words oddly revealing. For a regress does lurk only just around the corner, if we introduce the general thesis that modal statements about one sort of entity, must have their ground in relations among another sort.

Tooley begins his account with a hierarchical description of the world of universals and particulars.[9] Particulars are of order *zero*; properties of, and relations among particulars, are of order *one*. In general, a universal (property or relation) is of order $(k + 1)$ exactly if k is the highest order of entities to which it pertains. So for example, the property of *being malleable* is of order 1, and the property of *being a property of gold* is of order 2, but any relation between gold objects and the property of *being malleable* is of order 2 as well.[10] Let us call the latter *impurely* of order 2, saying that a universal is *purely* of order $k + 1$ if it pertains solely to entities of order k. Tooley moreover calls a universal *irreducibly* of order k if it cannot be analysed in terms of universals of lesser order. Finally, a relation R is called contingent exactly if there are entities which (*a*) can bear *R* to each other, but (*b*) can also fail to bear *R* to each other.

We perceive here at once what I called the first peril. I used 'pertain' rather vaguely. Tooley does not mean that a relation is, for example, of order *one* exactly if all its instances are particulars. For any relation, of any order, may fail to have instances at all. So we should understand this as: a relation of order *one* is a relation which can have particulars, but not universals, among its instances. This 'can' and 'cannot' may be the same as those which appear in the definition of contingency. But what are they? It is not at all clear that they could be either purely logical or grounded in higher order universals.

Be that as it may, we can now proceed to the definition of what I called necessitation relations.

> *R* is a *necessitation relation* exactly if it is a contingent relation, irreducibly and purely of some order >2, and the statement that *R* holds between certain universals of order 1 entails that a certain other (corresponding) relation holds between particular instances of those universals.

When that corresponding relation among particulars is exactly that all instances of the first universal are instances of the second, let us call *R* a *proper necessitation relationship*.[11]

See now how elegantly the introduced qualifications pre-empt the difficulties I raised for this sort of proposal before. This identification of necessitation relations does not catch extensional inclusion in its net, even if it is a real relation—for it is not irreducible. It is (analysable as) necessarily equivalent to a statement—of form *All A are B*—in which no order *two* universals appear at all; hence reducible in Tooley's sense. Similarly for the example: *A* bears *R* to *B* exactly if there have been or will be black swans and all instances of *A* are instances of *B*. The statement to the right of 'exactly if' is a necessary equivalent in which no universals of order *two* are even mentioned.

This fits in exactly with our story of the second peril, the dilemma relating the problems of identification and inference. Tooley has escaped the first horn by identifying necessitation relations (or as he calls them, nomological relations) by means of features that logically exclude from them any relation which is 'merely' extensional inclusion, whether alone or in conjunction with something else. But then—here comes the second horn—how can this identification be involved in, or even co-exist with a solution to the inference problem?

Perhaps this question puzzles you, for does not the identification of *R* as a proper necessitation relation involve as part of the definition, entailment of extensional inclusion? Yes; but that just forces the dilemma's second horn into the form: how could a relation *R* with the initially noted features (contingent, irreducibly and purely of order *two*) also entail extensional inclusion?

In my praise of how Tooley's account avoids the first horn of the dilemma, I assumed that extensional inclusion does not fit the definition of necessitation relationship, on the basis of a certain gloss I gave to 'analyse'. My prescription was that extensional inclusion is not irreducibly of order *two* because *A is extensionally included in B* is necessarily equivalent to *All instances of A are instances of B*, and no order-*two* universals are involved in the latter proposition. Nor would the situation get better if we threw in a conjunct or disjunct that does involve an order-*two* universal, since irreducibility requires that any correct analysis should be *entirely* in terms of order-*two* universals.

But in that case, the necessary equivalent exhibited by any correct analysis will give no logical clue to what is true at orders below *two*. The entailment asked for cannot be logical entailment, just as a fact that is purely about the parents cannot logically imply anything about the children. (Note that it would not be purely about the parents if it described them as parents, i.e. as people having children!) So the entailment cannot be a matter of logic. What is it then—necessary con-commitance? of a non-logical sort?

Now we have arrived again, after all, at the door behind which lurks the first peril. Unless we are to return to a necessitarian account—with universals as frills—we have besides logical necessity only the necessity that derives by definition from relations of a higher order. At the present point that would expand the word 'entail' in our definition of necessitation as follows:

> and the statement that *A* bears *R* to *B* entails that all *A* are *B*, in the following sense: there exists a(n impure) relation *R′* of order *three* which *R* bears to *A* and *B* and is such that, if any *X* bears *R* to *Y* and *R* bears *R′* to *X* and *Y* then it follows that all *A* are *B*,

for the case of proper necessitation. But note the word 'follows' in the last clause—it faces the same challenge as the earlier 'entails'. We will have the same impossibility of taking it to be the 'follows' of pure logic, and will be driven to a universal of order *four*. This is a typical 'Third Man' regress.

Don't give up! Not every regress is vicious. Couldn't we just accept that each law must be backed by an infinite hierarchy of *trans-order laws*, which are factual statements about impure higher order relations?

I think we could, in all consistency, if not in all good conscience. But we should not underestimate the regress. It will not be merely infinite, but transfinite; for the infinite hierarchy envisaged above still leaves one, as a whole, with the question of why *that* should entail that if A bears R to B then every A is a B. The hopes of finessing modal statements at the level of particulars, which Dretske held out so temptingly, are not dashed—but they do recede ever farther into transfinite distances, as we pursue them.

Another hope surely is dashed. *Distinguo*: not all regresses are vicious, in that the existence of an infinite regress does not always reduce the theory to absurdity. Regresses in explanation, however, are not virtuous: they leave us with something which may be consistent but is not an explanation. The explanatory pattern 'This is so, because it must be so, because it is a law that it is so' is destroyed if we say that it is only a sketch, the second 'because' needing the additional phrase '*and* if it is a law then it must be so, because it is a trans-order law that if something is a law then it is so *and* if it is a trans-order law that . . .'. A hierarchy need have no top, but an explanation without a bottom, an ungrounded explanation, is no explanation at all. The regress had to be stopped if there was to be an explanation of nomological statements in non-nomological terms. But it cannot be stopped.

3. DOES ARMSTRONG AVOID THE REGRESS?

David Armstrong's account of laws of nature is based on a theory of universals, previously developed in his *Universals and Scientific Realism*. The main question I want to address in this section is whether Armstrong's account avoids the debilitating regress we found above.[12] I will take for granted, therefore, that the relation(s) among universals which he introduces are independently identifiable—leaving the identification problem aside, to focus on the inference problem.[13]

To begin, let me repeat and add some terminology concerning universals. Armstrong's world has in it particulars and universals. The particulars come in two sorts, objects and states of affairs. A state of affairs will always involve some universal, whether monadic (property) or dyadic, triadic, . . . (relation). When I say particulars, I mean entities which are not universals, but which instantiate universals. It is countenanced that universals may instantiate other universals, and that there is a hierarchy of instantiation. So it is natural, as Armstrong does, to extend the terminology: call an *n*th order universal one which has instances only of $(n - 1)$th order, call it also an $(n + 1)$th order particular. Now call the particulars which are not universals, first-order particulars. When I say 'particular' without qualification I shall mean these first-order particulars. If *a*, *b* are particulars and *R* a relation, and *a* bears *R* to *b*, then there is a state of affairs, *a*'s bearing *R* to *b*—let us designate this state of affairs as *Rab*—in which these three are 'joined' or 'involved'. If *a* does not bear *R* to *b*, there is no such state of affairs—a state of affairs which does not obtain is not real.

To bring to light just a little of Armstrong's theory, let us look at how he handles the problem known as Bradley's Regress. In the state of affairs described above, the terms *R*, *a*, *b* are all names, one of a universal and two of particulars. These three entities are 'joined' in the state of affairs; but how? Is it that there is a certain three term relation *R′* which *R* bears to *a* and *b*? More generally, how are universals related to their instances? If a regress were accepted, it might or might not be vicious. After all, suppose *R* is the set of real numbers, and *X* one of its subsets. Then in set theory we could say $b \in X$ exactly if $<b, X> \in \{<y, z>: y \in R \,\&Z \subseteq R \,\&\, y \in Z\}$. Calling the latter set, a subset of $R \times P(R)$, by the name X', we continue with the equivalent $(<b, X>, X') \in X''$ for the membership relation restricted to $(R \times P(R)) \times P(R \times P(R))$ and so forth *ad infinitum*. But as I remarked before, that a regress does not lead to a contradiction, does not mean it is a satisfactory thing to have around. Armstrong adopts a view of the Aristotelian (moderate) realism type, in order to avoid Bradley's regress. If universals were substances (i.e. entities capable of independent existence), he says, the regress would have to be accepted. But they are not substances, though real: they are abstractions from states of affairs. One consequence is that there are no uninstantiated universals. We should not regard states of affairs as being constructed out of universals and particulars, which need some sort of ontological glue to hold them together.

We cannot embark on the general theory of universals here; we need to consider only those details crucial to Armstrong's account of laws. The important point here, for our purposes, was that his solution to the puzzle entails that there can be no uninstantiated universals. In this account the symbol '*N*' is used, prima facie in several roles; I shall use subscripts to indicate prima-facie differences and then state Armstrong's identifications. There is first of all the relation N_1 which states of affairs can bear to each other, as in

(1) N_1(*a*'s being *F*, *a*'s being *G*)

This N_1 is the relation of necessitation between states of affairs: formula (1) is a sentence which is true if and only if *a*'s being *F* necessitates *a*'s being *G*. But for this to be true, both related entities must be real; therefore (1) entails that these two states of affairs are real, hence

(2) *a* is *F* and *G*

However, both states of affairs could be real, while (1) is false, so N_1 is *not* an abstraction from the states of affairs of sort (2)—for then it would be the conjunctive universal (*F* and *G*).

This relation N_1 has sub-relations, in the sense that the property *being coloured* has sub-properties (determinants) *being red, being blue*. One sort is the relation of *necessitation in virtue of the relation*(*s*) *between F and G*, referred to as N_1 (*F*, *G*):

(3) N_1(*F*, *G*)(*a*'s being *F a*'s being *G*)

This is a particular case of (1), so (3) entails (1) and hence also (2), but again the converse entailments do not hold. In (3) we also have a sort of universalizability. Note that (3) says (*a*) that the one state of affairs necessitates the other, and (b) that this is in virtue solely of the relation between *F* and *G*. Hence what *a* is, does not matter. Of course we cannot at once generalize (3) to all objects, for (3) is not conditional; it entails that *a* is both *F* and *G*. Thus we should say that what *a* is, does not matter beyond the entailed fact that it is an instance of those universals whose relation is at issue. We conclude therefore that if (3) is true, then for any object *b* whatever, it is also true that

(4) if *b* is (*F* and *G*) then *N*(*F*, *G*)(*b*'s being *F*, *b*'s being *G*)

Now what is the relation between *F* and *G* by virtue of which *a*'s being *F* necessitates *a*'s being *G*? It is the relation of necessitation between universals, Armstrong's version of Dretske's → and Tooley's nomic necessitations. Let us call this N_2:

(5) $N_2(F, G)$

Thus if (3) is true then (4) and (5) must be true, and indeed, (3) must be true *because* (5) is true. This N_2, the target of course of the remark by David Lewis which I reported earlier, Armstrong takes as not calling for further illumination:

> the inexplicability of necessitation just has to be accepted. Necessitation, the way that one Form (universal) brings another along with it as Plato puts it in the *Phaedo* (104D–105) is a primitive, or near primitive, which we are forced to postulate (*What is a Law of Nature?*, 92).

'Forced to postulate' because the way the actual particulars (including states of affairs) are in our world cannot determine whether (3) is true.

We now come to the point where Armstrong may escape the threat of a regress of trans-order laws. To be specific: Armstrong proposes a surprising identification of a relation with a state of affairs. If this will lead to a *logical* inference from the existence of the universal *N*(*F*, *G*) to the conclusion that any *F* is a *G*, then the inference problem will have been solved without falling into the lawgiver's regress. But could that really work?

It appears that Armstrong thinks it does, for he writes that he hopes now to have arrived at a 'reasonably perspicuous view of the entailment' of all *F*'s being *G*'s by the statement that *F* necessitates *G*. He elaborates on this as follows:

> It is then clear that *if such a relation holds between the universals*, then it is automatic that *each* particular *F* determines that it is a *G*. *That is just the instantiation of the universal* (*N*(*F*,*G*)) *in particular cases*. The [premiss of the inference] represents the law, a state of affairs, which is simultaneously a relation. The [conclusion] represents the uniformity automatically resulting from the instantiation in its particulars. (*What is a Law of Nature?*, 97; italics in original)

But I shall try to show that, on careful analysis, the inferential gap appears only to have been wished away.

Armstrong proposes the postulate—surprising, but in accord with his theory of universals—that the universal $N_1(F, G)$—a relation between states of affairs—is identical with the state of affairs $N_2(F, G)$. In that case we can drop the subscripts and say that the state of affairs $N(F, G)(Fa, Ga)$ instantiates the state of affairs $N(F, G)$—in the way that the state of affairs *Rab* instantiates *R*, generally. The question whether $N(F, G)$ entails (If *Fb* then *Gb*), can then be approached by logical means. For suppose $N(F, G)$ is real, i.e. *F* bears *N* to *G*. Then it must have at least one instance; let it be described by (3) above. But in that case (4) is also true. We conclude therefore that if $N(F, G)$ is real, then for any object *b*, if *b* is both *F* and *G*, then $N(F, G)$ (*b*'s being *F*, *b*'s being *G*).

But we can go no further. What we have established is this: if there is a law $N(F, G)$, then *all* conjunctions of *F* and *G*, in any subject, will be because of this law. There will be no *F*'s which are only accidentally *G*. That shows us an interesting and undoubtedly welcome consequence. Although Armstrong does not give this argument, I must assume that he introduced the curious identification of N_1 and N_2 to reach some such benefit. However, this benefit is not great enough to get him out of the difficulty at issue. For what *cannot be deduced*, from the universal quantification of (4), is that all *F*'s are *G*'s. Any assertion to that effect must be made independently. Nothing less than a bare postulate will do, for there is no logical connection between relations among universals and relations among their instances.

Proofs and illustrations

For me, the above argument establishes that the inference problem remains unsolved. But let us look a little further into what might or might not be possible in some such theory of universals, if not Armstrong's own.

Armstrong reports that logicians are inclined to protest at this identification. Let us see why. In the preceding paragraphs I have more or less followed Armstrong's practice of using the same notation to stand for a *sentence* and also for the *noun* that denotes the state of affairs which is real if and only if the sentence is true. If we identify N_1 and N_2 as N, we have a *four-fold* ambiguity; '$N(F, G)$' can stand for

(*a*) the sentence '*F* necessitates *G*'
(*b*) the noun '*F*'s necessitating *G*'
(*c*) the predicate 'necessitates by virtue of (relation(s) between) F and G'
(*d*) the noun 'the relation of necessitating by virtue of (relation(s) between) *F* and *G*'

The identification is meant to give sense to the idea that the state of affairs *a*'s being *F* and *G* is (near enough) an instance of the law that *F* necessitates *G*. So the only identification needed is this:

the nouns (*b*) and (*d*) have the same referent

which is (for all I know about universals and states of affairs) consistent.

Worrying, however, is the occurrence of 'necessitates' in sentences (*a*) and (*c*)—and its further occurrence, in the guise of N_1, in sentence (*1*) above. Does this verb have the same meaning in all these cases? It would then stand for a single universal which is a relation (i) between universals like *F* and *G*, and (ii) between states of affairs like *Fa* and *Ga*. The former are first-order universals, hence second-order particulars, while the latter are first-order particulars. So what is *N*? Either it is not a third-order particular or the order-hierarchy is not simple. The alternatives posed here are not at all attractive. Suppose for instance Armstrong says *N* stands for a single universal, and that it need not be thought of as a disjunctive one, because the hierarchy is cumulative. Then he has to make sense of such assertions as that *N*(*F*, *Ga*) or that *N*(*Ga*, *F*), i.e. that his relation holds between a universal and a particular state of affairs. Perhaps he can say that such assertions are always false, but this would still presuppose that they make sense.

Armstrong has shown great determination not to multiply or complicate the diversity of his world by allowing disjunctive universals. If he resists the preceding line of thought however, he must say that 'necessitates' is ambiguous: it stands now for one relation and then for another (one second-order, one first-order) and not for a disjunction of the two. But that would destroy the identification.

Suppose on the other hand that Armstrong does not resist a new complication in this case, but says that *N* is a universal, which is however a *disjunction* of a first-order relation and a second-order relation. (This could be expressed without using the word 'disjunction', perhaps by calling N 'order transcendent' or whatever, but that would not really alter the case.) Then he has his identification. But the glory has gone out of it, since the relevant states of affairs are now instances of the relevant universal, just because the law has been reconstructed by swelling it so as to encompass those states of affairs. The real story, obscured by the notation, would still be this: $N_1(F, G)(Fa, Fb)$ holds exactly if $N_1(Fa, Fb)$ and $N_2(F, G)$ both hold. Let us now define $N(F, G)$ to hold exactly if $N_2(F, G)$ holds and also $N_1(F, G)(Fa, Fb)$ for all entities a such that a is F and G. Now drop the notation N_1, N_2 from the language—don't have names for N restricted to particulars, nor for N restricted to universals. Now you no longer have the language to raise the embarrassing question whether it could be that F necessitates G while some particular F is not a G. If you'd had the language, we would have had to answer either *Yes*, and the law does not imply the corresponding regularity; or *No*, there is a trans-order law that forbids it. And the regress would begin.

4. ARMSTRONG ON PROBABILISTIC LAWS

So far we have only discussed laws corresponding to universal regularities. How could universals and their relations account for irreducible probabilities?

As before, let us take the law of radioactive decay as example. This is the simplest; it involves only a single parameter: the atom still remains stable, or has decayed. We are not so ambitious here as to tackle the complex web of statistical correlations which gives quantum theory its truly non-classical air. Radioactive decay exhibits a form of indeterminism which is conceptually no different from Lucretius' unpredictable swerve. In addition, the decay law appeared first as deterministic law, of the rate at which such a substance as radium diminishes with time. After $1600k$ years, the remainder is only $(\frac{1}{2})^k$ of the original amount. Unlike Achilles' race with the tortoise, however, the process comes to an end, because each sample consists of a finite number of atoms.

But this reflection makes the original law inaccurate, for strictly speaking, there cannot be half of an odd number of atoms. What can the true law be for a single atom? The new, probabilistic law of radioactive decay is that each single atom has a probability (depending on a decay constant A), namely

(1) e^{-At}

of remaining stable for an interval of length t (regardless of the time at which you first encounter it in stable condition). The original half-life law is thus regularly violated by small numbers of atoms, and it can be violated also for substantial samples. The new law has as corollary, however, that for a large number of atoms, the probability of having more or less than $\frac{1}{2}$ left after 1600 years is very small. This small probability is the same sort of probability as (1), namely physical probability. Yet this corollary, if properly understood, should make it rationally incumbent upon us to attach only negligible personal likelihood to such violations.

The task of an account of laws is now two-fold: (*a*) to give an official meaning to such a probabilistic law, and to the objective probability involved in it, and (*b*) to do so in a way that warrants the guiding role of the objective probability for subjective expectation.

The second task was already the focus of several sections of the preceding chapter. The arguments there were general enough to plague any conception of objective probability which is logically unconnected with frequency or opinion. Such a conception we will also find here in the universals account of laws. But I do not propose the boring work of transposing those arguments, *mutatis mutandis*. I believe them to be devastating to any metaphysical reification of statistical models, and am content to leave the issue here.

The first task, on the other hand, is the focus of Armstrong's, and more recently Tooley's interest. They wish to bring probabilistic laws into the fold of their universals account of law. I will use the law of radioactive decay as a touchstone, at a certain point, for their success.

Armstrong begins[14] by asking us to consider an irreducibly probabilistic law to the effect that there is a probability P of an F being a G. One imagines that G may include something like: remaining stable for at least a year, or else, decaying into radon within a year. Adapting his earlier notation, he writes

(2) $((Pr{:}P)(F, G))$(*a*'s being F, *a*'s being G)

Read it, to being anyway, as *There is a probability P, in virtue of F and G, of an individual F being a G*. As with *N*, (*Pr*:*P*)(*F*, *G*) is a universal, a relation, which may hold between states of affairs, but of course only real ones. Suppose now that *a* is *F* but not *G*. Then (2) is not true, for in that case there is no such state of affairs as *a*'s being *G*. So (2), properly generalized, does not say something true about any *F*, but only about those which are both *F* and *G*. That is not what the original law-statement looked like. Nor does Armstrong wish to ameliorate this by having negative universals, or negative states of affairs, or propensities (i.e. properties like *having a chance P of becoming G* which an *F* could have whether or not it became *G*). Thus *a probabilistic law is a universal which is instantiated only in those cases in which the probability is 'realized'*.

Suppose there is such a law; what consequences does this have for the world? The real statistical distribution should show a 'good fit' to the theoretical distribution described in the law. The mean decay time of actual radium should show a good fit to law (1)—on its new, probabilistic interpretation—for example, also on Armstrong's construal. But I don't see why it should. We can divide the observed radium atoms into those which do and do not decay within one year. Those which do decay are such that their being radium atoms in a stable state bears (Pr:e^{-A}) (radium, decay within one year) to their decaying within one year. The other ones have no connection with that universal at all. Now how should one deduce anything about the proportions of these two classes or even about the probabilities of different proportions?

Open questions are not satisfactory stopping points, so let us leave the connection with actual frequency aside, and concentrate on probability alone. The reality of (*Pr*:*P*)(*F*, *G*) has one obvious consequence: a universal cannot be real without being instantiated, so there is at least one *F* which is *G*. Thus we have, for example:

(3) If it is a law that there is a probability of $\frac{3}{4}$ of an individual *F* being a *G*, and there is only one *F* then it is definitely a *G*.

This is worrying, but a one-*F* universe is perhaps so unusual that it can be ignored. However, suppose there are two *F*s, call them *a* and *b*. If we ignore the Principle of Instantiation, and assume this is the only relevant law that is real, we calculate: the probability that both are *G* equals $\frac{9}{16}$, the probability that *a* alone (or *b* alone) is G equals $\frac{3}{16}$, and that neither is G equals $\frac{1}{16}$. But the Principle rules out the last case. How does this affect the probabilities? We must give *zero* to the last case; the new probabilities of the other cases must be 'like the old' but add up to *one* again. This adjustment is called *conditionalization* (see Table 5.1). The new probabilities x, y, z must stand in the same proportions 9:3:3, and must add up to 1. The probability that a is G, for example, is now calculated by adding the probabilities of the first and second case: $(9/15) + (3/15) = (12/15) = 4/5$. In this case, we deduce, after a few steps:

(4) Given the law that there is a probability of $\frac{3}{4}$ of an individual F being a G, and a, b are the two only Fs, then the probability that a is a G equals $\frac{4}{5}$, and the probability that a is a G given that b is a G, is a bit less (namely, $\frac{3}{4}$ again).

So the trouble is not confined to a one-F universe; it is there as long as there is a finite number of F's. If the law says probability P, and there are n F's, then the probability that a given one will be G equals P divided by $(1 - (1 - P)^n)$. For very large n, this is indeed close to P, but the difference would show up in sufficiently sensitive experiments. Should we recommend this consequence to physicists, if they have ever to explain apparent systematic deviations from a probabilistic law?

TABLE 5.1. *Effect of the Principle of Instantiation*

G	not G		Probabilities	
a, b		$\frac{9}{16}$	x	$\frac{9}{15}$
a	b	$\frac{3}{16}$	y	$\frac{3}{15}$
b	a	$\frac{3}{16}$	z	$\frac{3}{15}$
	a, b	$\frac{1}{16}$	0	0

The second part of (4) is also striking. I made the calculation on the assumption that the objects a and b did not influence each other's being G or not G. This assumption stands—they could exist in different galaxies, say—but the difference between the probability of b being G *tout court*, and its probability given that a is G amounts to a statistical correlation. Now today's physics countenances such 'uncaused' correlations, though not ones arising simply from numbers present, so to say. A correlation without preceding interaction to account for it is always prima facie mysterious, and I note it as an interesting feature of Armstrong's account.

The first problem was generated by the fact that in Armstrong's theory, a universal cannot be real without having at least one instance. That same fact spells trouble independently for a law such as that of radioactive decay, which gives a distinct probability for each time-interval. Because Armstrong does not have negative universals, we should expect that only one of *remaining stable for interval t* and *decaying into radon within interval t* is a real universal. We had better consider both in turn.

If the *former*, we note that e^{-At} is positive for each t, no matter how small. So for each t there must be an instance: an atom that remains stable for interval t, starting from *now*. This means that either there is an atom which never decays, or else that there is an infinite series of atoms which remain stable respectively for at least one year, at least two years, at least three years, . . ., and so on. On the other hand, if the *latter*, there must be for each time t, an atom which decays before t (measuring from *now*). This means either that there is an atom which decays right now, or else that there are an infinite series of atoms which decay respectively before a year, a month, a day, an hour, . . ., and so on, has elapsed.

If, as we think, the amount of matter in the universe is finite, then two of these possibilities are ruled out at once. But the reasoning about *now* was quite general, and applies equally to every time. For there to be, at each instant, a radium atom which decays just then, would require an infinity of these atoms as well.[15] So only one possibility remains: there is at least one radium atom which will never decay.[16]

This is a striking empirical deduction. It shows that Armstrong's reconstruction of probabilistic laws is not mere word-play, but has empirical consequences, which were not present in the law as heretofore understood. I do not say verifiable—it is no use to apply for a grant, to find that sempiternal atom—but concrete and strikingly general. For the argument would apply to any law which delineates objective probabilities as a positive function of time.

These two problems resulted from the instantiation requirement; the third will be quite independent of that. To explain it, we must first look a little further into Armstrong's account. Armstrong proposes that we identify (*Pr*:*P*) as *a* subrelation of *N*, rewriting it as (*N*:*P*). How should this be interpreted? Armstrong has recently emphasized this reading:

> I hold that a probabilistic law gives *the probability of a necessitation in the particular case*. Necessitation is just the same old relation found in any actual case of (token) cause bringing about a (token) effect, whether governed by a deterministic law, a probabilistic law, or no law at all.[17]

But if the difficulties below prove too onerous, one should not lose sight of any possibility of retrenchment into another possible interpretation, such as that *N*—like temperature or propensity—has degrees, of which *P* provides a measure.

The main difference I see between the two interpretations is perhaps one of suggestion or connotation only. In an indeterministic universe, some individual events occur for no (sufficient) reason at all. If the law (*N*:*P*)(*F*, *G*) is real, and *b* is both *F* and *G*, there could prima facie be one of two cases. The first is that *a*'s being *F* necessitated ('brought along with it') its being *G*, in virtue of *F* and *G* and the relation (*N*:*P*) between them. The second is that *b*'s being *F* is here conjoined with its being *G* as well, but accidentally ('by pure chance'). Now the first reading ('probability of necessitation') suggests that this prima-facie division may have real examples on both sides of the divide. The second reading suggests that on the contrary, if *F* bears (*N*:*P*) to *G*, and *b* is both *F* and *G*, then *b*'s being *F* cannot just be conjoined with its being *G* but must have necessitated (to degree *P*) its being *G*. On the first reading there can be cases of something's being *F* and *G* which are not instances of the law, on the second not. But I think either reading could be strained so as to avoid either suggestion. We should therefore consider both possibilities 'in the abstract'.

Let us begin[18] with the first, 'official' reading, and suppose that the law (*N*:*P*)(*F*, *G*) is real with $P = \frac{3}{4}$, then there are three sorts of *F*: those which are not *G*, those whose being *F* necessitated their being *G*, and those which are *G* by pure chance. What is the probability that a given *F* is of the second sort? Well, if *P* is the probability of necessitation, then the correct answer should be *P*. What is the probability that a given *F* is of the third sort? I do not know, but by hypothesis it is not negligible. So the overall probability that a given *F* is a *G*, is non-negligibly greater than $\frac{3}{4}$. Thus again we have the consequence that if it is a law for *F*'s to be *G*'s with probability $\frac{3}{4}$, then the probability that an individual F is a G is greater than $\frac{3}{4}$. This time the consequence does not rest on Armstrong's special instantiation requirement.

Armstrong has replied to this problem as follows: the law does not give us the probability of *F*'s being *G*'s, once properly understood, but the probability of instantiation of the law. Thus the law is not wrong, if it gives that probability, correctly, even though the overall probability of *F*'s being *G*'s is greater. He adds to this 'How then can we tell which cases of the FG are which? "That is an epistemic matter" we realists reply. Perhaps one would not be able to tell' (ibid.). Testing of laws becomes a little difficult, of course. For anyone not quite so sanguine, it will be worth while to consider the alternative reading.

Suppose therefore that the third sort must be absent, due to some aspect of the meaning of '(*N*:*P*)'. Then if any *F* is *G*, it is of the second sort. Let us again ask: what is the probability that in the case of a given *F*, its being *F* bears (*N*:*P*)(*F*, *G*) to its being *G*? On the supposition that it is a *G*, the answer is 1; on the supposition that it is not *G*, it is 0; but what is it without suppositions? We know what the right answer should be, namely *P*; but what is it? The point is this: by making it analytic that there can be no difference between real and apparent instances of the law, we have relegated (*N*:*P*)(*F*, *G*) to a purely explanatory role. It is what makes an *F* a *G* if it is, and whose absence accounts for a given *F* not being a *G* if it is not. (It is not like a propensity, another denizen of the metaphysical deep, which *each* radium atom is supposed to have, given *each* the same probability of becoming radon within a year.) So we still need to know what is the probability of its presence, and this cannot be deduced from the meaning of '(*N*:*P*)' any more than God's existence can be deduced from the meaning of 'God'. It cannot be analytic that the objective probability, that an instance of (*N*:*P*)(*F*, *G*) will occur, equals *P*.

Thus we have three serious problems with Armstrong's universals account of probabilistic laws. The first and second derive from his special instantiation requirement, which other accounts do not share. The third derives from the specific reading he gives to the statement that a certain state of affairs instantiates a probabilistic law. But there appears a worse problem on the alternative reading.

Proofs and illustrations

Armstrong does consider a different approach, or a different sort of statistical law. Suppose, he says, it is a law that a certain proportion of *F*'s are *G*'s, at any given time, but individual *F*'s which are *G*'s 'do not differ in any nomically relevant way from the *F*'s which are not *G*'s. (This is what he could not say in connection with the approaches examined above.) The law would govern a class or aggregate. If the half-life law of radium were construed that way, we would say: if this bit of radium had not decayed, then another bit would have, so that it would still have been the case that exactly half the original radium would remain after 1600 years. But (as Richmond Thomason has pointed out in a paper about counterfactuals) we would most definitely not say that about actual radium. We would say that if this bit of radium has not decayed, then less than half the radium would have decayed. There would be no contradiction with the theory because the half-life of 1600 years only has an overwhelmingly high probability, not certainty. I suppose there could be another sort of physics in which the half-life law is a deterministic law, and (like in ours) individual radium atoms do not differ 'nomically'. If Armstrong's account fits that other law of radioactive decay better, that is scant comfort if it cannot fit ours.

5. A NEW ANSWER TO THE FUNDAMENTAL QUESTION ABOUT CHANCE?

In his recent book, Michael Tooley has improved and extended his account of laws (which we examined in section 2 above) to probabilistic laws as well.[19] His new account is designed to avoid the difficulties we found in Armstrong's. Moreover, Tooley has an answer to what I called the fundamental question about chance (see Chapter 4 above), as it can be posed for his counterpart to objective chance. This answer appeals to the concept of *logical probability*, and thus introduces a new element into that discussion.

The basic idea of Tooley's account of deterministic laws was that (*a*) it is a law that all *A* are *B* if and only if there is a nomological relation between *A* and *B*, and (*b*) nomological relations are those contingent, irreducible relations among universals whose holding necessarily implies certain corresponding universal statements about particulars.

The main difficulty we found was that there could not be any nomological relations, on an account of the sort Tooley wants to give. The necessary implication could not be a matter of pure logic, and any attempt to add missing premisss to the logic ('trans-order laws') leads to a debilitating regress. To introduce a sort of implication that is not purely a matter of logic leads us into a necessitarian account instead. This is not the sort of problem that can be solved by adding a postulate: you cannot postulate that one thing logically implies another when it does not, without making a logical mistake. You can't make an argument valid by adding the postulate that it is valid.

In the elaboration to probabilistic laws, Tooley in effect replaces logical implication by logical probability. This notion of logical probability is an old one: that there is a quantitative relation between proposition, which generalizes implication, and has the same logical status. Of course, if it is a matter of logic, then it must govern rational opinion, and the analogue to Miller's Principle will have the same status as: if *P* logically implies *Q*, then rational opinion cannot hold *P* more likely to be true then *Q*. This latter status is that we can show someone that, even by his own lights, he sabotages himself if he violates it. So if we think of Tooley's explication of probabilistic laws as introducing his notion of objective chance, then we can view him as answering the fundamental question about chance in two steps: if there is a law then there is a corresponding logical probability, and if there is a logical probability, then that must logically constrain rational opinion. We should look carefully at both steps.

Tooley begins with some very welcome criteria of adequacy. Suppose that all radium atoms decay within a trillion years; it could still be a law that they had a certain positive probability of remaining stable for longer. Suppose that there are only four *A* in the history of the universe, and that three are *B*; it could still have been a law that an *A* has a probability of 0.8 or $\frac{\pi}{4}$ of being a *B*. He proposes that for it to be a law that an *A* has probability *p* of being a *B*, requires the real existence of a certain relation between *A* and *B*, which he designates as:

A probabilifies *B* to degree *p*
or: *Law-Stat* (*B*, *A*, *p*)

To begin we must identify this relation, and we must do so in a way that will support the correct inferences. The inference he settles on as correct is this:

1. The argument from *Law-Stat* (B, A, p) and the additional premiss that x is an A, to the conclusion that x is a B, is logically valid to degree p

or, in terms of the quantified form of logical implication:

2. The logical probability of the proposition that x is a B, given that x is an A and that A probabilifies B to degree p, equals p.

He also discusses what other sorts of premisses can be added, without altering the logical probability; this I shall leave aside. We can see that 2 is *formally* like Miller's Principle, which connects objective chance and subjective probability. Now Tooley supports this solution to the inference problem, by identifying the relation among universals so as to secure 2:

3. Probabilification to degree p is that contingent, irreducible relation between universals such that 2 holds.

The question is now whether we can consistently postulate that there is such a relation as probabilification.

It would be boring to elaborate those difficulties with this idea, which we already encountered in section 2. No argument from spatial relations among trees to spatial relations among stones, to give yet another example, is logically valid without additional premisses relating trees and stones. To postulate that it is valid is a logical mistake. The same goes for logical validity to degree p, if there is a legitimate notion of that sort.

But perhaps that is too fast. Perhaps if we look closely at the notion of logical probability, there will be a new insight that can save us. For consider: the meaning of 'and' is surely not much more than its logical role; to understand this is to see, among other things, that as a matter of logic, (P and Q) implies P, and the probability—in *any* sense thereof—of (P and Q) can be no greater than that of P. As for 'and', so perhaps also for 'probabilifies'?

The notion of logical probability is unfortunately not nearly so clear as that of implication—and we need to set aside a great deal of scepticism to be able to discuss it seriously. I know how to identify a valid inference from one sentence to another: it is valid if merely understanding the words is sufficient to see that if the one is true, then so is the other. Now, how shall I identify validity to degree p? It is there if merely understanding the words is sufficient to see that, if the one is true then . . .. Then what? How can I complete this statement without using the word 'probability' again?

Carnap had a very clever reaction to this problem. Our understanding of 'probability' consists of (*a*) the rules for probability calculation, (*b*) the rule that if two sentences are entirely on a par as far as meaning is concerned, they have the same logical probability. To complete this identification it is required then to spell out what 'on a par' means, and to demonstrate (given such a spelling out) that the probabilities of all sentences are (thereby) uniquely determined. This completion was Carnap's programme.

Part (*b*) is clearly a 'symmetry' requirement, a reinstatement of the eighteenth-century principle of indifference, which fared so badly at the hands of late nineteenth-century writers. (See further my discussion of this history in Part III.) We can see how to begin here: if P and Q are logically equivalent sentences, then they are on a par. Also if two sentences are related by permutation of a single syntactic category they are on a par. This means for instance that if F and G are syntactically simple predicates of the same degree, then a sentence (... F ...) must receive the same logical probability as the corresponding sentence (... G ...). Carnap spelled out carefully all the invariants of syntax so as to explicate when two sentences are on a par.

However, even given all these requirements of invariance, the assignment of probabilities was not uniquely determined. Nor was the class of remaining probability functions sufficiently constrained, to make their common features informative. (See further *Proofs and Illustrations*.) Therefore the programme had failed: if Carnap's concept was correct, then there is no such thing as the logical probability.[20]

Carnap had a favourite probability function, called m*, and in his article Tooley referred to it as the correct logical probability function. How could this be warranted? Could we *postulate* that it is the correct one? Not in the sense that the above mathematical problem has a unique solution, when it does not. That is again like trying to postulate that an argument is valid, when it isn't. Could we postulate that eventually we will understand the notion better, and be able to add to Carnap's (*a*) and (*b*) certain other requirements, which will single out m* uniquely? What would *that* be—a postulate about the future of Western philosophy? Would the correctness about present philosophical views then depend in part on whether the military will reduce this world to ashes in the next century? Or would the postulate mean that we already have a richer concept of logical probability than either Carnap or anyone else has been able as yet to make explicit? That could be; but all of us being unable to tell, how shall we evaluate a philosophical position resting on this article of faith? If a philosophy requires an act of faith, of such a specific sort, what has it to say to those who do not share it? I am not accusing Tooley of having chosen any of these courses, but to be frank, I see no other course open to him.

Proofs and illustrations

Hindsight is easy, and always a bit *gênant*. But the problems for Carnap's early programme turned out to be both insuperable and elementary. The non-uniqueness of the measure rests on different considerations for finite and infinite vocabularies (or sets of properties and particulars), and I shall discuss it for these two cases separately.[21] Moreover, to show that the problem does not rest on Carnap's 'Humean' notion of what simple sentences say, I shall show how laws may be incorporated. I follow in part Tooley's proposals and in part John Collins's BA Thesis.[22]

Let language L have k simple sentences Q, R, ...; m one-place predicates F, G, ...; and n names a, b, c, ... and the machinery of standard first-order logic without identity. The simple sentences can be interpreted as laws. Call L *finite* if its vocabulary is finite, and otherwise *infinite* (in which case k, m, or n is not an integer but countable infinity). Let a TV be an assignment of T (*true*) or F (*false*) to each sentence, in accordance with (first-order) logic. Usually such a TV can be summed up rather briefly. Suppose for instance that $k = m = n = 1$. Then

Q	Fa	$(x)Fx$	$(x) \sim Fx$
T	F	T	T
T	T	F	F

depicts two TVs in summary.

Suppose that L is finite; then so is the number of its TV's. (This is in part because I have kept out identity, so we cannot count in this language.) A probability function P must assign a number between *zero* and *one* (inclusive) to each *TV*, and these numbers must add up to *one*. Then we *define*

$$P(A) = \text{Sum } \{P(t) : t \text{ is a } TV \text{ which assigns } T \text{ to } A\}$$
$$P(A|B) = P(A \text{ and } B)/P(B) \text{—defined if } P(B) \neq 0.$$

Now, what requirements can we put on P? The definition has already guaranteed that P will assign the same number to any two logically equivalent sentences, and that P will not violate the usual rules of probability calculation. So what remains is to determine what P must do with the individual *TV*'s.

This is the point where Carnap introduces the invariance requirements. It is clear that any specific meaning given to the vocabulary, or any factual information assumed, will break the sort of syntactic symmetries which these represent. For example if F and G stand for 'scarlet' and 'red', then no *TV* should give T to Fa and F to Ga. Similarly, if we already know that all F's are G's for some other reason. We now have two options. We can classify some of this information as really logical or verbal, and eliminate *TV*'s conflicting with it, before determining how P should treat (the remaining) *TV*'s. Or else we can ignore all such information to begin, define a perfectly 'informationless' P and then 'conditionalize' it on the information (along the lines of the second part of the above definition). The result, call it P', may be thought of as the mature logical probability, after assimilating meaning that goes beyond syntactic form. In this latter case it will not be so bad if P is not unique, as long as the mature P' *is unique*.

The two courses will lead to the same result, if the language is finitary and the idea is merely to delete 'bad' TV's. When the language is infinitary, *zero* probabilities begin to play a troublesome role, and the first course may work when the second won't. On the other hand, if we have the idea that some of the simple sentences speak about the probabilities of other sentences—i.e. if they express probabilistic laws—only the second course could work. For that we can make come out right only by insisting that the probabilities should be so distributed among the *TV*'s that the conditional probability—e.g. of b decaying in 5 minutes, given that b is a radium atom and the law gives probability e^{-5A} to the decay of radium atoms within 5 minutes—be correct. This cannot be done by deleting some *TV*'s, and it cannot itself be invariant under substitution of predicates (such as 'lead atom' for 'radium atom').

So let us proceed carefully in two steps. First we decide what an informationless probability P looks like. Then we will look for a 'mature' descendant P'' which reflects meaning and lawhood. If P itself is not unique, this need not worry us, as long as its 'mature' descendant is.

Recall the case of language L with $k = m = n = 1$. It does not have 2^4 *TV*'s, because a *TV* must give T to Fa if it gives T to $(x)Fx$, and F if it gives T to $(x) \sim Fx$. That leaves eight. Because the language is so small, no two of these are related to each other by a permutation of simple terms. If there is no other invariance requirement, that means these eight *TV*s can be assigned *any* probabilities summing to *one*. We could think of insisting on invariance if F is replaced by $\sim F$, or Q by $\sim Q$. That gives the grouping shown in Table 5.2.

TABLE 5.2

		Q	Fa	$(x)Fx$	$(x) \sim Fx$
Group I	(1)	T	T	T	F
	(2)	T	F	F	T
	(3)	F	T	T	F
	(4)	F	F	F	T
Group II	(5)	T	T	F	F
	(6)	T	F	F	F
	(7)	F	T	F	F
	(8)	F	F	F	F

Then every *TV* in Group I must be assigned the same value, and likewise every *TV* within Group II. But how the probability is allocated to the two groups is arbitrary. It is noteworthy that on either policy, Q and Fa will each receive probability $\frac{1}{2}$, which shows the extent to which the Principle of Indifference is operative. What is left indeterminate, clearly, is the probabilities of $(x)Fx$ and $(x) \sim Fx$. On the first policy, these could be anything; on the second they could also be any number between *zero* and *one*, but must be equal.

There is no way in which logical considerations could go further than this. You could say that all eight *TV*s must have the same probability. But this would mean, for example, that the probability of $(x)Fx$ rises from $\frac{2}{8}$ to $\frac{2}{4}$ conditional on the information that Fa.

This is a 100 per cent increase, which is just silly. If we thought that the number of names in the language reflected the number of things there are, then we should have said that $(x)Fx$ is *certain* given Fa. And if we didn't think that, we shouldn't be assuming that a can function as such a sensitive gauge of what *all* things are like. So the idea that all *TV*'s must be treated equally, can have no general appeal.[23]

Can we cut this down to a unique function P' by letting Q express a law that everything must be F? Indeed; that would in effect remove *TV* (2) from Group I and (5), (6) from Group II. But the liberty to distribute probability any way we like between the two groups still leaves many probability functions. Uniqueness would appear only if we added R to express the law that that $(x) \sim Fx$, and then tossed out any *TV* in which Q and R are both false. This would be a hypothesis of total determinism. We could also allow both Q and R to be false, while adding, say, S, T, ... to express probabilistic laws such that $P'(Fa|S)$ must equal, say, $\frac{3}{4}$. These restrictions leave the probability of $(x)Fx$ unconstrained again, however. To close the gap, another law could be introduced to fix the probability of $(x)Fx$ directly, and independently of that of Fa. That would be quite out of line with conception of a probabilistic law. In any case, that we could constrain a unique P' simply by dictating all the probabilities it must assign, is no news! The point of introducing logical probability into the account—that it is itself an independent and determinate logical notion, which provides a bridge between probabilistic laws and rational expectation—would here be lost altogether.

Let us look now at an infinite language; suppose specifically that there is an infinite set of names a, b, c, ... If F is a predicate, then each *TV* must give T or F to each of Fa, Fb, Fc, ... The number of *TV*'s is then the next infinite number: it has the power of the continuum. But from this it follows at once that most of them must receive probability *zero*. (For at most two can receive as much as $\frac{1}{2}$, at most three as much as $\frac{1}{3}$, ..., at most n as much as $\frac{1}{n}$. So at most countably many will receive a probability higher than *zero*.)[24] This is not debilitating: we can represent the *TV*'s by points on a line, and then use a distribution function over that line to determine positive probability for consistent finite sets of sentences. However, there are more than innumerably many such distribution functions. It does not help to say 'of course, you must use a constant distribution function'! For the representation of *TV*'s by points does not incorporate a non-arbitrary distance metric, between *TV*'s; hence it is largely arbitrary. (This is a point we encountered also in the discussion of chance in the preceding chapter; see Fig. 4.5.)

How far can the invariance requirements, imposed on probabilities assigned to single sentences, take us? Suppose we include the demand that simple sentences such as Q or predicates such as F be replaceable uniformly by their negations, without affecting logical probability. Then $P(Q) = P(\sim Q)$, hence each must be $\frac{1}{2}$; similarly for $P(Fa)$. Now $P(Fa) = P(Fa \,\&\, Fb) + P(Fa \,\&\, \sim Fb)$. If we try to keep probabilities positive (or non-infinitesimal) as long as possible, then $P(Fa \,\&\, Fb)$ will be less than $P(Fa)$. By a repetition of the argument, $P(Fa \,\&\, Fb \,\&\, Fc)$ will be less still; and so forth. Since $(x)Fx$ entails all those sentences, its probability will be less than each in consequences. This need not be zero, for the series could converge to a positive number.

Note well that replacing F by $\sim F$ does not turn $(Fa \,\&\, \sim Fb)$ into $(Fa \,\&\, Fb)$ but into $(\sim Fa \,\&\, Fb)$. The invariance requirement, even in this strong form incorporating negation, entails only:

$Fa \,\&\, Fb$ has same probability as $\sim Fa \,\&\, \sim Fb$
$Fa \,\&\, \sim Fb$ has same probability as $\sim Fa \,\&\, Fb$

but probability could be arbitrarily distributed between the two groups. A still stronger requirement, that any simple predicative sentence like *Fa* be everywhere replaceable by its negation, *salva probabilitate*, would make the two groups equal in probability. Then all four sentences receive $\frac{1}{4}$. By repeating the argument for $(Fa \,\&\, Fb \,\&\, Fc)$ and so forth, we would arrive at the conclusion that every *TV* has zero (or infinitesimal) probability—and the universal sentence $(x)Fx$ also. But just as above, there are independent reasons not to accept such a strong requirement as logically imperative. (The reasons are however that the programme would suffer, and not that the general command, to treat logically similar sentences similarly, is stopped here by an apprehended asymmetry. I am simply being as charitable to the programme as I can be.) If we did impose this very strong requirement, we would arrive at the unique function $m\dagger$ for the quantifier-free part of the language again, the one that Carnap rejected.

The plethora of distribution functions on a continuum makes P non-unique to a remarkable extent. What happens if we begin to interpret the single sentences, such as Q, as laws? If Q is the law that everything must be F, we can delete any TV which gives T to Q and F either to some sentence like Fa, or to $(x)Fx$. Then the probability of Fa, or of $(x)Fx$ given Q will be 1, while Q itself still has probability $\frac{1}{2}$. Now the negation invariance is broken: no reason to expect $P'(Fa| \sim Q)$ to be *zero*, so Fa will now have a higher probability than $\sim Fa$. This is rather curious in itself: the mere *possibility* that there is such a law, has raised the probability that something is F! But at this point we may drop even universal negation invariance, and let $P(Q)$ be anything. Uniqueness is not exactly nearer then.

If $(x)Fx$ had probability *zero*, the above manœuvre will not work, because the envisaged conditionalization of P on $(\sim Q \vee (x)Fx)$ to produce P' will have reduced the probability of Q to *zero* as well. In such a case, the deletion of TV's should occur first, and only then should the logical probability be determined, to the extent it can be. That is: the meaning of the laws has to be built into the language before the Indifference Principle is applied. That is certainly possible, and will then insure that $(x)Fx$, which had probability *zero*, now has the probability of Q (since the determination of $P((x)Fx| Q)$ is presumably unaffected). Again, since $P(Fa| \sim Q)$ is presumably also positive, the mere recognition of the possibility of a law (by designing the language to allow its expression) has raised the logical probability of Fa. The effect could be counteracted by incorporating other, contrary law statements. Thus the general conclusion is this: however it be done (as in the preceding paragraph or this one), the set of law-statements *expressible* in the language, significantly affects the logical prior probability of their instances.

For someone who views the correct language as having a law statement in it only if the law is true, this would be fine. But the logical design of the language cannot depend on a contingent truth. I do not see how this could seem fine to someone who would expect merely conceivable laws to be formulated in the language. Perhaps he could insist on a very carefully chosen set, to be expressible, so as to nullify this effect by their presence overall. But puzzling over this seems rather useless, given that the non-uniqueness we have found, establishes that there is no such thing as the logical probability.

6. WHAT THE RENAISSANCE SAID TO THE SCHOOLMEN

So far we have concentrated on the unsolvability of the identity problem and the inference problem, taken jointly. It is time to look at the universals accounts' most frequent claim: that laws so conceived truly explain.

This claim is most often advanced in the negative: universal regularities as such do not explain, but laws do, so a law must be or entail more than a regularity. So far, so good; but that does not yet establish that laws conceived as relations among universals do explain. The first major obstacle to the claim that they do is the failure to solve the inference problem: it simply does not seem that (irreducible higher-order) relations among universals can provide information about how particulars behave. While I'm anxious not to base criticisms on any specific theory of explanation, surely a minimal criterion is unmet here. But let us set all this aside, and see whether (if the information they give be granted to be as hoped) relations among universals can indeed truly explain, in the way that regularities cannot.

For the necessitarian accounts, possible-worlds style, the answer to the corresponding question was *No*, according to Foley's argument. For the law was there conceived as also a universal truth, though about worlds rather than about entities in a world. Now, if a mere universal truth does not have the wherewithal to explain, then the postulate of a universal truth about worlds cannot be as such the *terminus de jure* for explanation.

The form of this argument is *tu quoque*: you claim that *A* cannot be explained by *B* alone because *B* is a mere *X*, and you then explain *A* by explaining *B* by *C*—but *C* too is a mere *X*! Let us call this the *termination problem*: anyone who claims that something or other is not enough for explanation must enlighten us as to what is enough.

At first sight, the universals account fares better here. After all, it explains the universal truth about particulars by means of a singular truth about universals. But it was not the universal form of the universal regularity that made it incapable of explaining! The objection was that mere universality is not enough. We can't explain that this crow is curious by saying that all crows are curious *alone*; at best this will *point to* an explanation in terms of inheritable characteristics among birds. The failure of the possible-worlds account was that we don't receive the information 'All worlds physically possible relative to us are thus-and-so' on a background of beliefs that would lead us to go on in this fashion, 'Oh well, then there is probably a set of inheritable characteristics of crows in this, or all such worlds, such that . . .'. Or at least, the universal truth about worlds does not point to these missing pieces in our puzzle any more than the original universal truth about actual crows did.

Of course, the possible-worlds theorist says: but this news about crows in other worlds *means* that the regularity is *necessary*, that it is not an accident. By making it *mean* that, however, he robs the assertion of necessity of any force that the generalization about worlds lacks. After all, what is gained, except brevity, if one restates the same story by means of explicitly defined terms? A defined term is *only* an abbreviation, and nothing can be added by abbreviating.

Now again, the universals account looks as if it will fare better. It claims not to define, but to reveal the ground of, the necessity. The regularity in the particulars is made necessary, by the relations among the universals.

But now we face the dilemma: is this 'necessary' in 'made necessary' a matter of logic or not? In the first case, we have Molière's *virtus dormitiva* as pattern of explanation. In the second case, we land in the lawgivers' regress.

Molière was late, and only in fashion with his critique of the Schoolmen—a fashion harking back to the real struggle of the New Sciences against the Scholastic tradition in the Renaissance. Galileo still had to understand that tradition to fight it. Boyle and Newton already seem unaware of finer distinctions, but still appreciate the real gap between the two styles of explanation.

> That which I chiefly aim at, is to make it probable to you by experiments, that almost all sorts of qualities, most of which have been by the schools either left unexplicated, or generally referred to I know not what incomprehensible substantial forms, may be produced mechanically . . .[25]

Substantial forms—that means universals. But can't the Schoolman retort that the mechanics, describing what Boyle calls the 'mechanical affectations of matter', must fall into the same pattern of explanation as his own—if it is to explain at all?

To explore this question, just imagine a discussion in the Renaissance between a Schoolman (with a complex theory of natures, substantial forms, *complexio*, and occult qualities) and a new Mechanist (with a naïve theory of atoms of different shapes, with or without hooks and eyes). The Schoolman says that the mechanist account must eventually rely on the regularities concerning atoms, such as that their shapes remain the same with time, and there must be a *reason* for these regularities in nature. But the Mechanist can reply that whatever algebra of attributes, etc. the Schoolman can offer him, the inference, from the equations of that algebra to regularities in the behaviour of atoms, must rest on some further laws which relate attributes to particulars. For example, if A is part of B, it may follow that instances of A are instances of B, but not without an additional premiss which justifies, in effect, the suggested 'part-whole' terminology for the indicated relation between universals. Of course, if we define A to be part of B exactly if it is necessary for instances of A to be instances of B, the argument becomes valid. But then is the necessity appealed to in the definition itself grounded in some further reality? And if not, if there can be a necessity not further grounded in some further reality, he would like to return to his atoms, please, and say that their postulated regularities are not grounded in any further reality—it's just the way atoms are.

PART II
Belief as Rational but Lawless

INTRODUCTION

We have now seen that any philosophical account of laws needs a good deal in the way of metaphysics to do justice to the concept at all. We have also seen that, as a result, any such account founders on the two fundamental problems of identification and of inference. The extant accounts come to grief additionally in their attempts even to meet the most basic criteria relating to science and explanation. Their promises have all proved empty.

But there are still those traditional arguments, which conclude first that there *must be* laws of nature, and secondly, that we *must believe* that there are such laws. In modern terms, the threats are these: without laws of nature we can make no sense of science and its achievement, nor of rational expectation, and must succumb inevitably to the despair of scepticism.

In this Part I shall answer these epistemological arguments. More constructively, I shall propose a programme for epistemology and for philosophy of science which will allow them to flourish in the absence of laws of nature or belief therein.

6.
Inference to the Best Explanation: Salvation by Laws?

> As a man of science you're bound to accept the working hypothesis that explains the facts most plausibly.
>
> The Arch-Vicar of Belial, to Dr Poole, in Aldous Huxley, *Ape and Essence*.[1]

THE inference from the phenomena that puzzle us, to their best explanation, appears to have our instinctive assent. We see putative examples of it, in science and philosophy no less than in ordinary life and in literature.

It is exactly this pattern of inference, to the best explanation offered, that philosophers have drawn upon to claim confirmation of laws. They support this appeal in two ways: by pointing to the failures of traditional ideas of induction and by arguing that this inference pattern is the true rock on which epistemology must build. After examining their reasons, I shall argue instead that they would build on shifting sands. As long as the pattern of Inference to the Best Explanation—henceforth, IBE—is left vague, it seems to fit much rational activity. But when we scrutinize its credentials, we find it seriously wanting.[2] (For those more interested in IBE itself than in its connection with laws, sections 2 and 3 may be skipped.)

If both induction and IBE fail as rational basis for opinion and expectation of the future, traditional epistemology is indeed in serious difficulty. But rather than proclaim the death of epistemology, and submit either to an irenic relativism or to sceptical despair, I shall try to show in the next chapter that a new epistemology has been quietly growing within the ruins of the old (as well as show that it issues in a still more drastic critique of IBE).

1. ON THE FAILURES OF INDUCTION[3]

One contention, common to many writers, is that without some such concept as laws of nature, we can make no sense of rational expectation of the future. I have earlier presented this point in what I take to be its primordial form: if anyone says that there is no reason for the observed regularities, then he can have no reason to expect them to continue.

This assertion clearly denies the cogency of induction in a narrow sense (belief based on straight extrapolation from the data) while it holds out the hope of induction in a very broad sense (rationally formed expectation of the future). Let us here use the term 'induction' everywhere in its narrow sense: the procedure whose independent rationality friends of laws tend to deny. I may as well add at once that I agree with them on the critical point. My discussion will aim to underline their legitimate objections, but to show simultaneously how their critique is misdirected, and where it rests on dubitable premisses of their own.

Here is the ideal of induction: of a rule of calculation, that extrapolates from particular data to general (or at least ampliative) conclusions. Parts of the ideal are (*a*) that it is a *rule*, (*b*) that it is *rationally compelling*, and (*c*) that it is *objective* in the sense of being independent of the historical or psychological context in which the data appear, and finally, (*d*) that it is *ampliative*. If this ideal is correct, then support of general conclusions by the data is able to guide our opinion, without recourse to anything outside the data—such as laws, necessities, universals, or what have you.

Critique of this ideal is made no easier by the fact that this rule of induction does not exist. The rule was indeed baptized—presumably after conception, but before it was ever born. Sketches of rules of this sort have been presented, with a good deal of hand-waving, but none has ever been seriously advocated for long.[4] Every generation of philosophers, beginning with Aristotle, has seen that mere numerical extrapolation in any specific form, cannot be the rule described in our ideal. Criticisms brought forward in this century however, exhibit difficulties to plague every possible realization of the ideal. If the reader is already convinced of the inadequacy of induction in the narrow sense, there is no need to read the rest of this section.

What is extrapolation? Example of the alien die

Consider a die, which is to be tossed ten times, and the hypothesis that all ten tosses will come up *ace*. Let the evidence so far be that it has come up *ace* for the first seven tosses. Now, how could the rule of induction relate these data to the hypothesis? Should it tell us that all ten will be like the first seven? Rules of extrapolation can't be expected to do well if some relevant evidence is left unstated, and we do have other information about human dice. So suppose we found this die on an alien planet, and 'ace' is the name we give to one of the six sides of this geometric cube of unknown composition.

Of course, reader, you are still unwilling to suggest that the rule should tell us to infer that all tosses (or equivalently, the last three) will come up *ace*. After all, in this situation, *you* would not infer that. But perhaps the rule should be sophisticated beyond anything Bacon and Newton, the great advocates of this ideal, could imagine. Let it tell us the probability of the hypothesis, bestowed on it by the data. Then it could be asserted that a rational person must follow the rule of induction, in the sense that it provides him with the probability that the hypothesis is true, given the evidence.

This suggestion marks quite a shift, because it takes us from induction as extrapolation from mere numbers to something much more general. But every discussion of induction is forced to this. Suppose, for example, that instead we try to maintain the rule in as simple a form as possible, with as one corollary: if all instances have been favourable, and you have no other evidence, then believe that the next instance will be favourable as well. You will immediately insist, surely, that 'believe' must be qualified here, if you are ever to follow it. Believe with what confidence? Believe to what degree? Are a hundred instances not better evidence than ten, even if all have been favourable? Any such reaction replaces the simple rule with a more sophisticated one, of the order of probability assignments. Of course we should hasten to accept all worthy suggestions for improvement of the rule, rather than insist on beating a dead horse.

In both forms, the same problem about induction appears very clearly. It is that, being a *rule* it must have certain structural features—and as a result, its extrapolation from any data will be heavily influenced by what it does with small increases in data. But how it does that, must be either uninformative or arbitrary. That is the dilemma this ideal always foundered on.

To illustrate this, begin with the naïve 'straight rule': believe that the ratio of A to B overall equals the ratio observed so far. That tells us to believe that the sun will always rise. Unfortunately, it also tells us to expect all *aces* as soon as we have seen a single toss of the alien die, if it came up *ace*. We can't very well suggest that another ratio is any more plausible. So we must fiddle with the confidence: always believe what this rule tells you, but believe it weakly after one toss and strongly after many tosses.[5] Now what is needed is an exact prescription of what beliefs, degrees of belief, or confidence I should have at the outset; and an auxiliary rule about how this should change with the outcome of each new toss. The former could be perfect neutrality of opinion. (I do not assume it must be a precise subjective probability or anything like it.) But then the auxiliary rule must still say exactly how much non-neutrality there should be after one toss; and indeed, after $n + 1$ tosses (as a function of n, the previous outcomes, and the new outcome). Now you can look at the numbers and ratios as much as you like, but they will give you no clue at all to this auxiliary rule for massaging your confidence in the observed ratio. Myriads, continua, of such functions exist, and however little they diverge in the small, they lead to widely different consequences down the line. You can try to remain a little neutral among them: the more you do so, the *less arbitrary* will your rule of induction be, but also the *less informative*. Now, next problem: try and formulate a measure of balance between arbitrariness and informativeness. There you will again find a continuum of functions to choose from, and you will again confront the dilemma presented by the spectrum from capricious choice to trivializing neutrality.[6]

There are many other problems with the ideal of induction, even if this is (as I think) its fundamental flaw.[7] As to the empiricists who followed this banner *sans* device, their hope was placed in an empty promise. But does there indeed lie a better hope in the mobilization of laws to found rational expectation, as Dretske, Armstrong, and Tooley contend?

2. DRETSKE ON THE REMARKABLE CONFIRMATION OF LAWS

The scheme we inspected above might be called simple or bare induction; many have been the proposals to replace it by more complex schemes. One such is Dretske's proposal that we take note of the (supposed) remarkable tendency of laws to become well confirmed on the basis of very little evidence. He uses exactly the reflections and types of examples exhibited above, to sketch a rival picture of how we can rationally go beyond the evidence.

It appears, at first sight, that laws are beyond our epistemic grasp. The sun has risen every day; this appears to confirm the universal generalization that it always has and always will. But if we speculated that this was a matter of law, we would be asserting more: something beyond and in addition to the universal statement. The conforming instances support the generalization, but surely they do not support anything beyond that? Dretske tells us that this apparent problem is a pseudo-problem, resting on a mistaken empiricist epistemology. In fact, he claims, it is quite the other way round. This sort of evidence, of positive instances, does not at all support the universal statement, if taken in isolation. But it does support the hypothesis that it is a law that the phenomenon always occurs in that same way.

Dretske calls this conclusion 'mildly paradoxical' ('Laws of Nature', p. 267), but it seems more than that. Surely if it is a law that *A* then it is also true that *A*; hence I can become no less confident that it is so than that it is a law. The air of paradox is perhaps removed, if we take Dretske to be attacking the conception of evidential support that was implicit in the proffered argument. This conception appears to be the old ideal of purely numerical induction.

To tackle also more sophisticated epistemic schemes, Dretske makes the preliminary point that raising our probabilities may not amount to real confirmation.[8] Let us use the alien die, introduced in the preceding section, to illustrate his point. Suppose I begin with the initial assumption that the die is fair. Then my initial probability that it will come up *ace* all ten tosses, is very low—$(\frac{1}{6})^{10}$ which equals about six in a thousand million. After I have seen seven *aces* come up in in a row, while maintaining this assumption, the probability that the last three will come up, is still the same as it was: $(\frac{1}{6})^3$. But this is now also the new probability of the proposition that all ten come up *ace*. The probability of that proposition has therefore become 6^7 times—about 300 000 times—higher than it was. Our probability for the universal statement has increased dramatically—but we are in no better position to predict what comes next! Using a similar example about coin tossing, Dretske writes:

> But this, of course, isn't confirmation. Confirmation is not simply raising the probability that a hypothesis is true, it is raising the probability that the unexamined cases resemble (in the relevant respects) the examined cases. It is *this* probability that must be raised if genuine confirmation is to occur (and if a confirmed hypothesis is to be useful in *prediction*), and it is precisely this probability that is left unaffected by the instantial 'evidence' in the above examples. ('Laws of Nature', p. 258)
>
> The only way we can get a purchase on the unexamined cases is to introduce a hypothesis which, while *explaining* the data we already have, *implies* something about the data we do not have. (ibid. 259)

Let us criticize the example and its discussion immediately.

The moral is not correctly drawn, because it is only on the supposition of one explanatory hypothesis (e.g. fairness, or any other sort of *bias*) that the data can raise the probability of the universal statement without raising that of the remaining instances. If instead I profess some measure of ignorance about the bias of the die, then that ignorance becomes modified by the initial data, and my opinion about the unexamined cases changes right along with it. For example, if I had thought that the die was *either* fair *or* perfectly biased in favour of *ace*, then after seven *aces* I would have favoured the latter hypothesis considerably! I would accordingly think it more likely then than I did before that the last three would be *aces* too. So Dretske has generalized upon a special case.

The second point that had better be noticed is that these effects would appear in the same way if our background beliefs had nothing lawlike about them, as long as they relate to the instances in the same way. For consider another example. I am told that the ten coins I am about to be shown came either from Peter's pocket or from Paul's; that Peter's contained ten dimes and fifty nickels, while Paul's contained sixty dimes. The first seven to be put before me are dimes. Obviously, on the supposition that they all come from Peter's pocket, the hypothesis that all ten will be dimes has increased to the constant probability that the last three will be, namely $(\frac{1}{6})^3$. Without this supposition, but with the background belief that they are equally likely to come from Paul's as from Peter's, the probability of the last three being dimes has also increased, however. This example parallels the previous one perfectly, although here no laws of nature are involved at all. The non-lawlike hypothesis about pockets has the same effect. None of this has anything to do with the explanatory power that may or may not reside in the lawlikeness of background opinion, but only with what that opinion is (as expressed in terms of my personal probability).

Of course, someone else might be happy to say that the hypothesis, that all the coins come from Peter's pocket, explained why the first seven were dimes. But Dretske's contention appears to be that to be explanatory, the hypothesis has to be about something special, like laws or similar unordinary facts.

Dretske's alternative proposal

Suppose, however, that we do agree to this idea about what explanation requires. What rival to primitive numerical induction does Dretske want to propose? He proposes that we follow a rule of IBE.

If the first seven tosses yield *ace* this is best explained by the die's having a perfect bias in favour of *ace* (let us say). Should we now at once accept this hypothesis? But it was the best explanation already when we had just seen the first toss yield *ace* and for the same reason. Should I therefore have accepted the hypothesis of extreme bias already after one *ace*? Obviously not; so I am construing the proposal too naïvely. It cannot be that I'm simply to infer the truth of the best explanation. Rather, the all-or-nothing model of jumping from agnosticism to full belief must again be modified to accommodate, and trade in, degrees of belief or confidence. That conclusion is also evident in Dretske's discussion, though he talks of confidence rather than of probability:

> laws are the *sort* of thing that can become well established prior to an exhaustive enumeration of the instances to which they apply. This, of course, is what gives laws their predictive utility. Our confidence in them increases at a much more rapid rate than does the ratio of favourable examined cases to total number of cases. (ibid. 256)

This is not as easy to construe as it looks! Let us be careful, and see what meanings this passage can and cannot bear.

First, it cannot mean that our confidence in a proposition

increases rapidly with accumulating evidence, if we know (or believe) that it is a law. For then our confidence is already at a maximum.

Second, if it really were a law that all *A* are *B*, would that by itself make our confidence in it increase especially fast, regardless of what we believe at the outset? Surely not—the law of gravity may make things fall, but can't make people believe that things will fall.

Third, might Dretske mean that, in response to the same evidence, my confidence in the proposition *It is a law that all A are B* increases more rapidly than my confidence in *All A are B*? That could be. But since the former is supposed to entail the latter, it will catch up, and from there on must inevitably drive the latter ahead. For example, it seems likely to be a law that all radium decays, I must then regard it at least as likely that all radium does decay.

Besides these three possible meanings which Dretske cannot intend, how else could the assertions be construed? Perhaps the fault lies with our lack of imagination. Possibly Dretske is pointing to a rule, as yet unknown, which will make or revise our probabilities so as to give *bonus* marks to the hypotheses that explain observed phenomena. Then, if hypotheses to the effect that there exists a law are especially explanatory, they may get an especially high *bonus*. This suggestion cannot be dismissed, nor discussed probatively at the merely qualitative level. I shall discuss it in the next chapter, and argue that there cannot exist any such probabilistic rule of Inference to the Best Explanation, on pain of incoherence. We will also find there that other, more precise construals of Dretske's dictum leave it equally false.

3. ARMSTRONG'S JUSTIFICATION OF INDUCTION

At the beginning of his book, Armstrong wrote 'There is one truly eccentric view, brought to my attention by Peter Forrest. . . . This is the view that, although there are regularities in the world, there are no laws of nature.' Armstrong's response follows at once: 'Such a view, however, will have to face the question what good reason we can have to think that the world is regular. It will have to face The Problem of Induction. It will be argued . . . that [no such view] can escape inductive scepticism' (*What is a Law of Nature?*, p. 5). He makes good his promise in a later chapter, by a real *tour de force*. For there Armstrong purports not only to prove that belief in laws is needed, to avoid the sceptic's slough of despond, but to present us with a justification of induction—by an argument which does not depend on what the rules of induction are! I shall now analyse this carefully to show what he assumes along the way.[9] What this analysis will show, among other things, is that Armstrong's own argument *relies* on a previously assumed rule of Inference to the Best Explanation, and advances no independent support for it.

Armstrong begins with the explicit premiss (call it P_1) that 'ordinary inductive inference, ordinary inference from the observed to the unobserved, is . . . a rational form of inference'. On questions about what that form is, what rules may be being followed, he confesses himself largely agnostic. He defends the premiss along the lines of Moore, common sense against scepticism, saying that this premiss is part of the bedrock of our beliefs, indeed, that it 'has claims to be our most basic belief of all' (p. 54).

This premiss (P_1) is theoretically loaded despite the accompanying agnosticism on questions of form. It is undoubtedly true that we have expectations about the future, and opinions about the unobserved. It does not follow that we are engaged in ampliation—let alone some sort of ampliative *inference*, i.e. ampliation in accordance with rules. Perhaps we amend our opinions (*a*) by purely logical adjustment to the deliverances of new experience ('conditionalization', for example) and/or (*b*), some unpredictable free enterprise in the formation of new opinions, within certain limits required by rationality. The distance Armstrong slides here stretches from the Moorean common sense that we form rational expectations, to the philosophical modelling of this activity as *inference*.

Besides the explicit premiss, therefore, we have found as further premiss the statement that we believe the initial premiss, and that we either know or rationally believe it to be true. This extra premiss (call it P_2) is needed to understand the subsequent argument. Yet it is seriously questionable.

Suppose now that we do engage in some form of ampliative inference, which we believe to be rational. At this point in the argument, we need not yet know what that form is (one form that fits all ampliative inference is '*P*; therefore *Q*', but the 'ordinary inductive inference' presumably includes much less than everything fitting that form)—so let us call this form F.

Here follows the *first sub-argument*. Its conclusion is that it is a necessary truth that induction (i.e. in our present terms, ampliative inference of form F) is rational. This argument is based on P_2 rather than on P_1 and is an interesting variant on Peirce's argument for the reality of laws. P_2 says that we know or rationally believe induction to be rational. But that implies (via a premiss which I shall not number) that our belief that induction is rational must have a justification. That justification cannot be by induction or it would be circular. Nor can it be by deduction, since the relevant statement (i.e. P_1) is not a logical truth and any premiss from which P_1 could be deduced would face the same question as we have for P_1, thus leading to a regress. The only possible justification is therefore a claim to knowledge or rational belief not based on any sort of demonstration. That is a tenable claim only for a statement claimed to be known a priori. But (again via a premiss I shan't number) only necessary truths can be known a priori. Therefore P_1 is a necessary truth.

I am not sure that to be rational a belief must have a justification reaching back all the way to a priori truths. But if we allow, say, a priori truths plus the evidence 'of my own eyes', as basis for justification, the case for P_1 will not be significantly different. So Armstrong has 'established' that if induction is known to be rational, this must be a case of a priori knowledge. And if only necessary truths can be known or rationally believed without the sort of justification that P_1 is denied, then P_2 implies that P_1 is a necessary truth.

We come now to the *second sub-argument*. The conclusion we have reached is this: it is a necessary truth that ampliative induction of form F is rational. This fact (call it P_3) Armstrong insists, must be given an explanation. What F is will now finally make a difference. He proceeds as follows: he makes a proposal for what F is, demonstrates how P_3 can be true on the basis of this proposal, and then notes that the demonstration would fail if laws said nothing more than mere statements of regularity.

The proposal is that ampliative inference of form F is inference from the evidence, to laws that explain the evidence (call this P_4). He adds that this procedure is an instance of Inference to the Best Explanation, that this sort of inference (IBE) is rational (P_5), and that it is analytic (true by virtue of the meanings of the words) that IBE is rational (P_6). If we add that analytic statements are necessary truths, the explanation of (P_3) is complete. The footnote to be added about any 'regularity' view of laws is that it would make (P_4) false, because according to Armstrong regularities, unlike laws, do not *explain* the evidence which they fit or entail.

Note well that (P_4), (P_5), and (P_6) are not premisses of the overall argument. They are premisses of a sub-argument, whose correctness—once noted—is all that is asserted. Because it is correct, it gives us an explanation of (P_3). Since no other explanation of (P_3) is available, it is supposedly rational to believe the explanation offered, and hence the premisses on which it rests. *Here*, in this ultimate stage of the argument, we see a step made by means of inference to what explains. So IBE does function as premiss of the overall argument as well. What independent support could any of these premisses receive?

The defence of (P_4) must be that laws explain the phenomena which they fit or entail, and that laws provide the best among the explanations that can be given for such phenomena. This second part of the defence comes in a very cavalier little paragraph:

> It could be still wondered whether an appeal to laws is really the *best* explanation of [the phenomena]. To that we can reply with a challenge 'Produce a better, or equally good, explanation'. Perhaps the challenge can be met. We simply wait and see. (p. 59)

Would it be enough, to meet this challenge, to present some cases where the best available explanation of some phenomena does not consist in deriving them from laws? If so, there is enough literature for Armstrong to confront now; he need not wait.

The defence of the next premiss, (P_5)—namely, that IBE is rational—consists simply in the last premiss, (P_6): it is analytic, due to the meaning of the word 'rational', that IBE is rational. This conviction about IBE appears not only here, but throughout the book, and not surprisingly: IBE is the engine that drives Armstrong's metaphysical enterprise. It provides his view of science (p. 6: 'We may make an "inference to the best explanation from the predictive success of contemporary scientific theory to the conclusion that such theory mirrors at least some of the laws of nature . . .".'). He also regards IBE as being first of all a form of inference to be found pervasively in science and in ordinary life (p. 98: 'But I take it that inference to a good, with luck the best, explanation has force *even* in the sphere of metaphysical analysis'—(my italics)). To support (P_6) he does not see the need for more than rhetorical questions: 'If making such an inference is not rational, what is?' (p. 53); 'To infer to the best explanation is part of what it is to be rational. If that is not rational, what is?' (p. 59).

In sum, therefore, Armstrong has reached powerful conclusions on the basis of an assumption which is supported solely by a challenge to those who would doubt it.

In the next two sections I hope to meet Armstrong's challenge. I shall argue that inference to the best explanation cannot be a recipe for rational change of opinion. And then I shall try to answer the question, 'If that is not rational, what is?'

4. WHY I DO NOT BELIEVE IN INFERENCE TO THE BEST EXPLANATION[10]

There are many charges to be laid against the epistemological scheme of Inference to the Best Explanation. One is that it pretends to be something other than it is. Another is that it is supported by bad arguments. A third is that it conflicts with other forms of change of opinion, that we accept as rational.

Still, the verdict I shall urge is a gentle one. Someone who comes to hold a belief because he found it explanatory, is not *thereby* irrational.[11] He becomes irrational, however, if he adopts it as a rule to do so, and even more if he regards us as rationally compelled by it. The argument for this conclusion will be begun here and concluded in the next chapter.

What IBE really is

Inference to the Best Explanation is not what it pretends to be, if it pretends to fulfil the ideal of induction. As such its purport is to be a rule to form warranted new beliefs on the basis of the evidence, the evidence alone, in a purely objective manner. It purports to do this on the basis of an evaluation of hypotheses with respect to how well they explain the evidence, where explanation again is an objective relation between hypothesis and evidence alone.

It cannot be *that* for it is a rule that only selects the best among the historically given hypotheses. We can watch no contest of the theories we have so painfully struggled to formulate, with those no one has proposed. So our selection may well be the best of a bad lot. To believe is *at least* to consider more likely to be true, than not. So to believe the best explanation requires more than an evaluation of the given hypothesis. It requires a step beyond the comparative judgment that this hypothesis is better than its actual rivals. While the comparative judgment is indeed a 'weighing (in the light of) the evidence', the extra step—let us call it the ampliative step—is not. For me to take it that the best of set X will be more likely to be true than not, requires a prior belief that the truth is already more likely to be found in X, than not.

There are three possible reactions to this, each of which argues that IBE must be allowed *nevertheless* to play the role of leading to a new belief extrapolated from one's evidence. Clearly any such reaction must focus on the ampliative step, because the above objection is independent of the method of evaluation (of explanatoriness) that is used.

Reaction 1: Privilege

The first consists in a claim of privilege for our genius. Its idea is to glory in the belief that we are by nature predisposed to hit on the right range of hypotheses.[12]

We recognize here the medieval metaphysical principle of *adequatio mentis a rei.* Contemporary readers will not be happy to accept it as such, I think, and would hope for a justification. Such a justification could take two forms, allied respectively with naturalism and rationalism in epistemology.

The naturalistic response bases the conclusion on the fact of our adaptation to nature, our evolutionary success which must be due to a certain fitness. But in this particular case, the conclusion will not follow without a hypothesis of pre-adaptation, contrary to what is allowed by Darwinism.[13] The jungle red in tooth and claw does not select for internal virtues—not even ones that could increase the chance of adaptation or even survival beyond the short run. Our new theories cannot be more likely to be true, merely given that we were the ones to think of them and we have characteristics selected for in the past, because the success at issue is success in the future. The moths in industrial England became dark, not because they began to have more dark offspring but because the light ones were more vulnerable.

The rationalist response must be patterned after Descartes's argument for the correspondence of ideas to reality. Alvin Plantinga has suggested such a reason for privilege: given other beliefs about God, such as that we are made in his image, it is only reasonable to believe that we are specially adapted to hit on the truth when we come up with our (admittedly limited) guesses at explanation. Plantinga applies this even to belief in propositions and other abstract entities. But it takes more than a generally agreed concept of God to get this far. For even if he created us naturally able to perceive the truth about what is important for us in his eyes (perhaps to discern love from lust, or charity from hypocrisy, in ourselves), this may not extend to speculations about demons, quarks, or universals. *Privilege* is consistent, but seems incapable of either naturalistic or rationalist support.

Although it does not count as an argument, I should also point out that Privilege is entirely at odds with Empiricism. By this I mean the position that experience is our one and only source of information. Clearly this leaves open a great deal—experience may be very rich in its possible varieties; on the other hand, the information it brings us may not come as if in the voice of an angel, but in dubious and defeasible form. However that may be, the position sets one clear limit: if we do have innate or instinctive or inborn expectations, we'll be just lucky if they lead us aright, and not like lemmings into the sea. However basic or natural our inclination may be toward, for example, more satisfying explanations, that inclination itself cannot be relevant information about their content's truth.

Reaction 2: Force majeure

The *second* reaction pleads *force majeure*: it is to try and provide arguments to the effect that we *must* choose among the historically given significant hypotheses. To guide this choice is the task of any rule of right reason. In other words, it is not because we have special beliefs (such as that it will be a good thing to choose from a certain batch of hypotheses), but because we must choose from that batch, that we make the choice.

The *force majeure* reaction is, I think, doomed to fail. Circumstances may force us to act on the best alternative open to us.

They cannot force us to believe that it is, *ipso facto*, a good alternative.

Perhaps it will be objected that the action reveals the belief, because the two are logically connected. And in a general way that is certainly true. But it is exactly in situations of forced choice that action reveals very little about belief. Think of William James's example of a walker in the mountains who has the choice of jumping over a crevasse, or remaining for the night. Suppose that both a fall and exposure mean almost certain death. The prize is equal too: life itself. She jumps. Does this reveal that it was very likely to her that she would get across? Not at all, for even a very low chance would be reasonable to take in this case. Or if the Princess must open one of N doors, the Tigers skulking behind all but one, we can only conclude about the door she opened, that it seemed to her no less probable than $1/N$ that it would lead to freedom. And if there might be a Tiger behind each, and she is forced to choose nevertheless, we cannot even conclude that.

In the case of science we certainly observe theory choice. But just what belief is revealed there? Let us look carefully at the practice, and see what belief it entails, if any.

Scientists designing a research programme, bet their career and life's satisfactions on certain theoretical directions and experimental innovations. Here they are forced to choose between historically given theoretical bases. They are forced by their own decision to be scientists, to opt for the best *available* theory, by their own light. What beliefs are involved in this, can be gauged to some extent from their goals and values. Does this scientist feel that his life will have been wasted if he has spent it working on a false theory? Then he must feel that Descartes's and Newton's lives were wasted. Or does he feel that his life will have been worth while if he has contributed to progress of science, even if the contribution consisted in showing the limits and inadequacies of the theories he began with and the discovery of some new phenomenon that every future science must save? In the latter case his choice between theories, as basis for research, does not reveal any tendency to belief in their truth.

Reaction 3: Retrenchment

The *third* reaction is to *retrench*: 'Inference to the Best Explanation' was a misnomer, and the rule properly understood leads to a revision of judgement much more modest than inference to the truth of the favoured hypothesis. The charge should be that I have construed the rule of inference to the best explanation too naïvely. Despite its name, it is not the rule to infer the truth of the best available explanation. That is only a code for the real rule, which is to allocate our personal probabilities with due respect to explanation. Explanatory power is a mark of truth, not infallible, but a characteristic symptom.

This retrenchment can take two forms. The *first form* is that the special features which make for explanation among empirically unrefuted theories, make them (more) likely to be true.[14] The *second form* is that the notion of rationality itself requires these features to function as relevant factors in the rules for rational response to the evidence. I will take up both forms—the first in the remainder of this section, and the second in the next chapter. Let us note beforehand that the first must lean on *intrinsic explanatoriness*, which can be discerned prior to empirical observations, and the second specifically on *explanatory success* after the observational results come in. What the criteria are for either, we shall leave up to the retrencher.

Retrenchment, form 1

Is the best explanation we have, likely to be true? Here is my argument to the contrary.

I believe, and so do you, that there are many theories, perhaps never yet formulated but in accordance with all evidence so far, which explain at least as well as the best we have now. Since these theories can disagree in so many ways about statements that go beyond our evidence to date, it is clear that most of them by far must be false. I know nothing about our best explanation, relevant to its truth-value, except that it belongs to this class. So I must treat it as a random member of this class, most of which is false. Hence it must seem very improbable to me that it is true.

You may challenge this in two ways. You may say that we do have further knowledge of our own best explanation, relevant to its truth-value, beyond how well it explains. I'm afraid that this will bring you back to the reaction of Privilege, to glory in the assumption of our natural or historical superiority. Or you may say that I have construed the reference class too broadly. That is fair. The class of rivals to be considered should be on a par with our own, in ways that could affect proportions of truth. So we cannot include in it two theories, each having the same disagreement with ours at point *X*, but then disagreeing with each other at point *Y*, on which ours has nothing to say. So for each statement ours makes, beyond the evidence we have already, we can include only one theory disagreeing with that statement, for every way to disagree with it. But of course, there is only one true way to agree or disagree at this point. So the conclusion, that most of this class is false, still stands.

David Armstrong, replying to a version of this argument, writes 'I take it that van Fraassen is having a bit of fun here.'[15] Yes, I had better own up immediately: I think I know what is wrong with the above argument. But my diagnosis is part and parcel of an approach to epistemology (to be explained in the next chapter), in which rules like IBE have no status (*qua* rules) at all. As a critique of IBE, on its own ground, the above argument stands.

One suspicious feature of my above argument is that it needed no premisss about what features exactly do make a hypothesis a good explanation. Let us consider a contrary argument offered by J. J. C. Smart, which does focus on one such feature, simplicity. I think that both argument and counterargument will be rather typical of how any such debate could go (for any choice of explanatory feature). Smart begins as follows:

> My argument depends on giving a non-negligible a priori probability to the proposition that the universe is simple. . . .
>
> Let p = the observational facts are as if there are electrons, etc.
> q = the universe is simple
> r = there really are electrons, etc.
>
> We can agree . . . that $P(p) > P(pr)$, as of course we have to! But I want to say $P(pr/q) > P(p\bar{r}/q)$.[16]

That is, the probability *on the supposition that the universe is simple*, is greater for our best explanation which entails that the phenomena will continue as before, than it is for a denial of that explanation which agrees on those phenomena.

The reason for this judgement, on Smart's part, must hinge on a connection between explanation and simplicity. So it does. It is exactly because the explaining hypothesis is simple (as it must be, to qualify as explanation) that the supposed simplicity of the universe makes the hypothesis more probable than its denial (under the further supposition that the explanation is right about all the phenomena, even those to come). And it seems at first sight plausible to say that a supposed simplicity in nature, will make simple theories more likely to be true.

But this plausibility derives from an equivocation. If the simplicity of the universe can be made into a concrete notion by specifying objective structural features that make for simplicity, then I can see how Smart may have arrived at the opinion that the universe is (probably) simple. For there can be evidence for any objective structural feature. But if the universe's simplicity means the relational property, that it lends itself to manageable description by us (given our limitations and capacities) I cannot see that. The successes we've had are all successes among the descriptions we could give of nearby parts of the universe, and of the sort which our descriptive abilities allow. Suppose it is true that the frog can distinguish only the grossest differences between objects at rest, but can notice even small moving objects. Then his success in catching insects flying by is no index of how many potential prey and potential enemies sit there quietly watching him.

Could simplicity in the first sense, which we might have reason to surmise, make more probable the simpler-for-us among the accounts we can give? That depends in part on how much simplicity of theories has to do with simplicity of the world as described by those theories. But suppose there is an intimate connection—unlike, say, in literature where the simplicity and economy of form in poetry does not limit it to simplicity of subject, in comparison with prose. Then still the allocation of probabilities is effectively prevented by a very modest consideration. Simplicity is global. A part of a structure, which is very simple overall, may be exceedingly complex considered in isolation. Here is a simply described set: that of all descendants of Geoffrey of Monmouth, alive today. Now, try to describe it purely in terms of features recognizable today (by geographical location, blood type, what have you) without reference to the past. Considered as part of a historical structure, the set is easily delineated; viewed short-sightedly in the twentieth century it is not. Similarly for the simplicity of the universe as a whole, and those aspects and parts of it on which our sciences focus. The simplicity a situation has in virtue of being part of a simple universe, does not make more likely any simple putative description of it by itself alone.

Retrenchment, form 2
I already raised the possibility of a sophisticated, probabilistic version of the rule briefly above. Combining the ideas of personal probability and living by rules, the new rule of IBE would be a recipe for adjusting our personal probabilities while respecting the *explanatory* (as well as predictive) success of hypotheses. This will be investigated in the next chapter, after we have taken a much closer look at the representation of opinion.

5. WHAT GOOD IS THERE IN THIS RULE?

There must clearly be a solid basis for the appeal and renown that IBE has enjoyed over the years. It is important to ferret this out. But as with any subject, we must carefully separate the inflated claims of philosophical exponents from the grains of common sense which gave those claims their initial appeal. Eventually we must also show that the common-sense part is equally respected in our own account.

If I already believe that the truth about something is likely to be found among certain alternatives, and if I want to choose one of them, then I shall certainly choose the one I consider the best. *That* is a core of common sense which no one will deny.

But how far is it from this common sense to IBE, conceived as cornerstone of epistemology? This rule cannot supply the initial context of belief or opinion within which alone it can become applicable. Therefore it cannot be what 'grounds' rational opinion.

That is only the first point. Next we can see that even if the rule is applicable, we might very well not wish to apply it. Suppose it seems likely to me that one of the first six horses will win the race, and that of these, horse No. 1 is the best. It does not follow at all that I shall then wish to predict that horse No. 1 will win, for this might mean no higher probability than $\frac{1}{5}$ for its winning. Similarly, if I turn away then, and just at the end of the race a great cheer goes up, the best explanation for me of this cheer will be that horse No. 1 has just won. And still I shall be no readier to say that this is what has just happened. If a *force majeure* makes me predict, then I shall indeed say 'No. 1'—but this will certainly not reveal then that I believe it.

A rule which we would often, when it is applicable, prefer not to apply, is not a rule we are following! Common sense will often prevent us from applying IBE; and it could not very well do that if it were the epistemic categorical imperative. Even more important, perhaps, is this: even if the rule is applicable, and we make our choice in accordance with it, we may not be following the rule. This sounds paradoxical, but think: must a choice among hypotheses, even if unconstrained, necessarily be a choice to believe?

In general, the common-sense choice will be in the context of an opinion to the effect that this batch of hypotheses have a balance of certain virtues, and are well fitted to serving our present aims. Likelihood of truth will presumably be among those virtues, or among those aims, but it need not be alone. Informativeness with respect to topics of interest may be another. *When we choose the best, therefore, the choice must be interpreted in terms of the basis for choice.* If likelihood of truth is not the sole basis, then the choice—choice to accept—must not be equated with choice to believe.

How little comfort common sense gives to philosophical fancy! What has happened to extrapolation of general truth from evidence alone? No more than the metaphysician do I think that common sense brings its own clarity with it. So I propose next to discuss the whole subject again in a higher—and more constructive—key. Specifically, I will consider versions of IBE that make full use of the conceptual resources of probability. And yet, its fortunes will not improve thereby.

7. Towards a New Epistemology

> . . . it is by no means clear that students of the sciences . . . would have any methodology left if abduction is abandoned. If the fact that a theory provides the best available explanation for some important phenomenon is not a justification for believing that the theory is at least approximately true, then it is hard to see how intellectual inquiry could proceed.
>
> Richard Boyd[1]

So far I have only offered a critique of traditional epistemology with its ampliative rules of induction and inference to the best explanation (abduction). But this mainstream does not constitute the only tradition in epistemology. The seventeenth century gave us besides Descartes and Newton also Blaise Pascal, and from his less systematic writings there sprung a stream that in the succeeding three centuries has become a powerful river: the underground epistemology of *probabilism*.

After introducing its basic ideas, I shall show that it leads us into a much more radical and far-reaching critique than we have seen so far. The rule of IBE, and indeed the whole species of ampliative rules, is incoherent. I shall deliver this critique at the outset through a proxy, a foil, one particular sort of probabilist: the orthodox Bayesian. The rigours of *his* views are however also considerably more than his arguments can demonstrate. I will go on then to propose an epistemology that is certainly still probabilist, but offers a reconciliation with traditional epistemology. It does give room to practices of ampliation beyond the evidence. (In the next chapter, we shall see that it also allows us finally to illuminate how objective chance should guide our personal expectation—remember the fundamental question, concerning chance, which previous chapters had to leave unanswered?—and how it may enter the opinion that guides rational decisions.) To end I shall argue that, despite the ominous warnings of the past, our new epistemology is driven into neither sceptical despair, nor feeble relativism, nor metaphysical realism.

1. PASCAL: THE VALUE OF A HOPE

Blaise Pascal is well known for his Wager in which the stake is one's whole life and the possible gain, eternity. The argument is subtler than its caricatures allow, though certainly not cogent *enfin*. But it was the first example of a truly revolutionary way of thinking. Its basic principle was stated succinctly in the great logic text prepared by his colleagues at Port-Royal:

> To judge what one must do to obtain a good or avoid an evil, it is necessary to consider not only the good and the evil in itself, but also the probability that it happens or does not happen; and to view geometrically the proportion that all these things have together.[2]

This passage presents the new paradigm of rational judgement. The person who judges and decides must determine the 'value of the hope'—as Christiaan Huygens so engagingly called it in his monograph of 1656/7—involved in each possible alternative before him. How will he determine this value?

It was Huygens who codified the principles precisely. His standard phrase is

> To receive a contract which . . . is worth . . .

The contract is described by stating probabilities and possible pay-offs. Such a contract pays this or that depending on what turns out to be the case (the ship comes in, the ship goes down with all hands, . . .). Let the contract pay x if A is the case and nothing otherwise. Denote such a contract by $[A]^x_0$. Then the value of the hope is found as follows:

> the value of $[A]^x_0$ equals: x times the probability that A is the case.

The cost of entering the contract (e.g. the price of a lottery ticket) can be listed as part of the pay-offs. Thus if the contract pays 10 if A be true, and zero otherwise, but costs 4 to enter, we can denote it as

$$[A]^{10-4}_{0-4} = [A]^{6}_{-4}$$

and its value equals then

6 times the probability that A, plus (-4) times the probability that not A.

Of course this is equivalent to taking a 'book' of two simple contracts, namely $[A]_0^6$ and $[\text{not } A]_0^{-4}$.

With this basic insight into how one calculates the value of the hope in mundane projects, Pascal approached the project in which the stake is one's life, and the hope eternal bliss. The insights continued to be developed on a mundane level, by for example Bayes and the Bernoullis in the next century, Jevons and De Morgan in the nineteenth, Ramsey and De Finetti in the twentieth century. While its details were being worked out, philosophy mainly ignored it as mathematical gamesmanship or materialistic technology of the mind. It is neither. Pascal and those who followed him showed us how to reconceive all the problems of epistemology.

2. THE PROBABILISTIC TURN

So Pascal and the *Port-Royal Logic* taught us that the opinions which enter our practical deliberations, just like the evaluations of good and evil, are not simple. They are not merely: *yes* this will happen; *no* that will not. We must think in terms of probability.

But, it is at once objected, we do not: few people ever learn to grasp such words, and fewer to apply the concept. This is false—not in what it says (that is true enough), but in what it insinuates. For our opinion does take on many subtle forms; this is so for soldier, sailor, fisherman, beggarman, poor man, thief, as well as for scientist, scholar, gentleman, crook. No one's opinions can be adequately described in simple *yes* and *no* terms. The concept of probability, if brought into this context, is indeed technical; its value must be that it provides a systematic description of the opinions framed by everyone without it. Here are forms in which, for example, fishermen will express their opinion about the coming weather. The first few forms are *qualitative*, then come *comparative* then *quantitative*; and next we see that all the preceding can take on *conditional* form as well.

1. Rain seems very likely to me.
2. It seems likely to me that rain will end by evening.
3. Rain seems much likelier to me than snow.

4. Rain seems as likely as not.
5. Rain seems twice as likely as not.
6. Rain seems likely if it turns cold. (Rain seems likely to me on the supposition that it will turn cold.)
7. Snow seems likelier than rain if it turns cold.
8. Snow seems to me as likely as not if it stays cloudy tonight.

I do not maintain that these examples yield a complete typology of judgement. But I do think that these are all quite common forms of judgement.

We can introduce the notion of probability here in a preliminary way by suggesting a *translation* of all such judgements into probability terms. The translation relates 'likely' and 'likelier' to 'probability' in the way that 'tall' and 'taller' are related to 'height'. Thus we have, for example,

1*. My probability for rain is very high.
2*. My probability for rain is much higher than my probability for snow.
4*. My probability for rain is the same as for no rain.
5*. My probability for rain is twice that for no rain.
6*. My probability for rain, conditional on the supposition that it turns cold, is high.

I don't need to do more than give you these examples, to show clearly what the recipe is for translation. We can now also use obvious abbreviations, and a scale from *zero* to *one* for the probabilities. That allows us to abbreviate 4* and 5* very effectively:

4*. $P(\text{it will rain}) = P(\text{it will not rain})$
i.e. $P(\text{it will rain}) = \frac{1}{2}$
5*. $P(\text{it will rain}) = 2P(\text{it will not rain})$
i.e. $P(\text{it will rain}) = \frac{2}{3}$

Finally numbers! As my examples make very clear, it is *not* supposed that I have somewhere in my head a numerically precise probability function. I may not have any judgement, or only a vague judgement, on some questions. The probability terminology is only being used to provide us with a single, systematic way of characterizing all ten forms of judgement. To complete the notational convention, let us use the upright bar to indicate conditionality

8**. $P(\text{it will snow} \mid \text{it stays cloudy tonight}) = \frac{1}{2}$

Distinguish this carefully from the bogus translation:

8?. If it stays cloudy tonight, then P(it will snow) = $\frac{1}{2}$

We will not be tempted to 8? if we realize that P(—) = — is first and foremost the form in which I *express* my opinion, rather than a form in which I state what it is.[3] With 8?, we would license the wrong inferences. For the following two arguments are exactly similar *fallacies*.

A. Peter wants to lose his left hand if he must lose a hand at all.
In fact, Peter must lose a hand.
Therefore Peter wants to lose his left hand.

B. My probability for snow, if it turns cold, equals $\frac{1}{2}$.
In fact it will turn cold.
Therefore my probability for snow equals $\frac{1}{2}$.

The first example was suggested by Richmond Thomason to help elucidate conditional statements in moral deliberation: *modus ponens* does not apply! The same is true of conditional probability.

To conclude then: an initial more adequate typology of judgements of opinion characterizes them as attitudes expressed by *It seems to me that* and its modes. A second, at least equally adequate typology characterizes them in terms of *my personal probability* and its modes.

3. LOGIC OF JUDGEMENT

We come now to the logic of judgement, that is, of expressions of opinion as here conceived.[4] The familiar problem of logic is the one studied by Aristotle: find patterns of reasoning in which, *if* the premisses are true, *then* the conclusion is true. In such a case, we say the conclusion *follows from* the premisses. But is this the only important sort of consequence relation?

Consider the similar case of commands: 'Peter, give Paul a horse for his birthday!' and 'Peter, give Paul some present for his birthday!'. Does the second follow from the first? How to draw an analogy? 'True' does not seem to apply—we need an analogous concept.

Here is a simple proposal.[5] Let us call the command *satisfied* if the commanded state of affairs does occur (for whatever reason). And say that the second command *follows* from the first exactly if it is satisfied in any possible state of affairs (or world) in which the first is satisfied.

We ignore here altogether the question of truth-conditions for the statement 'He commanded that *C*'. Instead, in the logic of commands we focus on what a command 'entails', i.e. requires, in order to be fulfilled.

Let us now consider opinion. There too we have two sorts of discourse, though it often takes the same verbal form

(*a*) Expression of my opinion 'It seems likely to me that it will rain'

(*b*) Statement of autobiographical fact '(I am in the state of opinion in which) it seems likely to me that it will rain'

(*c*) Statement of biographical fact 'It seems likely to Bas van Fraassen that it will rain'

(*b*) and (*c*) are not significantly different, but (*a*) is.

One approach to the logic of judgement would be to ask for truth conditions for (*b*) and (*c*), i.e. to ask which biographical statements entail each other. In fact, this is the approach followed in most of the literature on the propositional attitudes, starting with Hintikka's *Knowledge and Belief*. But it seems inevitably to focus on a trivial question. Since most of us are often confused and unclear, very few opinions follow, in this sense, from other opinions. The second approach is to address (*a*), in terms of the *point* of such discourse. The point of commands is to make things happen, and so we get a minimal handle on their internal logical connections by looking at when they are satisfied. The point of having opinions is to have an (internal) guide to the conduct of life—to successful conduct, by one's own lights. If (*a*) were followed by

(*d*) 'It seems also unlikely to me that there will be any precipitation'

we can see at once that successful conduct of one's life guided by (*a*) cannot involve guidance by (*d*).

This second approach—linking the logic of judgements of opinion with the very point of having opinions at all—seems like a good idea; but how shall we carry it out?

Any act or decision can be evaluated in two ways. If we evaluate it beforehand, we ask how *reasonable* it is, and afterward, we ask to what extent it was *vindicated*. The two cannot be the same since the agent cannot have knowledge beforehand of the exact outcome and consequences of his action—vindication or the lack thereof lies as yet beyond his ken. But there must be a connection, since the point of deciding or acting lies in the outcome (broadly construed). Therefore a minimal criterion of reasonableness is that *you should not sabotage your possibilities of vindication beforehand.*

If your aim, in giving commands to Peter, was that he should give a present to Paul, then you are vindicated if he does. Suppose you give him the two commands, to give Paul a horse and to give Paul nothing. Then you have given commands which cannot be jointly satisfied—so the vindication will necessarily leave something to be desired. Similarly, if your aim, in making factual descriptive statements, is to give true information, then you are vindicated if your statements turn out to be true. Should you make several mutually incompatible statements, they *cannot* be jointly true, so your vindication will *necessarily* be less than total.

In both cases, vindication was sabotaged by the choices made to begin. Thus it is easy to see that reasonableness is closely connected, at a certain level, with vindication—even though reasonableness cannot guarantee vindication. And the examples also show how the criterion of reasonableness in terms of not sabotaging the possibilities of vindication, has something to do with logic.

We must therefore, as next step, find criteria of vindication for judgements of opinion. There are two simple ones. The first is *calibration*. Consider a weather forecaster who says each morning something like 'My probability for rain today equals x per cent'. Sometimes x equals 10, sometimes 80, etc. How good a forecaster is he? We call him *perfectly calibrated* if the following is the case, for every numeral substituted for 'x':

> The proportion of rainy days among those on which he says 'My probability for rain today equals x per cent' equals x per cent.

This criterion can also be applied to other sorts of statements. It is then easy to see how he could sabotage his possibility of perfect calibration, for example, if every day he says

My probability for rain today equals x
My probability for precipitation today equals y

and he does not typically make y at least as great as x. That would be a *logical* fault.[6]

The second criterion for the vindication of opinion is gain, of whatever sort you value. Suppose that you buy a number of wagers or contracts, evaluated in terms of your opinion by Pascal's paradigm. On each you either win or lose money when the time comes to settle up. Vindication = net gain. If your opinions were such that we know beforehand that you *cannot* have a net gain, or that you *must* have a net loss, then you sabotaged beforehand your very possibility of vindication. We call this a Dutch Book situation:

> A Dutch Book is a set ('book') of wagers such that under all circumstances, the total pay-off is negative.

An example would be $[A]^{8}_{-10}$ and $[\text{not } A]^{8}_{-10}$ which under all circumstances has a total payoff of -2. A person's state of opinion is called *incoherent* if the value of the hope is, for him, positive for both these wagers. For such a person, the possibility of vindication for his opinion is sabotaged from the start.

Now what is logic? Exactly what it was for Aristotle, but transposed to other things besides factual/descriptive statements as well. Specifically, in the case of judgements of opinion, we want to know:

(1) What combinations of judgements constitute an incoherent state of opinion?
(2) Which judgements follow from a given set of judgements?

For (1) the short answer can be given as above, either in terms of calibration or in terms of net pay-offs on wagers. The longer answer—the outcome of the study of the logic of these judgements—will give us secondary criteria in terms of the *form* of the judgements alone. That is the probability calculus; its proper discussion is postponed till a later chapter (but see *Proofs and Illustrations*).

For (2) the short answer will be:

(2′) Judgement *A follows* from judgements *B*, *C*, . . . (*on pain of incoherence*) if any judgement incompatible with *A* is also incompatible with the set *B*, *C*, . . .

where we define incompatibility as follows:

(3) The judgement X is *incompatible* with the set of judgements B, C, . . . exactly if any state of opinion characterized by judgements X, B, C, . . . is incoherent

There are some reasons for dissatisfaction with these approaches to the logic of judgement. Calibration makes sense primarily for judgements which are essentially repeatable. Wagers require pay-off points—consider that the proposition *We shall find out, by time t, that A* is very different from the proposition A itself. But perhaps it is a further ideal of reason that the logic of judgement should be no different from what it would be if we gambled with angels (truthful, guileless, and fair) who are always able to reveal the relevant facts at any time.

Proofs and illustrations[7]

Here follows the *Two-Minute Dutch Book Theorem*; it shows that a person who is not vulnerable to Dutch Books is one whose degrees of belief obey the probability calculus. That calculus can be summarized in two axioms:

I. $0 = P(A \text{ and not } A) \leqslant P(A) \leqslant P(A \text{ or not } A) = 1$
II. $P(A \text{ and } B) + P(A \text{ or } B) = P(A) + P(B)$

where the assignment is to the propositions expressed (not the sentences). We assume Pascal's and Huygens's principle that the value of contract $[A]^x_0$ equals x times $P(A)$. But we will only need (zero-one) cases in which $x = 1$; in *these* cases, the value equals the probability.

Table 7.1 is a tabulation of pay-offs on various (zero-one) contracts, in each logically possible case. It is clear that if you pay more than 1 for a contract of form $[—]^1_0$, you will lose money; if you sell $[A \text{ or not } A]^1_0$ for less than 1, you will lose money, and so forth. Thus the Dutch Book justification of axiom I can be read off from these 'truth-tables', just because we are dealing with wagers for which probability = value.

TABLE 7.1. *Justification of axiom I*

Possible case	*A*	not *A*	*A* and not *A*	*A* or not *A*
(1) *A* true	1	0	0	1
(2) *A* false	0	1	0	1
	0 ⩽ pay-off ⩽ 1		always = 0	always = 1

For axiom II (Table 7.2) we imagine two people: Peter sells Paul the bets $[A]_0^1$, and $[B]_0^1$ and Paul sells to Peter the bets $[A \text{ and } B]_0^1$ and $[A \text{ or } B]_0^1$. What will the pay-offs be? It is clear that in each possible case, Paul and Peter received the same income. If either had paid more for his contracts than the other, he would have made a poor deal—and could have realized this beforehand.

TABLE 7.2. *Justification of axiom II*

Possible cases	*A*	*B*	*A* and *B*	*A* or *B*
(1) *A* true, *B* true	1	1	1	1
(2) *A* true, *B* false	1	0	0	1
(3) *A* false, *B* true	0	1	0	1
(4) *A* false, *B* false	0	0	0	0
	what Paul receives		what Peter receives	

Therefore the total value of the book of wagers bought by Paul—namely, $P(A)+P(B)$—equals the total value of the other book of wagers—that is, $P(A \text{ and } B)+P(A \text{ or } B)$—according to anyone whose opinion does not advise either Paul or Peter to make a deal which necessarily loses money.

4. INFERENCE TO THE BEST EXPLANATION IS INCOHERENT

After our critique of IBE in the preceding chapter, there was just one possibility left. Behind the naïve rule of IBE there might lie a recipe for adjusting our personal probabilities, in response to new experience, under the aegis of explanatory success. Since no such recipe has been precisely formulated, we are again (as so often with ideas about confirmation) looking at the gleam in some hopeful's eye. But we can try to see what it is like to do *without* any such rule—shall we be disadvantaged? What about such common examples as: I see dirty dishes and, although other explanations are possible, infer the best, namely that someone has eaten? Obviously these must be reconstrued: I make no inference at all, it was already highly probable to me that someone has been eating, given that there are dirty dishes. But nothing much can be proved in this anecdotal fashion.

We can also try to ask what *any* such rule would bring. The answer is that it would make us incoherent. This is the striking and powerful critique by which probabilism attacks the mainstream tradition.

My argument will be rather long, although the main technical aspects are left till the later chapters on probability theory. First I will briefly explain how the simplest sort of probabilist—the *orthodox Bayesian*—amends his opinion in response to experience. He uses no ampliative rule, but only logic. Still he is not reduced to epistemic helplessness! Returning to the last chapter's example of the Alien Die, we'll look at a Bayesian statistical model to see how quickly and sensitively his predictions change in response to evidence. Second, I propose to study the perils and fortunes of a particular person, Bayesian Peter, who becomes converted to the use of some sort of probabilistic IBE. He quickly discovers that he is led into incoherence. The argument is general: it applies to *any* ampliative rule in epistemology.

Purely logical updating

To depict your state of opinion, you can use a model which I call the Muddy Venn Diagram. Just represent the propositions you have an opinion about by areas on the usual Venn diagram, and represent your personal probability by means of some quantity of mud heaped on them (see Fig. 7.1). Call the total mud present one unit always; the proportion of mud on an area equals the probability of the proposition represented by that area. This model is easier to grasp than any axioms or rules, and does just as well. (There are in fact deep theorems to show that we have here in essence the most general model of probabilities, provided the mud can be fine enough.) Suppose now that I go for a walk in the garden, and come away absolutely convinced that a flying saucer has landed there. I reconstruct this as follows: I had originally a certain state

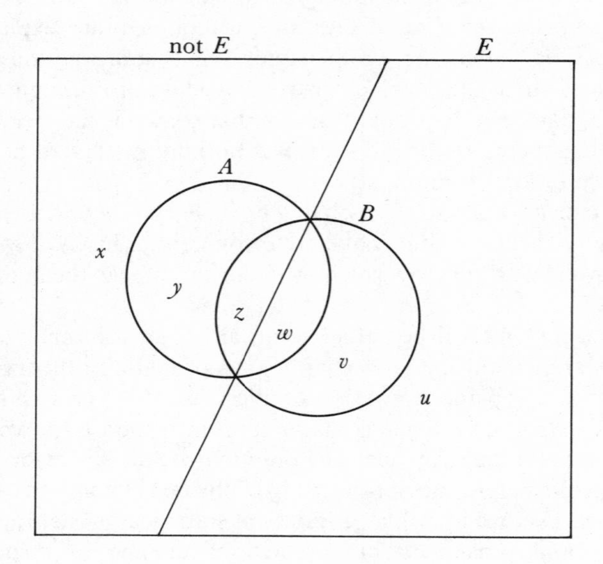

FIG. 7.1. The Muddy Venn diagram

of opinion, but accepted the constraint to become certain of this new proposition *E*, and adjusted my opinion accordingly. There is a rule for such adjustment, *Simple Conditionalization*.

This rule is easily explained in terms of the Muddy Venn Diagram. You simply wipe away all the mud on the area representing not-*E* (which has become certainly false for you). This has two effects: it raises the probability of *E* to 1, and it leaves all the odds between propositions which imply *E* (the only ones which remain, represented by areas inside that of *E*) exactly the same as they were.

To the orthodox Bayesian, this is the sole form of response to experience. That is also why he or she is called 'Bayesian', for to say that he updates by simple conditionalization may also be expressed as 'he updates by Bayes's theorem', which is really the same thing. Now to show how these simple principles work in practice, we'll look at the sort of statistical model that simulates an orthodox Bayesian in action.

A statistical model: the alien die revisited

For an example of a statistical model, let us continue with the alien die. The specific model I have introduces a factor X of bias, which can take N different forms: $X(1)$, ..., $X(N)$. If the die has bias $X(I)$, then the probability of *ace* on any one proper toss equals I/N. The one bias whose role we will specially inspect is the perfect bias $X(N)$, which gives *ace* the probability $(N/N) = 1$. In this model, each bias has the same initial probability $1/N$. Now we toss the die properly and independently J times, and it comes up *ace* each time. To update the model, one applies Bayes's theorem, and this gives each bias a new, posterior probability. (See *Proofs and Illustrations.*) Figure 7.2 shows the probabilities of the hypotheses of bias $X(I)$ initially, after four *aces* and after ten *aces*. Notice that the model as a whole has a certain bias, for it allows that the die

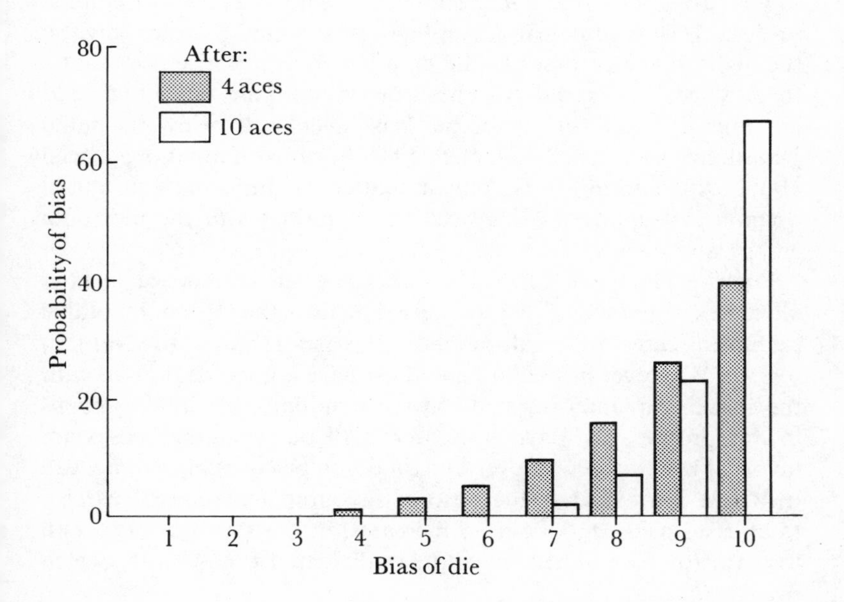

FIG. 7.2. The Alien Die

may have perfect bias $X(N)$, but not $X(0)$. Such bar graphs are a little crude: our diagram depicts probabilities of each *bias* hypothesis, in the model with $N = 10$, initially, after four *aces* and after ten *aces*. The probabilities depicted in Table 7.3 are approximate. Only a few of the hypotheses of bias are displayed here, but the probability is very small (less than 1 per cent) after four *aces* for I less than 4, and after ten *aces* for I less than 7.

TABLE 7.3. *Posterior probability of bias*

number of tosses	4	10
	%	%
Probability of bias I, given that all were aces:		
Bias $I = 8$	16	7
Bias $I = 9$	26	23
Bias $I = 10$	39.5	67

The probability that J *aces* should come up in a row, conditional on bias $X(I)$, is of course much higher for high I than for low. So the high bias hypotheses fit the data better. But notice how subtly the evidence works: the hypothesis of bias $X(8)$ has much increased its standing after four *aces*, but has fallen well below its initial plausibility after ten. None of this has to do with anything special about explanation—it is only a matter of adjusting our initial opinion in response to how well our hypotheses *fit* the new data empirically.

Dretske made the curious remark that our confidence in laws 'increases at a much more rapid rate than does the ratio of favorable examined cases to total number of cases' ('Laws of Nature', p. 256). Whatever he meant here, must have a great deal to do with the special explanatoriness of laws, for nothing like that happens in this progress by Bayes's theorem. All our examined cases are favourable here, so whatever our total number of trials will be, call it M, we can calculate the ratio of favourable examined cases to total cases as J/M. This ratio increases, in the step from $(J-1)$th toss to Jth toss at the rate (J/M) divided by $(J-1)/M$, which equals $J/(J-1)$.

This has nothing to do with model or hypothesis; the rate is 10/9 between our 9th and 10th toss as long as every toss comes up *ace*. What about our confidence in hypothesis of bias $X(N)$? At what rate does it increase? That depends on the model—i.e. on the rival hypotheses in the competition. Some results are shown in

Table 7.4. The comparison is the exact opposite of what Dretske suggests. These ratios all tend to one in the limit, but confidence in the 'little law' of bias does so more slowly.

TABLE 7.4. *Rates of increase*

	Increase in probability of $X(N)$		Increase in proportion of favourable cases
	$N = 2$	$N = 10$	
Toss 3 to 4	1.0588	1.194	(4/ 3) = 1.333
Toss 9 to 10	1.00098	1.05556	(10/ 9) = 1.111
Toss 14 to 15	1.0000305	1.0273	(15/14) = 1.07
Toss 19 to 20	1.0000009	1.01478	(20/19) = 1.05

Nor do the posterior probabilities climb more quickly in other ways. Our diagram shows these for $N = 10$, and does not suggest any more favourable interpretation of Dretske's remark. If we have envisaged ten tosses, then after four *aces*, the ratio of favourable examined to total equals 40 per cent. Meanwhile the probability of the 'little law' of total bias has climbed only to 39.5 per cent. After ten *aces*, if the total is some arbitrary number M, the former ratio climbs from $1/M$ to $10/M$, a tenfold increase, while the probability of total bias climbs from 10 per cent to 67 per cent, only a 6.7 fold increase.

All this may only underline the fact, of course, that by reading a textbook in probability or statistics or even experimental design, we won't get any inkling at all of the special confirmational role of lawhood and lawlike explanation!

Yet the simple, logical update these texts teach does lead to powerful prediction. What probability do we give to, say, the fifth toss coming up *ace*? At the beginning of the process, the answer is simply the average of all the probabilities bestowed by bias. After four *aces* already, it is a weighted average of these probabilities, for our hypotheses no longer have equal standing.

Probability of ace	$N = 2$	$N = 10$
Initial probability	0.75	0.55
After 4 *aces*	0.97	0.87
After 10 *aces*	0.9995	0.955

Let us take note of these effects or predictions, for we shall need to compare a person who does not infer to the best explanation with one who does.

The perils of Bayesian Peter

The person who proceeds as above is—within these confines, though perhaps not elsewhere—a perfect Bayesian agent. Let us call him Peter. Imagine now that he meets a travelling Preacher, who convinces him that the hypotheses of bias are like little or specialized laws, which also have explanatory power. The hypotheses of high bias not only predict four *aces* in a row with high probability; and not only turn out to be 'right', to some degree, if that happens; but will offer us (after the fact) a good explanation of why we saw nothing but *aces*. Our Peter feels intuitive agreement welling up in his breast. But then the Preacher goes on to say: in view of this explanatory success, you should raise your credence in the more explanatory hypotheses.

'What?' exclaims Peter. 'More than I would anyway?' 'Yes', says the Preacher. 'Our forefathers all inferred to the best explanation, and in daily commerce, our humbler brothers still do. We who have seen the light of probability should not disdain their insight, but give due respect to explanatory success.' This Preacher then proposes a rule which fixes a posterior probability for each hypothesis of bias, depending not only on the initial probabilities and the series of outcomes, but also on a feature which he calls explanatory success. (He also, of course, gives an auxiliary rule for gauging that sort of success.) We need not inquire into details. Suffice it to say that his proposal is non-trivial, and really gives *bonus* probabilities to certain hypotheses. Of course, the most explanatory hypothesis, for all *aces* is $X(N)$ and so it definitely gets a *bonus*.

The Preacher goes away

The Preacher goes away, but not before our heretofore Bayesian adopts the rule. After the Preacher goes, Peter contemplates further tosses. If tossing yields all *aces*, he will give hypothesis of bias $X(N)$ a higher probability than he would have otherwise.

At this point, however, Peter remembers a test designed by

De Finetti, which appears to be relevant.[8] Imagine that you are asked to announce your probability q for an event—*ace* again on the next toss, for example—and will be penalized q^2 monetary units if it does not happen and $(1 - q)^2$ if it does. (Your penalty will be the square of your probability for what does *not* happen.) What announcement should you best make? Imagine most specifically that the test will be carried out after a run of four initial *aces*, and the question is whether the next toss will also yield *ace*.

To be quite concrete, suppose that Peter has the ($N = 10$) model, so his initial probability that such a test would be followed by a fifth *ace*, equals 0.87. But now he has also adopted the Preacher's special rule, so he foresees that after a run of four *aces*, he will redistribute his probabilities still further in the recommended way. Thus his probability at the time of this test (if it occurs) for yet another *ace* will be higher than 0.87—say it will be 0.9. The difficulty this presents for him is that De Finetti suggests that you must better calculate your expected penalty, before deciding what answers you will give in such a case. And from his *present* vantage point he calculates by his present, initial probabilities, before the rule will have exercised its leavening influence. He calculates as follows:

Expectation value of the penalty:
(*a*) If I say '0.9'—$0.87(0.1)^2 + 0.13(0.9)^2 = 0.114$
(*b*) If I say '0.87'—$0.87(0.13)^2 + 0.13(0.87)^2 = 0.1131$

So it would be better for him to lie! Or, to put it less charitably, his present expectation for acting on his newly adopted rule is worse than for not acting on it. (See further *Proofs and illustrations*.)

Our erstwhile Bayesian backslides, and gives up on the new rule.

The Preacher returns

The Preacher returns and hears the story. What do we expect the Preacher to say? Of course he says: 'Your problem was that you did not give up enough of your pagan ways. This probability of 0.87 had already become quite irrelevant to any question of what would happen after an initial run of four *aces*. If you had followed the rule properly, you would not only have foreseen that your new probability afterward would become 0.9. You would also *now* have had as conditional probabilities, those probabilities that you knew you would get to have, after seeing such an initial run of *aces*. Thus the leavening effect of explanatory power in our lives, although activated properly only after seeing explanatory success, reaches back from the future into the present as it were, and revises our present conditional opinions.'

The backsliding convert, docile as can be, is reconverted, and the Preacher leaves again. Shortly afterward, one of Peter's old Bayesian friends comes to see him, with a dynamic coherence argument due to David Lewis.[9] Rather than explain the argument, this friend proposes some bets. Feeling that certain subjects are a little touchy just now, so soon after the reconversion, he is careful not to propose any conditional bets. What he proposes instead for proposition E (the initial four tosses show *ace*) and H (the fifth toss shows *ace*) are the following three bets:[10]

Bet I pays $10 000 if E is true and H false
Bet II pays $1300 if E is false
Bet III pays $300 if E is true

Each pays *zero* in all other cases. Together they evaluate these bets, on the basis of their shared initial probabilities (model $N = 10$). These are

Initial probability that E is true equals the average of $(0.1)^4$, ..., $(0.9)^4$; that is, 0.25333
Initial probability that E is false is 0.74667
Initial probability that E is true and H false is the average of $(0.1)^4(0.9)$, ..., $(0.9)^4(0.1)$, 0; that is 0.032505
Fair cost of bet I equals $325.05
of bet II equals $970.67
of bet III equals $76.00
where fair value is calculated as pay-off times initial probability. The total cost of all three bets equals thus $1371.72. Keep your eye on this number!

Peter buys all three bets from his friend. But this friend is a schemer. It is easy to see what he will do after the fourth toss. Suppose that not all four have come up *ace*. Then Peter has lost bets I and III, but has won bet II. The friend pays Peter $1300 accordingly, but has then a net gain of $71.72.

If, on the other hand, E has come true, then Peter has already won III and lost II. So the friend pays him $300. Bet I has now become equivalent to:

Bet IV pays $10 000 if *H* is false

which the friend now proposes to buy from Peter. At this point, Peter who has been following the Preacher's rule, has new probability 0.9 for *ace* on the next toss. So his probability for *H* being false is only 0.1. Thus he agrees to sell bet IV to his friend for $1000.

Now no money will change hands after the fifth toss, for *in effect* this friend has bought back bet I. He has paid out only $1000 + $300 = $1300, and has therefore a net gain, once more, of $71.72.

What is really neat about this game is that even Peter could have figured out beforehand what would happen, if he was going to act on his new probabilities. He would have foreseen that by trading bets at fair value, *by his own lights*, he would be sure to lose $71.72 to his friend, come what may. Thus, by adopting the preacher's rule, Peter has become *incoherent*—for even by his own lights, he is sabotaging himself.

What is the moral of this story? Certainly, in part, that we should not listen to anyone who preaches a probabilistic version of Inference to the Best Explanation, whatever the details. Any such rule, once adopted as a rule, makes us incoherent. We need no rule to tell us to use Bayes's theorem, when applicable—that is only logic.[11] What leeway is left, beyond Bayesian orthodoxy, we must explore next.

Proofs and illustrations

Thank heavens for little computers! The rules of calculation are rather simple, but the examples require a lot of arithmetic. In the model, the initial probability $P(I)$ of bias $X(I)$ equals $(1/N)$. The conditional probability of *aces* in the first J (independent tosses), given bias $X(I)$ equals $(I/N)^J$. The orthodox conditional probability of bias $X(K)$, given this outcome equals then

$$P(\text{bias } X(K) \mid J \text{ aces}) = \frac{P(\text{bias } X(K) \text{ and } J \text{ aces})}{P(J \text{ aces})}$$

$$= \frac{P(J \text{ aces} \mid \text{bias } X(K))P(K)}{\Sigma P(I)\, P(J \text{ aces} \mid \text{bias } X(I))}$$

$$= \frac{(1/N)\, P(J \text{ aces} \mid \text{bias } X(K))}{(1/N)\, \Sigma P(J \text{ aces} \mid \text{bias } X(I))}$$

$$= \frac{P(J \text{ aces} \mid \text{bias } X(K))}{\Sigma P(J \text{ aces} \mid \text{bias } X(I))}$$

which is a form of Bayes's Theorem.

De Finetti's test, which I also discussed, uses the Brier score, perhaps the best known of the scoring rules. General studies of this subject contain many fascinating results.[12] The function $p(1-q)^2 + (1-p)q^2$ has the derivative $-2p(1-q) + 2(1-p)q = -2(p-q)$. Thus the function reaches its minimum for $0 \leqslant p, q \leqslant 1$ at $-2(p-q) = 0$, i.e. at $p = q$.

5. Sketch for an Epistemology

So far my intent has been destructive. I have also used a foil: the orthodox Bayesian, who sees conditionalization—mere logical assimilation of the evidence—as the sole motive force in our epistemic life. It is high time that I should be more constructive, and in my own voice. But since an entire epistemology would require a book, which I am not now in any position to write, I can only give a small sketch.[13] What I hope for is some reconciliation of the diverse intuitions of Bayesians and traditionalists, within a rather liberal probabilism. The old we might call *defensive epistemology*, for it concentrates on justification, warrant for, and defence of one's belief. That aspect will have to be left behind, I think, but there is another aspect, a certain allowance for true epistemic *decisions* that can be retained.

Rational reliance on prior opinion

When Bertrand Russell wrote *The Problems of Philosophy* in 1912, his opening paragraph sounded almost exactly like Descartes: 'Is there any knowledge in the world which is so certain that no reasonable man could doubt it?'[14] And his initial answer, in terms of sense-data, sounded like the traditional British version of secure foundations for knowledge and belief. For by page 19, the initial question has been transformed into 'Granted that we are certain of our own sense-data, have we any reason for regarding them as signs of the existence of something else . . .?' Any attempt to meet Descartes's problem in this way, head-on, courts the eminent danger of forced scepticism about our knowledge of the external world. But in the answers Russell was prepared to give we can see very clearly how the problem had become transformed since the seventeenth century:

Of course it is not by argument that we originally come by our belief in an independent external world. We find this belief ready in ourselves as soon as we begin to reflect.... Since this belief does not lead to any difficulties, . . . there seems no good reason for rejecting it . . . we cannot have *reason* to reject a belief except on the ground of some other belief. (p. 25)

As his colleagues on the Continent were beginning to say: *we are already outside in the external world*! The whole burden of rationality has shifted from justification of our opinion to the rationality of *change* of opinion.

This does *not* mean: we have a general opinion to the effect that what people find themselves believing at the outset (of, say, a philosophical exercise such as this) is universally likely to be true. It means rather that rationality cannot require the impossible. We believe that our beliefs are true, and our opinions reliable. We would be irrational if we did not normally have this attitude toward our own opinion. As soon as we stop believing that a hitherto held belief is not true, we must renounce it—on pain of inconsistency![15]

Two conceptions of reason

In his further discussions, Russell clearly shows adherence to one concept of rationality, whereas I would advocate another. The difference is analogous to that between (or so Justice Oliver Wendell Holmes wrote) the Prussian and the English concept of law. In the former, everything is forbidden which is not explicitly permitted, and in the latter, everything permitted that is not explicitly forbidden. When Russell is still preoccupied with reasons and justification, he heeds the call of what we might analogously call the Prussian concept of rationality: what is rational to believe is exactly what one is rationally compelled to believe. I would opt instead for the dual: what it is rational to believe includes anything that one is not rationally compelled to disbelieve. And similarly for ways of change: the rational ways to change your opinion include any that remain within the bounds of rationality—which may be very wide. *Rationality is only bridled irrationality.*

So: we are outside in the world already, full of opinions and expectation. In response to experience we amend our opinion and form new expectations, but with critical deliberation. This deliberation, it would seem, can also issue in the rejection or change of previous opinions, or adoption of new ones that go well beyond the evidence of our senses. Where are the bounds of rationality?

The two main answers correspond to the above duality. The *first* is that there are certain rules of right reason, canons of logic in a wide sense, which must be followed, and which dictate the only allowable accommodation of new deliverances of experience. It is the view to which at least lip-service was paid by the Newtonian writing about induction, and is strictly held by the orthodox Bayesian. The second view is that we are creatively inventing new hypotheses and theories, and that we are not only prone but rational to embrace them, after viewing favourable evidence, while they still go greatly beyond any accommodated evidence-cum-previous-opinion we could ever have hoped to assemble. This view of rational free enterprise of the spirit (which acutely needs the permissive concept of rationality) was fittingly advocated by the American pragmatists.

Thus William James, upon reading W. K. Clifford's 'The Ethics of Belief', reacted strongly against the latter's view that it is wrong always, everywhere, and for everyone to believe anything on insufficient evidence.[16] James claimed instead that in forming and shaping our opinion, we pursue two main aims: to believe truth and to avoid error. The extent to which we pursue these two aims, that draw us in different directions, must to some extent be a matter of choice: 'he who says "Better go without belief forever than believe a lie!" merely shows his own preponderant private horror of becoming a dupe' (p. 100). In other places, James recognizes still further aims that might rightfully compete for attention. Thus in 'The Sentiment of Rationality' he describes the desire for theoretical parsimony, as the philosophical passion *par excellence*. But, strong and worthy as this sentiment—now perhaps expressed more usually in such terms as 'explanation'—is, it is only one of many subordinate desiderata: 'The interest of theoretic rationality, the relief of identification, is but one of a thousand human purposes. When others rear their heads, it must pick up its little bundle and retire till its turn recurs' (p. 5). James thus clearly represented one traditional view in epistemology, which recognized rationally allowed leaps, far beyond the solid security of previous opinion-plus-evidence so far. Could we rationally have, in any other way, come to our present high opinion of the theories of Darwin, Einstein, and Bohr, whose empirical implications stretch throughout all dimensions of world history, past and future?

Rule-following and rationality

Could there not be a third view, that we have rules which (*a*) we are rationally compelled to follow, which (*b*) leave nothing to our choice when we proceed rationally, and yet (*c*) give us new expectations that are not logically implied by our old opinions leavened by new experience?

The answer, we know from the perils of Baysian Peter, is *No*. Here it was crucial to have the correct representation of opinion. It is only in the context of *probabilism* that we can fruitfully discuss the insidious connection between rule-following and security. The subject is inevitably technical, and many aspects must be left to Part III. The arguments demonstrate that rules for the accommodation of evidence, operating solely on previous opinion and the described deliverances of experience, are severely restricted in form. When those deliverances are simply propositions taken as evidence, the sole qualifying rule is that of *conditionalization*.[17] This is the form of accommodation of evidence, via Bayes's theorem, which I discussed in the preceding section. There we saw arguments concerning coherence which make short shrift of rival rules. These arguments, due to De Finetti and David Lewis, show that any such rule which goes beyond the mere logical accommodation of evidence—as singled out by the above noted symmetry arguments—makes its adherents incoherent in the following precise sense: by their own lights, they sabotage themselves through the commitment to follow that rule.

These are hard words; who can hear them? Have I not now left room only for that first view, that rationality compels us to follow rules which are strictly non-ampliative with respect to our previous opinion-cum-new evidence?

No; the conclusion is only that *if and when we commit ourselves to a rule* for the revision of opinion, it must be non-ampliative. This is the rigid connection between security and rule-following to which I alluded. But who says that we must commit ourselves to such rules beforehand? David Lewis has a reason. He asserts: A rational person who envisages the different possible episodes that further experience may bring him will also know (if we abstract from practical limitations) what he will believe in each case, at the end. That seems to be equivalent, in this context, to the assertion that a rational person is committed beforehand to a recipe for belief-revision. How else could he know what he would come to believe after any possible course of further experience, except by being so committed?

But must we accept this assertion about the rational person? What of the person who says: 'I can envisage all of these possible episodes, one and only one of which will come to pass—I do not know now exactly what opinions and expectations I will form in response, I shall in most respects make up my mind then and there, hopeful and confident that I shall proceed both rationally and creatively'—is he not rational? It will be no use to object to him that if he expected to proceed rationally, then he would now expect to proceed by rules. For that is the very point at issue, the point disputed. I conclude that rationality does not require conditionalization, nor does it require any commitment to follow a rule devised beforehand.[18]

Normal and revolutionary episodes

It is true that I shall *normally* expect to do no more than accommodate new evidence, in the sole logical fashion proper to such accommodation. There remains always the option to leaven this process with theoretical innovation and courageous embrace of new hypotheses that have gained my admiration.

When I declare innovation to be rational, the fear will be that I have effectively destroyed rationality altogether, by pre-empting all criticism. I think not, for how could the way in which we fashion our world-picture be so different from all other enterprise? Imagine our friend Peter turning up one day with 'Yesterday I saw a flying saucer, so I have understood the error of my ways and now I firmly believe in reincarnation.' The suspicion is presumably that it will be no use to point out to him the considerable difference between evidence and conclusion even on his own account. After all he can claim first to have merely assimilated his startling experiences, and then to have embraced an innovative new explanation thereof. But this objection begs a question. When I say that rationality does not require conditionalization—does not require new opinion to be logically forced by new experience and prior opinion together—I have not implied that standards of criticism do not exist, but only that they are not a matter of logic.[19] In opting for voluntarism or pragmatism in epistemology, one implicitly allows the relevance of just those critical standards that apply to other sorts of enterprise. Was our friend Peter courageous, reckless, feckless, foolish, or sensible; did he slavishly follow received tradition or sensibly treat it with respect; did he avoid criticism or worse yet give in to sceptical despair? These are all questions we could also ask when someone builds a house, starts a career, joins a revolution, paints, fights, loves, . . .; why not also when a person fashions, adapts, and upholds a view about the world and our place therein?

It is not very satisfactory to respond to such a serious danger as total epistemic relativism with only a suggestion, a historical reference, and some rhetorical questions. I shall return to this subject in the last section of this chapter. This sketch for an epistemology is admittedly programmatic, though based on the joint convictions that the traditional epistemology lies in ruins, and that within its ruins, a new epistemology has already grown up, to vigour if not to maturity. In any case, my critique of Inference to the Best Explanation, and other soi-disant canons of right reason, had to be placed in proper context: you had a right to know the face behind the critic's voice. So: I am a probabilist, though not a Bayesian. Like the Bayesian I hold that rational persons with the same evidence can still disagree in their opinion generally; but I do not accept the Bayesian recipes for opinion change as rationally compelling. I do accept the Bayesian extension of the canons of logic to all forms of opinion and opinion change.[20] But I think that logic only apparently constrains—a little more logic sets us free.

Proofs and illustrations

My main argument for voluntarism in epistemology is not among those given in this chapter, but remains the one in my 'Belief and the Will'. I shall not repeat it here, but I do want to outline several paths to the same conclusion.

The reconciliation in epistemology which I have proposed in this section, is made possible by the leeway introduced into both traditions by the view that, in rationality as in morality, what is not required is still not always forbidden. That means of course that rational change of opinion is not a case of rule-following. It includes an element of free choice, even if this choice will *normally* be to merely 'update' one's opinion in accordance with the rules of probability kinematics. The factors relevant to the free choices that interrupt the periods of normal progress—I phrase this deliberately so as to remind us of Kuhn—are in the domain of pragmatics.

But I do not claim to be the first to introduce this element. Carnap himself introduced it in his later work, when he discussed the role of caution in choice of inductive methods: neither those methods nor any other rules guided that role.[21] Perhaps Carnap did not appreciate how much this changed the overall programme, or the epistemic world-picture—which we still associate with the conceit of Carnap's robot designed to learn from experience—behind his programme. Isaac Levi, in his *Gambling with Truth* introduced the same element in the guise of what he called the risk-quotient, which is entirely up to the agent.[22] This echo of pragmatism is heard throughout Levi's writings.

6. BETWEEN REALISM AND SCEPTICAL DESPAIR

The epistemological position I have now sketched, was clearly framed in opposition to metaphysical realism and foundationalism. But that brings with it the danger of a slide into other extremes. Those extremes are scepticism and relativism.

Here are the dangers we face. Irenic relativism holds: there is (1) no objective criterion or rightness for opinion, and(/or) (2) no non-trivial criterion of rationality—anything goes, there is no truth except truth-for-you. Scepticism too comes two-fold: even if there is such a thing as objective rightness or truth, there is (3) no possibility of attaining it, and (/or) (4) no possibility even of rational opinion. Our best strategy will be, I think, to take up these claims in the order (1), (3), (4), (2). The aim will be to show that the combination of probabilism and voluntarism I have sketched cannot be fairly convicted of either relativism or scepticism.

To begin then with the question of objective truth and right opinion. Certainly our opinion is right or wrong, and this depends on what the world (the facts we make judgements about) is like. Of course, this must be properly construed. The vaguer my judgements are, the less decisive the criterion can be. If I judge that it will rain today, then I am right exactly if it does rain, and wrong otherwise. If I judge that it is three times as likely to rain today as not, then what? Do not confuse this with the judgement that the chance of rain is $\frac{3}{4}$. *That* is true exactly if there is such a thing as objective chance, and it equals $\frac{3}{4}$—an entirely different question, which falls under the first heading of 'true/false' judgement. No, if my judgements are of the 'seems . . . likely . . .' form, the criterion is very insensitive if applied to single judgements, and becomes sensitive only if applied to a substantial body of judgements. Such a body could be, for example, all the judgements that are mine, concerning subjects S_1, S_2, . . ., over the period of time. . . . For example, if every morning I make such a judgement of personal probability for rain, then I am as right as I can possibly be with respect to class X of days, if for each proportion r, the proportion of rain on days in X on which I announced my personal probability as r, indeed equals r. A measure can be introduced to describe lesser perfection.[23]

It may be said that this criterion does not go very far. Very well; this is exactly how far opinion sticks its neck out with respect to the facts. To this extent only can it be objectively in or out of tune with the facts. We have other criteria as well for evaluating opinion, but the point at issue was whether there is one, non-trivially applicable, objective criterion of rightness—and there is. Along the way we have rejected not only claim (1) but also claim (3) of scepticism for we can certainly attain this rightness, if only by accident. Let us turn now to the dangers to rationality.

David Armstrong, writing in answer to my suggestion that his views on induction and inference rest in part on neglect of the Bayesian position, says:

> This position seems to me Humean scepticism all over again, but with the probability calculus added to deductive logic. As such, I take it to be part of the problem rather than a serious attempt to solve it. As a Moorean I

think that if bedrock commonsense clashes with philosophy, then we have strong reason to think that philosophical argument is wrong rather than commonsense.[24]

It does not seem to me that a Bayesian can be called a sceptic, for he gives such great weight to previous opinion. The sceptic argues that (I) there can be no independent justification to believe what tradition and ourselves of yesterday tell us, *and* (II) that it is irrational to maintain unjustified opinion. The Bayesian agrees with (I). But it is clear that the Bayesian disagrees with (II). In addition, the sceptic argues that (III) there is no independent justification for any *ampliative* extrapolation of the evidence plus previous opinion to the future, and that (IV) it is irrational to do so without justification. We must grant that the orthodox Bayesian does agree with both these theses. The extent of his scepticism is not debilitating, however. His disagreement with thesis (II) allows him to live a happy and useful life by conscientiously updating the opinions gained at his mother's knees, in response to his own experience thereafter.

Whether Armstrong is right or wrong to think that common sense disagrees, he is certainly right that Moore disagreed with the conjunction of (III) and (IV). I too disagree, though again with the second conjunct only. Thus I disagree with the sceptic even more than the Bayesian does, for I disagree with both (II) and (IV); and I am sceptical only to the extent, James would say, fitting to an empiricist philosopher.[25] For we can and do see the truth about many things: ourselves, others, trees and animals, clouds and rivers—in the immediacy of experience. The traditions handed on in our literature tell us much more, and (I hold) are to be treated critically but with respect. My scepticism is with the general theories and explanations constantly handed out about all this, and my disdain is reserved for the illusory peace that satisfying (!) explanations bring. It is possible to remain an empiricist without sliding into scepticism, exactly by rejecting the sceptics' pious demands for justification where none is to be had.

So I will not be called a sceptic; but what remains? Once we give up the possibility of ultimate justification, and allow as rational both the reliance on previous opinion (after critical scrutiny; whatever it be, if coherent) and rule-compelled ampliation, do not all criteria of rationality collapse? Shall we not be reduced to the irenic relativism that equates any virtue with the delights of the pigsty? As pertinent example, imagine a friend of laws of nature who insists that, by my lights, the reality of laws must be as credence-worthy as any other hypothesis which goes beyond my evidence.

Like the orthodox Bayesian, though not to the same extent, I regard it as rational and normal to rely on our previous opinion. But I do not have the belief that any rational epistemic progress, in response to the same experience, would have led to our actual opinion as its unique outcome. Relativists light happily upon this, in full agreement. But then they present it as a reason to discount our opinion so far, and to grant it no weight. For surely (they argue) it is an effective critique of present conclusions to show that by equally rational means we could have arrived at their contraries?

I do not think it is. Recall that voluntarism entered at two points. The first concerned the status of judgement. A judgement of opinion, such as 'It seems likely to me that we have evolved from lesser organisms', is not itself an autobiographical statement of fact. It does not state or describe, but avow: it expresses a propositional attitude. To make it is to take a stand. To adopt an attitude or a stance is akin to commitment, intention. Of course, this can then be criticized in various ways. But what sort of criticism is it to note that I could, without flaunting the bounds of rationality, have taken a different stand? How much pause should that give me?

The fact is, in taking this stand, I know *that* already. Since it is akin to commitment, let us look there for analogy. Suppose, to use images of various decades in living memory, that I am—or, speaking as member of a group, *we* are—committed that racism or poverty shall end, that the homeless shall be fed, that war shall be no more. Any such commitment *may* be subject to a telling critique. In taking the stance, we avow that we have faced all reasons to the contrary. But we do not pretend that, if we had historically, intellectually, or emotionally evolved in some different way, we might not have come to a contrary commitment. We accuse this alternative history (if predicated on the same basic facts) of being mistaken or wrong (just like our opponents) but not of being irrational. To do that, would be to deny the element of free choice involved in our own decision. However comforting that might be—how nice to think we could not rationally have concluded differently!—it would still be intellectual bad faith. And none of this detracts one bit from our actual commitment.

So I reject this reasoning that so often supports relativism. But because I have rejected it without retreat to a pretence of secure foundations, the relativist may think that I still end up on his or her side. That is a mistake. Just because rationality is a concept of permission rather than compulsion, and it does not place us under the sway of substantive rules, it may be tempting to think that 'anything goes'. But this is not so.

To take one good example: belief in laws of nature does not 'go'. It is true that if some philosophers believe in the reality of laws, they are not *ipso facto* irrational. But it does not follow that they are in a position to persuade us or even give us good reason to follow suit. They may be, for us, in exactly the same position as the sceptics: their conclusions, if we accepted them, would drastically realign our opinions—but those conclusions are based on premisses which they can neither substantiate nor make appealing to us.

Indeed, if we look to see how they could try to persuade us, we shall find them blocked at every turn. On the view I have described, to persuade us they must show either that the reality of laws is part of a tradition we have now and which deserves to fare well under critical scrutiny, or else that it provides us with theoretical innovation of great value and promise. In the preceding chapters we have seen attempts to do this, properly based on prior attempts to explicate the subject under discussion. These attempts all failed, if my criticisms were right.

Summing up this way, I am not simply placing the onus on the opposition. For if it could have been shown that the reality of laws was indeed inherent in traditional, prevalent opinion—left intact under scrutiny—that would have sufficed. What I do not accept is that someone should simply look inside himself, express his or her own spontaneous inclination to assent to the reality of this or that, and then be taken to have conveyed the burden of considered common sense. I would not even accept that if the subject were considerably less abstruse or culturally conditioned than this one. To demand more than personal testimony is not a mere shift of onus.

What interwoven judgements of fact and value would the rational friend of laws have to evoke in us, before we could view the reality of laws as confirmed? It cannot be merely a matter of evidence to fit a model's curves, for the fit will be equally good, regardless of whether its more arcane parts correspond to something real, or not. That is why the friends of laws emphasize explanation so much. As they correctly point out, to have a cherished model that fits is not *ipso facto* to have a belief that explains. The way we cherish a model may consist, after all, in our confidence merely that it will continue to save all observable phenomena and experimental results. Thus we must turn to the claim that the reality of laws explains the phenomena, and to the value attached to this feat.

But what reason is there to think that the laws, if they be real, explain the phenomena they cover? It cannot simply be that a law is the sort of thing that is *supposed* to explain. If we make it definitional or analytic that laws explain, or explain if they be real, then we have automatically removed their explanatoriness from the list of reasons for their reality. (Imagine the dialogue: 'Bachelors are single men—that is analytic.' 'Yes, but are there any?' 'You had better believe it—they couldn't very well remain single if they didn't exist, could they?') So the question: *Do laws explain?* has to be a substantive question, which must be answered, with substantive reasons for the given reply. But it is a question which we have no way of answering, without a previous account of what laws are. Again, I do not merely shift the onus if I say that a rational person cannot reasonably accept the reality of laws until there is an account that makes them acceptable. For this account is needed to answer the substantive question which arises at this point. If it remains unanswered, there is nothing but the hollow sound of words to accept or reject.

For each of the accounts given, and all the variant accounts which would fall under the rubric of our discussion, the conclusions are these. We must disagree that laws, *so conceived*, are inherent in the opinion which critically scrutinized tradition has bestowed on us already. We must disagree that the philosophical account of science, in which laws figure, can withstand scrutiny at that point. We must disagree that the laws, *so conceived* (on any of those accounts), explain—even by very minimal and common requirements upon explanation, to be met before we admire the grammar of laws and causation. Nor do they connect successfully with the pursuits of science. Nor does their exploration give us new insight into the aim and structure of science. Neither do they succeed in giving us a new epistemological scheme to make intelligible how evidence supports science, or even how we may coherently adjust our opinion to experience day by day. And finally, all the accounts suffer from internal difficulties—summarized as the problems of identification and of inference—so severe as to undermine even what partial satisfaction they could give us on these counts.

The great seventeenth-century writers on science, gave the concept of law a prima-facie claim on our loyalties. The claim has not been borne out, the unanswered questions have proved unanswerable, the promises of explanatory gain even have proved empty, and the original insight has evaporated before us. The fact that belief in laws of nature is not *ipso facto* irrational does not ameliorate any of that. There is no significant sense in which 'anything goes'.

8.
What If There Are No Laws? A Manifesto

> There is one truly eccentric view. This is the view that, although there are regularities in the world, there are no laws of nature.
>
> D. Armstrong, *What Is A Law of Nature?*

AFTER our look at the answers to *What is a law of nature?* the 'truly eccentric' view does not seem quite so eccentric any more. When all attempted answers to a question run into such difficulties, should we not consider—as Kant urged—a denial of its presuppositions? If we say that the regularities are all there is, shall we be so badly off? In the preceding chapters I have touched on my own views from time to time, on such subjects as objective chance and rational change of opinion, to intimate that more agnostic, anti-realist alternatives exist. I shall now frankly advocate the philosophical view that there are no laws of nature. After cataloguing the difficulties this view must face, I shall outline a perspective on science which can support it.

1. PROBLEMS THAT LAWS PURPORT TO SOLVE

The moon orbits the earth. Why does it do so? What reason can there be for expecting it to continue to do so? If you place a cut dandelion in a glass of inky water, the flower turns green. Why does it turn green? What reason can there be for believing that this will happen every time we try it?

I have no inclination at all to make light of such questions. But I do think that they have been read by philosophers in a special way, which we are not bound to take equally serious. I think this even more emphatically when the philosopher draws attention to the modal language in which such questions are underlined: 'The fact is that the dandelion has got to turn green, if it doesn't die first, if you give it ink in its water—there are no two ways about it!' True—but the inquisitive child who says this, will move from wide-eyed astonishment to wide-eyed appreciation if we teach him some elementary plant-physiology. It is only the philosopher who may, at the end of such a lesson, complain that we have told him only more facts, described how and not explained why. If we cannot be satisfied by more advanced biology lessons, if in the end we have a desire for understanding which no mere facts can satisfy, if finally there remains

> Infinite passion/and the pain
> Of finite hearts that yearn

I will assert frankly that we shall take stones for bread if we think to still this longing with possible worlds, universals, and laws.

Accounts of law purport to give us (1) a theory of explanation, (2) a theory of confirmation, (3) an explication of necessity, and most of all, (4) a way of understanding science, its aim and structure. If we are to hold that there are no laws of nature, we must provide these very things, without recourse to the reality of laws. The most important task is to provide (4), a rival key to what science is and does, though each of the four must be handled.

Of course we all know what dissatisfaction will be expressed about any way of carrying out this task, without recourse to laws, universals, propensities, necessities *in re*, and the like. The dissatisfaction will appear in such persistent questions as 'But how do you account for the difference between what is *really* possible and what is impossible?' or 'But why do only some (true) theories *really* explain?' There is no onus on us to answer such questions, once we have thoroughly discredited their presuppositions. They will always persist in the sense that our language, the way we speak, allows their formulation. That point, however, must be met by semantics and pragmatics, by an analysis of the logic of 'why', 'must', and 'possible'. It need not be met by the assertion that for a certain sort of model, constructed in semantics for modal discourse, every model element corresponds to some element of reality. To give an analogy: whether or not some form of moral relativism is tenable, it cannot very well be defeated by the mere question of what the difference is between *really* moral and immoral, or why only some reasons *really* exonerate.

2. THE FOUR TASKS AND THE SEMANTIC APPROACH TO SCIENCE

I shall give a short sketch here of a programme—which I see as being carried out co-operatively by many philosophers, although they form no school and have numerous disagreements—to give us an account of explanation, confirmation, necessity, and science. Only a sketch is possible: we have here, after all, four of the main perennial pursuits of Western philosophy. The sketch will also be skewed toward my own work and views, often not shared by philosophers I shall cite. The enterprise I am trying to describe is a cluster of attempts to make sense of our topics of concern without reliance on metaphysics, at least of the pre-Kantian variety.

Explanation[1]

The question why the sky is blue is different from any yes/no question, or which or what question. That may be for two reasons: it may be because it asks for a different sort of information, or because it has a different logic. The first alternative is taken by those philosophers who say that explanation—which is surely the right name for what why-questions request—must include information about laws, causal connections, powers or propensities, real possibilities and necessities in nature, and so forth. The second alternative allows us two further options, to say that a why-question is not a request for information at all, or else that it requests the same sort of information as ordinary questions—ordinary factual information.

The option that an explanation does not consist in information is perhaps that taken by those who locate explanatory value in theoretical unification. Such a view has been proposed in different forms by, for example, Michael Friedman, Clark Glymour, and Philip Kitcher. My own view is of the last sort—I hold that why-questions have a different logic (in a broad sense) from those other information requests, but do request merely ordinary information. Crucial to the logic of 'why' and 'because' is a certain context-dependence, which entails that the same why-interrogative sentence requests different information in different contexts. When I listed the question why the sky is blue, you undoubtedly took it in a certain way, quite contrary to the singer of the Broadway ditty

Dites-moi pourquoi
le ciel est bleu?
. . .
Est-ce-que parce que
je vous aime?

Scientific explanations are simply, in my view, those which draw upon science for the adduced information. That is not always what the questioner wants. (Nor is it true that the request must be for information which is either from science or from some rival to science.) But even when he or she does want scientific information, the context can make it a request for very different bits of information—preceding events, standing conditions, material composition, or functional role, for example.

This approach to the subject of explanation—I do not pretend that the resultant account is complete, especially not in the matter of criteria of evaluation—has no need to posit available information on a special sort of subject such as laws or causes. Certainly a person may ask a why-question which, in context, amounts to a request for information about what laws of nature there are. Any person may bring any presupposition to any question he wishes to ask—and asking is free. But there is no need for us to construe the usual sort of requests for scientific explanation in that way.

The debates concerning explanation continue, and metaphysics intrudes especially through the notions of causation and necessity. I shall not discuss explanation further in this book.

Confirmation

This is a very deceptive and misleading term, and I think we should definitely discard it. The etymology of 'confirmation' builds into it a directive to find a certain sort of theory belonging to the tradition of 'defensive' epistemology. For it asks how evidence establishes, confers warrant, makes firm, gives support. We should not allow our approach to epistemological questions to be dictated in this way. But of course we still have the task now conventionally referred to under this heading: to explicate rational response to evidence, rational constraints on belief and opinion.

Can this explication proceed without reliance on metaphysics? Descartes insisted that epistemology should precede metaphysics.

He had a point: what good is it to offer an account of what the world is like, if belief in that account is the sole support offered for the reliability of any means to test it? The point turns into a dilemma: either we can reach conclusions about reliability of belief without presupposing any metaphysics—or the whole enterprise described here is mistaken in intent. Descartes himself took of course the first horn, and purported to demonstrate an indubitable basis of knowledge. Indubitable does not imply true—but what more could you ask? In retrospect, however, his demonstration was not presuppositionless. Even less so were the British empiricist attempts to find a secure foundation in the data of sense. As Reichenbach pointed out forcefully, they tricked themselves into trying to supply what, as they had told the rationalists, could not be had.[2]

I do not want to embark here on a treatise in epistemology. Much less do I want to look for a presuppositionless approach in philosophy of science. The task we have is only to develop an account in which rational opinion and change of opinion is possible without certain kinds of presupposition. For example, in contact with science, we change our opinions and expectations about observables, physical objects, and events. Can we do so rationally without a belief in laws of nature and other sorts of things outside that domain of observable, physical entities?

In the preceding two chapters I have addressed that question. I shall leave it there, except that I shall try to substantiate my claims about probability in Chapters 12 and 13.

Necessity

The ghost of modality, to echo Herman Weyl, is not easily laid.[3] Nor should it be banished entirely: it haunts our language to good purpose. But this is one case in which the linguistic turn of philosophy in this century, has done philosophy of science a great service. For philosophy of language has given us the wherewithal to approach the explication of necessity arrived with no ontological weapons or metaphysical charms. I will return to this in the last section of this chapter.

Approach to science

The particular approach to science that I favour is the *semantic view of theories*, which has been advocated by a number of recent writers, in various forms. See *The Semantic Conception of Theories and Scientific Realism* by F. Suppe, which describes the prehistory (Weyl, Beth, Suppes), the development in the 1960s and 1970s (Suppe, van Fraassen, Giere), relations to other rivals of the syntactic/axiomatic approach, and Suppe's own recent work.[4] Or one may go more directly to pages 221–30 of F. Suppe (ed.), *The Structure of Scientific Theories*;[5] Chapter 5, 'Theories', of R. Giere, *Understanding Scientific Reasoning*;[6] or Chapter 3, 'Models and Theories', of his more recent book *Explaining Science*.[7] Still other sources are Chapter 3 section 4 of my *An Introduction to the Philosophy of Space and Time*[8] and Patrick Suppes's, 'What Is A Scientific Theory?'[9]

The most exciting recent development has been the adoption and development of this approach by certain philosophers of biology. See E. A. Lloyd, *The Structure and Confirmation of Evolutionary Theory*,[10] and P. Thompson, *The Structure of Biological Theories*,[11] and also the writings of Jon Beattie.[12]

I will present the semantic approach in the context of my anti-realism, which others may not share, but the semantic approach is not wedded to that. In that approach, the focus is on models rather than on axioms or theory formulation—we may share this focus and still disagree on how much, in a good model, is meant to correspond to reality.

According to the semantic view, to present a theory is to present a family of models. This family may be described in many ways, by means of different statements in different languages, and no linguistic formulation has any privileged status. Specifically, no importance attaches as such to axiomatization, and a theory may not even be axiomatizable in any non-trivial sense.

Laws do appear in this view—but only laws of models, basic principles of the theory, fundamental equations. Some principles are indeed deeper or more fundamental than others. Pre-eminent among these are the symmetries of the models, intimately connected with the conservation laws, but ubiquitous in their influence on theory construction. Our diagnosis is *not* that the more fundamental parts of a theory are those which reflect a special and different aspect of reality, such as laws of nature! It is only the content of the theory, the information it contains (and not its structure), which is meant to have the proper or relevant *adequatio ad rem*.

To outline an approach is not yet to take, even less to complete, the step-by-step journey it promises to guide. If the semantic view has worth, it must be that the unending task of the philosophical analysis of science can progress under its aegis. The chapters that follow this one are meant to contribute to that task with an inquiry into the role of symmetry in model and theory construction. I hope to continue this in a book on quantum mechanics. The books and papers I listed in the footnotes all contribute to this project. Only through such labour can it become clear (as I hope it will) that the benefits metaphysics so airily promised and failed to bring can be had without metaphysics.

3. REALISM AND EMPIRICISM

Philosophy of science attempts to answer the question *What is science?* in just the sense in which philosophy of art, philosophy of law, and philosophy of religion answer the similar question about their subject. But of course such a question can be construed in different ways. For better or for worse our tradition has focused on the scientific theory rather than on scientific activity itself. We have concentrated on the product, rather than on the aim, conditions, and process of production, to draw an analogy which already points in its terminology to the product as most salient feature. Yet all aspects of scientific activity must be illumined if the whole is to become intelligible. I shall therefore devote this section to the aim of science, and to the proper form for cognitive attitudes toward scientific theories.

The activity of constructing, testing, and refining of scientific theories—that is, the production of theories to be accepted within the scientific community and offered to the public—what is the aim of this activity?

I do not refer here to either the motives of individual scientists for participating, or the motives of the body civic for granting funds or other support. Nor do I ask for some theoretically postulated 'fundamental project' which would explain this activity. It is part of the straightforward description of any activity, communal or individual, large-scale or small, to describe the end that is pursued as one of its defining conditions. In the most general terms, the end pursued is success. The question is what counts as success, what are the criteria of success in this particular case?

We cannot answer our specific questions here without some reflection on what sort of thing this product, the scientific theory, is. A scientific theory must be the sort of thing that we can *accept* or *reject* and *believe* or *disbelieve*; accepting a theory implies the opinion that it is successful; science aims to give us acceptable theories. To put it more generally, a theory is an object for epistemic or at least doxastic attitudes—the attitudes expressed in assertions of knowledge and opinion. A typical object for such attitudes is a proposition, or a set of propositions, or more generally a body of putative information about what the world is like, what the facts are.

If anyone wishes to be an instrumentalist, he has to deny the appearances which I have just described. An instrumentalist would have to say that the apparent expression of a doxastic attitude toward a theory is elliptical; 'to believe theory *T*' he would have to construe as 'to believe that theory *T* has certain qualities'. There is indeed such a view about the structure of theories, due especially to Sneed, Moulines, and Stegmüller. They contrast their 'non-statement view' and the traditional 'statement view'.[13] The latter, as the name suggests, is that a theory is just a very complex statement. Two statements may be logically equivalent, in which case they 'say the same thing'. Thus two axiomatizations of a theory, being logically equivalent, may be called two statements or formulations of the same theory. Such a traditional view has some tensions in it. The very locution 'two statements may be formulations of the same theory' suggests that a theory is not a statement at all, but something which is merely formulated or expressed by means of a statement. Secondly, as soon as we think of a theory as something that must be linguistically expressed, we encounter the meta-linguistic paradoxes that plague our conceptions of the linguistic. The 'non-statement view' eliminates these tensions by insisting that a theory is not even the sort of thing which can properly be said to be true or false.

This looks like a high price to pay. Don't we believe, assert, deny, doubt, and disagree about theories? And do such propositional attitudes not presuppose at least that a theory is the sort of thing which can be true or false? On the non-statement view these locutions must all be reinterpreted. For example, to believe a theory really means to believe that the theory bears a certain relation to empirical reality. That relation must be carefully described of course, but can be given the slogan formulation: it saves the phenomena.

In my opinion, this goes too far. For it seems clear that we can discuss two separate questions: what does the theory say the world is like? *and* what does the theory say the phenomena are like? Since the phenomena are just the observable part of the world, and since it is logically contingent whether or not there are other parts, it follows that these questions are not the same. Indeed, the second question is part of the first, in the sense that a complete answer to the latter is a partial answer to the former. The 'non-statement view' appears to deny the intelligibility of the bigger question—but the question seems intelligible.

There were indeed reasons for Stegmüller and his collaborators to go that far along the road to logical positivism and instrumentalism. But those reasons are properly accommodated, I think, by the semantic view of theories, although that is a form of what Stegmüller calls the statement view.

I shall not follow the path that leads to a non-statement view. Let me state here, as a first tenet, that the theory itself is what is believed, partly believed, or disbelieved.

At this point we can readily see that there is a very simple possible answer to all our questions, the answer we call *scientific realism*. This philosophy says that a theory is the sort of thing which is either true or false; and that the criterion of success is truth. As corollaries we have that acceptance of a theory as successful is, or involves, the belief that it is true; and that the aim of science is to give us (literally) true theories about what the world is like.

That answer would of course have to be qualified in various ways to allow for our epistemic finitude and the consequent tentativeness of reasonable doxastic attitudes. Thus we should add that although it cannot generally be *known* whether or not the criterion of success has been met, we may reasonably have a high degree of belief that it has been met, or that it is met approximately (i.e. met exactly by one member of a set of 'small variants' of the theory), and this imparts similar qualifications to acceptance in practice. And we should add furthermore of course that empiricism precludes dogmatism, that is, whatever doxastic attitude we adopt, we stand ready to revise in face of further evidence. These are all qualifications of a sort that anyone must acknowledge, and should therefore really go without saying. They do not detract from the appealing and as it were pristine clarity of the scientific realist position.

I did not want to discuss the structure of theories before bringing this position into the open, and confronting it with alternatives. For it is very important, to my mind, to see that an analysis of theories—even one that is quite traditional with respect to what theories are—does not presuppose realism. Let us keep assuming with the scientific realist, that theories are the sort of thing which can be true or false, that they say what the world is like. What they say may be true or false, but it is nevertheless literally meaningful information, in the neutral sense in which the truth value is 'bracketed'.

We have come now to the position I advocate. With the realist I take it that a theory is the sort of thing that can be true or false, that can describe reality correctly or incorrectly, and that we may believe or disbelieve. All that is part of the semantic view of theories. It is needed to maintain the semantic account of implication, inference, and logical structure.

There are a number of reasons why I advocate an alternative to scientific realism. One point is that reasons for acceptance include many which, *ceteris paribus*, detract from the likelihood of truth. In constructing and evaluating theories, we follow our desires for information as well as our desire for truth. For belief, however, all but the desire for truth must be 'ulterior motives'. *Since therefore there are reasons for acceptance which are not reasons for belief, I conclude that acceptance is not belief.* It is to me an elementary logical point that a more informative theory cannot be more likely to be true—and attempts to describe inductive or evidential support through features that require information (such as 'Inference to the Best Explanation') must either contradict themselves or equivocate.[14]

It is still a long way from this point to a concrete alternative to scientific realism. Once we have driven the wedge between acceptance and belief, however, we can reconsider possible ways to make sense of science. Let me just end by stating my own anti-realist position, which I call *constructive empiricism*. It says that the aim of science is not truth as such but only *empirical adequacy*, that is, truth with respect to the observable phenomena. Acceptance of a theory involves as belief only that the theory is empirically adequate (but acceptance involves more than belief). To put it yet another way: acceptance is acceptance as successful, and involves the opinion that the theory is successful—but the criterion of success is not truth in every respect, but only truth with respect to what is actual and observable.

While truth as such is therefore, according to me, irrelevant to success for theories, it is still a category that applies to scientific theories. Indeed, the *content* of a theory is what it says the world is like; and this is either true or false. The applicability of this notion of truth-value remains here, as everywhere, the basis of all logical analysis. When we come to a specific theory, there is an immediate philosophical question, which concerns the content alone: *how could the world possibly be the way this theory says it is?*

This is for me the foundational question *par excellence*. And it is a question whose discussion presupposes no adherence to scientific realism, nor a choice between its alternatives. This is the area in philosophy of science where realists and anti-realists can meet and speak with perfect neutrality.

Proofs and illustrations

According to constructive empiricism, acceptance of a theory involves a certain amount of agnosticism, or suspension of belief. (As far as science is concerned, of course; an individual scientist may additionally believe in the reality of entities behind the phenomena. Similarly a chess player may wear flowers or hum a madrigal while playing.) But can this sort of cognitive attitude be accommodated by probabilism?

The argument that it cannot goes like this. Consider, for example, the hypothesis that there are quarks (and not merely that the theory saves the phenomena). The scientist has initially some degree of belief that this is true. As evidence comes in, that degree can be raised, to any higher degree. That is a logical point: if some proposition X has positive probability, conditionalizing on other propositions can enhance the probability of X.

The mistake in this argument is to assume that agnosticism is represented by a low probability. That confuses lack or suspension of opinion with opinion of a certain sort. To represent agnosticism, we must take seriously the vagueness of opinion, and note that it can be totally vague. The next question is of course how to represent vague opinion. The way this is done by both Levi and Jeffrey is to assimilate vagueness to ambivalence of a certain type. If my state of opinion is totally precise, it is represented by a single probability function. But otherwise the representation is a set of such functions, each of which is more precise than I am—they each represent my opinion equally well, but they will go too far. What is common to them, that is what my opinion is. For example suppose I think *A* exactly as likely as not, but am totally agnostic about *B*. Then every function in that representing class gives $\frac{1}{2}$ to *A*, but for every number $0 \leqslant x \leqslant 1$, one of these functions gives x to *B*.

What is the effect of new evidence? If hypothesis *H* implies *E*, then the vagueness of *H* can cover at most the interval $[0, P(\text{E})]$. So if *E* then becomes certain, that upper limit disappears. For the most thorough agnostic concerning *H* is vague on its probability from zero to the probability of its consequences, and remains so when he conditionalizes on any evidence.

4. ACCEPTANCE OF A STATISTICAL THEORY

The preceding discussion of theory acceptance ignores the presence of statistical theories—even irreducibly statistical theories—in science. I was initially of the opinion that the discussion would be completed by the analysis of empirical adequacy for such cases.[15] The notion would have to be quantified, along the lines of the concept of goodness of fit in orthodox statistics. But now I think that the limits of the view I offered of theory acceptance lie in its focus on belief. Belief must be replaced by the nuances of gradated opinion, modelled as personal probability.

The question becomes then: if we accept a theory, how do the probabilities it offers guide our personal expectation? The answer I shall now begin to elaborate is: in the form of Miller's Principle. That is what constitutes acceptance. For someone who totally believes the theory, that guidance will involve all theoretical probabilities. For the scientist *qua* scientist (as described by empiricism) only the theory's probabilities for observable phenomena will play this guiding role. 'Objective chance' is in either case the honorary epithet we give to the probabilities in theories we accept.[16]

In Chapter 4 I stated the fundamental question about objective chance: why and how should it constrain rational expectation? The 'how' is answered by Miller's Principle and its generalizations. But then, if anyone gives an account of chance, he must answer the 'why', and supply the warrant for that principle.

Ramifications of Miller's Principle

Before going any further, we should take a hard look at Miller's Principle itself. Any attempt to give it proper credentials must take full cognizance of its more curious features as well as of its basic appeal.

This principle, named after David Miller, was the first premiss of Miller's Paradox.[17] The value of this paradox—which I shall not discuss here in its own right—is that it makes clear how not to use Miller's Principle. The following three statements may look like proper instances:

(1) $P(A|\text{chance }(A) = 0.3) = 0.3$
(2) $P[A|\text{chance }(A) = \text{chance }(\textit{not }A)] = \text{chance }(\textit{not }A)$
(3) $P[A|\text{chance }(A) = \text{chance }(A)] = \text{chance }(A)$

but only the first is correct. It is easy enough to see for example that if (3) were correct, our personal probability P would have to coincide with chance everywhere! But whatever chance is, we are often ignorant of it, and ignorance is not irrationality.

It must also be noted that adding something to a supposition, in probabilistic contexts, can alter things drastically.[18] Thus the probability that a person speaks Swedish well, given that he has Swedish parents is high—but not on the supposition that he was born of Swedish parents resident in America. Similarly

(4) $P(A|A \textit{ and } \text{chance }(A) = x) = x$

is incorrect unless $x = 1$. We see therefore that we cannot in general add to the antecedent, without upsetting the connection. But since chance is meant to sum up everything *so far* that is relevant to what will be true, we know at least that we can add anything already settled. In other words, what is settled now as being true or false, can be trivially assigned chance *one* or *zero*. Thus if I know or fully believe that B is already settled one way or the other, it can't affect chance now. But this requires no addition to the principle because if I fully believe that *B* has chance = 1, then I fully believe that *B*, and so *B* can be added for me to any antecedent.[19]

The intuition that Miller's Principle is a requirement of rationality firmly links its credentials to a certain view of ourselves—namely that we are finite, temporally conditioned rational beings. We have no crystal balls, and no way to gather information about the future which goes beyond the facts which have become settled to date. If we thought instead that Miller's Principle must apply to all possible and conceivable rational beings, we would have to conclude that omniscience implies determinism:

(*a*) $P_t(A)$ is 1 if *A* is true and 0 otherwise, for all propositions *A*
(*b*) $P_t(\text{chance}_t(A) = x_A) = 1$ for a unique real number x_A—from (*a*) and the logic of 'chance'
(*c*) $P_t(A) = x_A$—from (*b*) and Miller's Principle
(*d*) x_A is 1 or 0—from (*a*) and (*c*)

The supposition (*a*) means that this conceived person has full belief ($P_t = 1$) exactly for each true proposition (*omniscience*), and the conclusion is that $\text{chance}_t(A)$ is a certain number x_A, which must then be zero or one (*determinism*). The proper conclusion to draw, of course, is the one announced already above, that Miller's Principle is a requirement of rationality for *us*. Here we are conceived as beings whose source of information at this moment is inherently limited to factors which are already represented—in summary fashion—in objective chance.

Even so, we must furthermore observe a possible restriction on the domain of propositions for which chance is well defined. It may not make sense to ask what is the objective chance—as opposed to our subjective estimate—that there exist, for example, abstract entities.[20] Similarly perhaps for propositions about the global structure of space-time, or propositions to the effect that, for example, there exist quarks, or that forces are real. There may be no clear a priori ruling on what the domain of chance encompasses.

It should now be fairly clear why we cannot expect any a priori proof of the principle. After all we do not expect to find such a proof *either* of the claim that omniscience implies determinism *or* of the assertion that our sources of information are temporally limited in a certain way. Nor can we envisage a definition of chance that would make that notion fully objective and yet link it *logically* to what we ourselves are like.

At the same time, it is just as clear that if we placed no restriction of objectivity on the notion of chance, we would have a veritable plethora of ways to satisfy Miller's Principle.[21] Let B be any proposition you like, and define for your own subjective probability function P:

$$x_A = P(A|B)$$
$$y_A = P(A|\text{not } B)$$
$$\text{chance }(A) = \begin{cases} x_A \text{ if } B \text{ is true} \\ y_A \text{ if } B \text{ is false} \end{cases}$$

Now you can deduce that $P(A|\text{ chance}(A) = z) = z$ for all numbers z for which the antecedent is not ruled out. Similarly for any exhaustive logical partition B_1, B_2 ... of possibilities, that goes beyond the coarse division into B and *not B*. But this reconstruction is unacceptable, for it has the consequence that chance itself is subjective.

Hard-headed 'non-supervenience' metaphysics of chance leaves us entirely incapable of drawing connections between chance and rationality—but subjective chance is an oxymoron. So what remains?

Chance and science

Let us draw the lines as follows. Chance is the probability which science purports to give us. Scientific statements of probability are descriptive of objective chance in just the way that other scientific statements are descriptive of forces, fields, the space-time metric, or fitness and hereditability. But what does *acceptance* of this science involve?

Because I do not equate belief and acceptance, for scientific theories, the problem separates into two distinct parts for me. The first is to explain what a probabilistic theory says about the world. The brief answer can indeed be that it says that the objective chances (objective probabilities, physical probabilities) are thus or so. This is only a 'verbal' answer; we should be told in detail how to understand the models of a probabilistic theory. This had better entail that their probabilities are, in a very straightforward sense, the theoretical counterparts of relative frequency. In my opinion, this first part of the problem is solved by the modal frequency interpretation of physical probabilities.[22]

The first part was to state what such a theory *says*; the second is to state what it means to *accept* such a theory. We have now come to the point announced at the beginning of this section: an explanation of what theory acceptance is, that also covers statistical theories. I will explain this now, and then I will also show again that the improvement does not help the metaphysician. The clue will be the analogy in the roles of theories and experts.

Expert functions

Artificial intelligence supplies us today with 'expert systems' and before that we also turned to experts, or turned to theories in much the same way. If I say that I regard Peter as an expert about snuff-boxes, and mean this in its strongest, unqualified sense, that entails:

(1) I believe whatever Peter says, or will say, about snuff-boxes.

But perhaps Peter does not just express yes–no opinions, but also judgements such as that one thing seems more or less likely to him than another. That is, this expert opinion comes to us as Peter's personal probability. And my opinions too have gradations of this sort—I have my own personal probability. Now I regard Peter as an unqualified expert on snuff-boxes exactly if:

(2) My personal probability for a proposition about snuff-boxes, given that Peter's probability for it equals x, equals x.

This expresses my state of opinion in a very rich, sophisticated fashion, which has many implications. It constrains the structure of my opinion and reasoning in subtle ways even when I'm ignorant of Peter's opinion—for (2) is schematic and conditional.

We notice of course that (2) has the formal structure of Miller's Principle. Chaim Gaifman, who wrote the most advanced formal treatment we have of this subject, introduced this idea of expertise as guiding clue.[23] We can define:

(3) If P is my personal probability function, then q is an *expert function for me concerning* family F of propositions exactly if $P(A \mid q(A) = x) = x$ for all propositions A in family F.

To say that q is such an expert function for me is merely a partial description of my state of opinion. Often, indeed in the interesting cases, this is not nearly all there is to it. To describe me further in relevant respects, we may have to add that I am committed to continue regarding q in this fashion, or that I have certain additional opinions concerning the source of q, or so on. It is especially important to describe the limits of expertise; for example, for a certain other family F', $P(A|q(A) = x \text{ and } Y) = x$ for all A in F and all Y in F'; but if Y is outside F', it may provide evidence overriding the expert. In the case of irreducible physical probability, we conceive it as not overridden by any proposition which is solely about the past or present.

Acceptance of a probabilistic theory

My attitude to a theory may be, in part, that it provides me with an expert function. This is exactly similar, except that the probability is now not a person's, but given by a theory. We must be careful not to regiment this idea too much. Our epistemic attitudes toward theories admit of the same subtleties as the attitude we have to our fellow creatures. I might say for example that I accept contemporary physics, and if pressed will elaborate in many ways what my acceptance involves. One thing it might involve is that I take this physics, without qualification, as expert on the probabilities of observable events—meant in the sense explained above. This is what I take to be the proper construal of acceptance, when the theory is viewed as irreducibly probabilistic.

Have I now really rejected the concept of objective chance, and any probability except the measure of opinion? Not at all. When physics says that a radium atom has a 50 per cent probability of decaying within 1600 years, it says something about what the world is like, and nothing about opinion. But to accept a theory may not always involve believing everything it says. The attitude might instead involve regarding this probability (or rather the radioactive decay function $p(\textit{stable for period } t) = e^{-At}$) as expert, without going any further. To draw an analogy: I could regard Peter as expert on snuff-boxes even if he invariably presented himself as repeating the opinion of an angel heard in a dream. The *meaning* of what he (or a theory) says cannot automatically give him (or it) expert status; conversely, to regard him as expert about something implies nothing about the meaning or status of what else he says. So the *concept* of chance remains that of an objective feature of the world, described by science, but acceptance of this science need not involve the belief that this feature is real. Even if one did have such beliefs, however, they could not justify acceptance in this sense—that is what we saw in Chapter 4.

Against metaphysics

What would the metaphysician wish to add to our account? He will surely object that I have ignored reasons: he will say that it is not reasonable to take a person, or theory, as expert unless you do so on the basis of substantial reasons. Of course, these reasons must be other beliefs. Thus if (1), I believe whatever Peter says about snuff-boxes, this is reasonable only if I believe

(1″) Everything Peter believes about snuff-boxes is true

which is a fact about the relation between Peter and the real world.

Of course I agree to this. That is obvious: anyone who says (1) and believes something at odds with (1′) is inconsistent. That is merely a logical point. Similarly, if q is an expert function for me, I would be incoherent if I also had the opinion that q is unreliable in some definite way.[24] That too is a logical point.

The metaphysician will insist that the logical point is not enough. His own story about his acceptance of contemporary science, he says is this:

I. My personal probability for A, given that the objective chance of A equals x, equals x.
II. Whenever science assigns a probability to A, that is meant to be the objective chance of A; and I believe that science is true.
III. The probabilities given by science, constitute expertise for me in the sense of (3) above.

This, he says, is a reasoned position to adopt, for III follows from I and II; while III by itself, without the warrant of I, would be unreasoned and unreasonable.

But by telling the story this way, he has placed himself in an inescapable dilemma. For I shows that objective chance constitutes expertise for him in the sense of (3). Now he must either justify *that*—the problem we discussed before, which has no solution in his case—or else he must join ranks with us. That is, if he cannot justify I independently, then he must say that it is permitted to accept something as expert without that kind of justification. We are then equally allowed to take this permission, and to say III without I, hence without the burden of metaphysics. The right conclusion is that the correctness of the stance which is pictured either in III or in I, cannot be equated with a certain fact about the world.[25]

5. PROBABILISTIC THEORIES AND DECISION-MAKING

Earlier I had characterized acceptance of a scientific theory as involving not full belief, but only belief in its empirical adequacy. I added that in addition to this belief (the epistemic aspect) acceptance also has pragmatic aspects, such as, for example, commitment to a certain research programme. Now I have suggested as amendment: in acceptance, our epistemic attitude is that of taking the probabilities entailed by the theory, for certain propositions, as an expert function.

But there are at this point a number of unanswered questions. Is Miller's Principle sufficient to characterize the attitude of taking q as an expert function? What does this attitude amount to if I accept a theory which gives only conditional probabilities? What role, exactly, does all this play in decision-making—is Huygens's 'value of the hope' still the sole factor needed there? I shall try to answer these and other questions by exploiting the generalization of Miller's Principle by David Lewis, myself, Brian Skyrms, and most especially Chaim Gaifman's theory of expert functions.

Generalization to conditional probabilities[26]

Suppose the whole of my expert's opinion, on the relevant subject, can be summed up as the probability function p. Then of course my attitude is that all my probabilities concerning that subject coincide with function p. If P stands for my probability function restricted to just that subject, we have accordingly the generalization

1. $P(\cdot|q = p) = p$

introduced by Brian Skyrms. This is in functional notation; spelled out it means that $P(A|q = p) = p(A)$ for every proposition A in their common domain.

Skyrms has also given the very simple argument that therefore, I also appropriate my expert's conditional probabilities in the same way. To simplify for a moment, suppose that I fully believe his personal probability function q to be one of the finite number p_1, ..., p_N. That is

2. $P(q = p_1 \text{ or } ... \text{ or } q = p_N) = 1$

Notice that this disjunction is also exclusive. Therefore P is simply the mixture (weighted average) of the appropriate conditional probabilities:

3. $P(\cdot) = P(q = p_1)P(\cdot|q = p_1) + ... + P(q = p_N)P(\cdot|q = p_N)$

But by Miller's Principle the ith conditional probability is just p_i. So P is a mixture of those p_i.

Now suppose that $p_i(A|B) = x$ for i = 1, ..., N. This carries over to P, i.e. $P(A|B)$ equals x as well.[27] That is just arithmetic about mixtures. (Suppose $(a/b) = (c/d) = e$ then $(xa + yc)/(xb + yd) = e$ too.) Generalizing on this argument, we get the conditional version of Miller's Principle:

$$P(A|B \;\&\; q(A|B) = x) = x$$

which is especially important because scientific theories often give us only conditional probabilities, for various possible initial conditions. As we shall see below, however, this nice fact about conditional probabilities cannot be extrapolated to other significant features.

Panels of experts and mixtures of states

The above reasoning can be regarded equally from different points of view. I began with: suppose we fully believe that our expert function q is the same as one member p_i of some finite set ($i = 1$, ..., N). If my opinion is correctly represented by a probability function too, I have some probability w_i for each of these alternatives.

Formally this is exactly like this situation: I have a panel of N experts, call them X_i ($i = 1, ..., N$) and each has his own probability function p_i to advocate. But I do not regard them all equally (especially since they disagree), so I give X_i the weight w_i. Then I treat the mixture $q = \Sigma w_i p_i$ as my expert function (= the opinion of the panel as a whole as regarded by me).

But formally it is also like this situation: I have a single expert, Petra, but she has only conditional opinions $p_i = p(.|Y_i)$ for a certain partition ($i = 1, ..., N$) of all the possibilities. Then I myself supply also a probability w_i that Y_i is the actual possibility. Again the mixture $q = \Sigma w_i p_i$ is the 'overall' verdict, which is the expert function.

Finally, of course, Petra is just like a theory I accept at least to the extent that I take guidance from its probabilities conditional on that partition. All these analogous (formally equivalent) ways of presenting the situation are illuminating; let us explore them in more detail.

If I have a panel of experts, I must 'weigh' their opinions somehow. The two analogies above both indicate that the weight I assign each is in effect my personal probability for this expert turning out to be 'right about everything'. That is not a proposition easy to spell out, and can perhaps be maintained only by introducing a very artificial sort of proposition. It does not need to be spelled out, however, for the way in which I combine these different experts' opinions constitutes exactly my attitude toward this panel, the precise sense in which I accept it as my panel of experts.

When I accept a theory it may provide me with models which are themselves like panels of experts. In such a model, a system may be capable of different *states* each of which determines the probabilities we are interested in. But we may not be able to attribute a specific state, so we have a probability distribution over the possible states. Sometimes the former are called *pure* states, and the probability distributions over them *mixtures* or *mixed* states. The analogy is then: a mixture of pure states corresponds to a weighted panel of experts.

There is now a very important question which can be asked both for a panel of experts and for a mixture of pure states. It is the question: to what extent does unanimity of the components (i.e. the experts, the pure states) carry over to the whole? What is the effect when it does not carry over?

Some cases are easy. Let our expert function q be a mixture of individual expert opinions $p_1, ..., p_n$. If $p_i(A) = x$ for each $i = 1, ..., n$

then obviously $q(A) = x$ too. Indeed, if x is 1 or 0 we can also assert the converse: q assigns that value to A only if all the p_i do. We can also generalize it as: if all the p_i assign values to A in the interval D, then so does q. And in the preceding subsection we saw in addition the striking result that if all the p_i unanimously give conditional probability x to A given B, then $q(A|B) = x$ also.

But unanimity about odds or about stochastic independence does not carry over. And that makes for trouble, as we shall see. It is easy enough to illustrate the point. Suppose I am about to toss a coin twice. Able thinks it is a fair coin. Baker thinks it is a trick coin that always comes up *heads*. But both think that successive tosses are independent of each other. They are my panel of experts and I give their opinion weight ($\frac{1}{2}$) each. Are the tosses independent for me? Obviously I may regard them as physically independent (or not); the real question is whether q(Heads on 2nd toss|Heads on 1st toss) = q (Heads on 2nd toss), in view of the fact that this is so for both Able's probability function and Baker's. The answer is *No*. The reason is easy to see: if the coin comes up heads on the first toss, that increases the probability that it is a trick coin.

Simpson's Paradox

This spells trouble for decision-making. Indeed we have come to one limit of Pascal's and Huygens's paradigm—the value of the hope, for me, may not be a sufficient clue to what I should do.

The trouble does not yet occur if my only decisions are of the sort: what do you bet that the second toss yields *heads* if the first one does? There I simply go by the value of my own hope, and I should. But now consider this more complicated case. Able and Baker are also medical experts, and they give me certain probabilities. Table 8.1 shows a typical sort of example (of *Simpson's Paradox*, as it is often called) in which a proportional weighting of two experts appears to reverse their judgement. The percentages given in this example are accurate to within 0.01 per cent. The great disparity between the experts is presumably due in part because they place me in a different reference class, for which they have separately been collecting data. That is the traditional way to think of it.

TABLE 8.1. *Simpson's Paradox*

	Able (weight 0.6) %	Baker (weight 0.4) %	Me %
P (cancer)	75	12.5	50
P (smoking)	33.33	75	50
P (cancer and smoking)	33.33	12.5	25
P (cancer|smoking)	100	16.66	50

In this example both experts assert a positive correlation of smoking and cancer; for each *P*(cancer|smoking) is greater than *P*(cancer). Thus both would tell me (unless they have some further relevant opinions not presented here) that it will be more prudent not to smoke. But when I look at my own probabilities—derived from theirs via their weights—I see no correlation. Thus if I merely look at the value of the hope for me, there is no prudential reason not to smoke.

In general, if I take Able and Baker as my panel of medical experts, I would surely expect to follow any recommendation they both make. At the very least, the whole above example *could* fit a situation in which the prudent thing for me to do is not to smoke. Thus the naïve decision rule—maximize your subjective expected value—is now seriously in doubt.

One issue may be set aside immediately. This is not just one of the 1001 boring varieties of Newcomb's paradox. It is true that as told above the experts' calculations involve their respective probabilities that I *will* smoke. That is at least very curious if the issue is for me to decide whether or not to smoke. Did their estimates take into account the fact that I will try to reach a decision in the light of these very estimates? Is my smoking behaviour itself being treated as a symptom? But the story need not be told this way at all. Suppose the above figures are derived exactly from studies in which inhalation of smoke was random and randomly independent of volition. The proposition *Smoking* then means simply that I will inhale a certain amount of smoke (daily, on the average) and has nothing to do with my decision as such.

The fact is that accepting a panel—or for that matter, a scientific theory of similar structure—as expert predictor, requires us to reconsider procedures for decision-making.

The limits of empiricism?

We must now diagnose this problem, and see what effect it has on the question of what is involved in acceptance of a scientific theory. The initial diagnosis I propose is this. When the decision rule is simply to maximize expectation, nothing is said about any proposition which does not figure explicitly in the statement of the problem. In the above example, for instance, I am supposed to decide whether or not to smoke, on the basis of probabilities and values assigned just to the little algebra of sixteen propositions generated by logical combinations of *smoking* and *cancer*. Let us call that the *base algebra*. In reality, propositions outside the base algebra must be taken into account. For the solution to remain empiricist, the proper decisions reached must still supervene on statistical predictions for observable events.

It is clear that we must distinguish between two sorts of panels of experts: those for prediction and decision, and those for prediction alone. There is indeed a simple formal distinction that corresponds. Logically a panel of experts for prediction can be replaced equivalently by a single expert. The latter has as his probability function just the relevant weighted average of the panel. That is, the panel's 'joint' prediction equals his individual prediction. And the panel's opinions are respected to the extent that any absolute or conditional probability judgement on which they all agree, is upheld also so by that single 'resultant' expert.

But if the panel is regarded as expert for decision as well, no such replacement is logically possible. The criterion would be: a single expert who shares all those probability and preference (based on expected value) judgements on which all the members of the panel are unanimous. The relevant results are that in all but the most trivial cases, anyone who met this criterion would duplicate one member of the panel, and therefore would not be averaging the panel's opinions or preferences where they disagree.[28] This is true even if all members of the panel assign the same preference values everywhere.

Let us put this in terms of a theory and states as well. A given pure state will assign probabilities, for example, to propositions that say that some physical parameter will take on some value. Hence this parameter has an expectation value in each such state. A mixture of such pure states is also a state, and assigns a weighted average of those probabilities to the same propositions. Again we can calculate expectation values. Now it may be true of two parameters that *in each of the pure states*, one has a higher expectation value than the other, while it does *not in the mixed state*. The same point can be made about correlations.

It seems clear therefore that something more than statistical relations on the base algebra has practical significance. All the developers of 'causal decision theory' reached this conclusion in some form.[29] Some of their formulations appear to be compatible with empiricism (e.g. Skyrms). But others have concluded: there must be unobservable and/or modal facts (e.g. about causation) described by propositions outside the base algebra which are playing a role. That conclusion comes about in this way. There is a single formula for a modified calculation of expectation, which refers to an (unspecified) partition of propositions. It is natural to ask: is there also a single way to specify which partition that is? And then it is natural to make up an answer to that question. Since the answer would be quickly refuted if it had empirical content, it is given in terms of some notion which does not have such content. Then the way everything fits nicely together is cited as support for the idea that this empirically uninformative notion must stand for something real.

But this is a plausible way to proceed only if a certain intermediate conclusion is warranted, namely that we must turn to propositions about the unobservable, because there are no relevant differentiating empirical predictions.

Empiricism vindicated

The flaw I see in the 'genuine causes' version of causal decision theory, is that it insists on a single characterization, valid for all cases, of that something extra—beyond statistical relations on the base algebra—which has practical significance. No general concept is available for such generality, except for certain traditional philosophical concepts such as cause or causal relevance. If we see no correlation between S and C taken by themselves, and then we notice that they are positively correlated conditional on T as well as on $-T$, what shall we do? Well, that depends on what T is. Depending on that we shall either conclude or not conclude that manipulation of the incidence of S will affect the incidence of C.

So there are two cases. What general name can we offer to label these regardless of what S, C, and T are? Enter metaphysics.

But how would we really reach a decision in a specific case? I submit that the two cases would be separated by distinct empirical predictions, on a larger algebra of propositions. It is true that we cannot rule out that we would reverse our decision if we were to investigate a still larger algebra—but that is just the usual risk of the incompleteness of science. To illustrate this we can look at a specific example, in which the relevant factors, if real, are even postulated to be unobservable.

The statistician R. A. Fisher described the following theory about smoking and cancer. It says that there is indeed a positive correlation between cancer and smoking, but this is because some people have a physical factor that predisposes to both smoking and cancer, and absence of the factor predisposes one to not smoking and not having cancer. Let us add that this factor (whether gene or whatever) is strictly unobservable. We have here a theory which poses two possible 'pure' states {factor present, factor absent}. Each has its own probabilities for statements generated by the descriptions {-has factor, -does not have factor, -has cancer, -smokes}. By suitable mixing via the probability of presence of the factor, we derive the usual statistics for smoking and cancer, which show the positive correlation. Yet a person who accepts this theory should say: it does not matter whether you smoke or not, as far as cancer is concerned.

Should we conclude that a person who accepts this theory, and arrived at that practical conclusion, must believe in the presence of that unobservable factor in a certain portion of the population? That would have to be the conclusion, if he or she could not point to predictions concerning observable phenomena which differentiates Fisher's theory from the usual ones.

But of course one could point to such predictions, and they would be just the ones that everyone would cite to justify the practical advice. For the theory predicts that if all smoking materials are removed or destroyed, the incidence of cancer will be no different. It also predicts that among people who may be subject to involuntary smoke inhalation, there will be no correlation between cancer and smoke intake.

Similarly, the choice merely to accept a panel as one whose unanimous recommendations should also guide practical action, is not blind. It can be based on an empirical evaluation of the reliability of their predictions for a relevant *further* set of propositions.

Proofs and illustrations

A concrete example revisited

I have in my hands the usual coin chosen from a magician's supply, and am about to start tossing. For my predictions, I rely on two experts, Able and Baker, and I believe that one of them is right. But my probability that Able is right equals 0.6 and that Baker is right 0.4. Able believes the coin is fair, Baker believes it has probability 0.9 of coming up heads. We do the experiment many times, each time fixing upon a number N of tosses, and writing down whether or not they all come up heads. Then we note the proportion of cases in which they did (see Table 8.2). If Able was right, the sequence X_N should closely fit the sequence A_N, and if Baker was right, it should closely fit B_N. In neither case will it closely fit my probabilities M_N. Knowing that, have I not placed myself in a bad position?

TABLE 8.2. *A panel of experts*

	Able (weight 0.6)	Baker (weight 0.4)	Me	Actual %
P(run of 1 head)	0.5	0.9	0.7	X_1
P(run of 2 heads)	0.25	0.81	0.474	X_2
P(run of 3 heads)	0.125	0.729	0.367	X_3
P(run of N heads)	A_N	B_N	M_N	X_N

But wait! This experiment (let us call it EXP1) was done many times with the same coin. In that case my weighing (0.6/0.4) was not well-calibrated, so no wonder that I'm not well-calibrated on the question of sequences of heads. If instead we sometimes choose a new coin from a 60/40 mixture of fair and (9/1) biased coins, then the actual proportions *will* closely match my probabilities M_N. And Able's and Baker's will both show a poor fit in this second experimental setup (call it EXP2).

How should this influence my behaviour and decisions? No matter which experiment we are doing, of these two, if I am asked to bet on a single run of N tosses, that they will all come up heads,

I should base it on the same probability M_N. That is point *one*. But point *two* is that there are bets I could be offered which I will assess differently in these two cases. Suppose I am asked to bet on such propositions as

(1) In the total (very large) number of runs, the proportion of heads among all the outcomes will be approximately 0.9.
(2) In the total (very large) number of runs, the proportion of occurrences of heads on the 10th throw after 9 heads, equals approximately the proportion of heads among all tosses

Now, if we are in experimental setup EXP1, I shall base my bet for (1) on my probability 0.4 that Baker is right. But if instead we are in experimental setup EXP2, I shall base it on my own probability function.

Similarly, in EXP1 I shall make my bet for (2) exactly the same as each of the expert's bets—I shall take their joint advice and regard the outcome as virtually certain. But in EXP2, I shall look at my own probability function, and regard (2) as virtually certain to be false.

It is clear then that I'm treating my experts differently in these two cases, although I derived my probabilities from theirs via the generalized form of Miller's Principle.

This difference in attitude on my part is not at all due to some primitive inclination or insight, that allocates the decision-advice role to some prediction expert panels and not to others. The difference in decision procedure is solidly linked to differences in empirical predictions outside the simple discourse generated by the statements; {the next run of N tosses will have K heads and N–K tails}.

6. THE LANGUAGE(S) OF SCIENCE

One motive of the semantic approach was to remove the surfeit of philosophy of language. Suppes's initial motto—*mathematics, not metamathematics for philosophy of science*—signals this desire. But of course there are relevant issues about language. I shall here address three. The first concerns the objection that the semantic approach trades on a distinction without a difference. The second is about *which* language to address at all, when analysing a theory.

And the third is the issue of necessity—possibility, modality—which I take to be one place in philosophy where the linguistic turn did more good than harm.

I must admit that I too was, to begin, overly impressed by certain successes of modern logic. One reviewer was able to quote my remark in 1970, that the interrelations between the syntactic and semantic characterizations of a theory 'make implausible any claim of philosophical superiority for either approach'.[30] So how can the semantic approach offer any advantages?

To begin with a point of logic, when a theory is presented by defining the class of its models, that class of structures cannot generally be identified with an elementary class of models of any first-order language.[31] In a trivial sense, everything is axiomatizable, because a thing must be described in order to be discussed at all. But for logicians, 'axiomatizable' is not a vacuous term, and a scientific theory need not be axiomatizable in their sense—or, as they say, the family of models may not be an elementary class.[32] As soon as we go from mathematics to metamathematics, we reach a level of formalization where many mathematical distinctions cannot be captured—except of course by fiat as when we speak of 'standard' or 'intended' models. The moment we do so, we are using a method of description not accessible to the syntactic mode.

On the practical side we must mention the enormous distance between actual research on the foundations of science and syntactically capturable axiomatics. While this disparity will not affect philosophical points which hinge only on what is possible 'in principle', it may certainly affect the real possibility of understanding and clarification. When the advantages of the semantic approach are denied, there seems most often to be a slippery slide back and forth between the practical and the theoretical. First we say: the focus on models puts us in closer touch with actual work in the foundations of the sciences. That is a 'practical' advantage, aligning the philosopher with the scientist and mathematician. It is then immediately retorted that from a logical point of view, focus on models is not essentially different from focus on deductive theories. We reply that a scientific theory is just the typical exception to even the most liberal extension of axiomatizability, because its family of models won't form an elementary class. Then the retort is that scientists have never heard of these metamathematical notions, and philosophy of science is done a disservice by such logical *Spitzfindigkeit*. Too true!

Given this initial appreciation of the situation, shall we address ourselves to language at all? Let us look for a moment at language and the study of language in a general way. Russell made familiar to us the idea of an underlying *ideal language*. This is the skeleton, natural language being the complete living organic body built on this skeleton, the flesh being of course rather accidental, idiosyncratic, and moulded by the local ecology. The skeleton, finally, is the language of logic; and for Russell the question was initially only whether *Principia Mathematica* needed to be augmented with some extra symbols to describe the skeleton fully.

Against this we must advance the conception of natural language as not being constituted by any one realization of any such logical skeleton. Logic has now provided us with a great many skeletons. Linguists have uncovered fragments of language in use for which some of these constructed logical skeletons provided more or less satisfactory models. Natural language consists in the resources we have for playing many different possible language games. Languages studied in logic texts are models, rather shallow models, of some of these specific language games, some of these fragments. To think that there must in principle exist a language in the sense of the objects described by logic, which is an adequate model for natural language taken as a whole, may be strictly analogous to the idea that there must exist a set which is the universe of set theory.[33]

So if we now apply our logical methods in the philosophy of science we should, as elsewhere, set ourselves the task of modelling interesting fragments of language specially relevant to scientific discourse. These fragments may be large or small.

In my opinion, it would be a poor choice to try and describe a whole language in which a given theory can be formulated. The reason is that descriptions of structure in terms of satisfaction of sentences is much less informative than direct mathematical description. It is the choice, explicit or implicit, to be found in almost all linguistically oriented philosophical studies of science. It was the implicit choice behind, certainly, almost all logical positivist philosophy of science.

At the other extreme, we may choose a very small fragment, such as what I have called the fragment of *elementary statements*. Originally I characterized these as statements which attribute some value to a measurable physical magnitude. The syntactic form was therefore trivial—it is always something like '*m* has value *r*'—and therefore the semantic study alone has some significance. Under pressure of various problems in the foundations of quantum mechanics, I broadened my conception of elementary statements in two ways. First I admitted as possibly distinct statements the attributions of ranges (or Borel sets) of values. Second, I admitted as possibly distinct the attribution of states of certain sorts, on the one hand, from that of values to measurable magnitudes on the other.[34]

There are points between these two extremes. I would point here especially to certain forms of natural discourse that are prevalent in the informal presentation of scientific theory, but which have a long history of philosophical perplexities. The main examples are causality and physical modality. From an empiricist point of view, there are besides relations among actual matters of fact, only relations among words and ideas. Yet causal and modal locutions appear to introduce relations among possibilities, relations of the actual to the possible.

This subject is especially important to us now, since the challenge in the subject of laws lies so firmly in the intuitions concerning what must and can be, or can happen, in nature. But we must in any case, in our century, confront the subject of modality. Since irreducible probability is now a fact of life in physics, and probability is a modality, there is no escaping this problem. Yet, if we wish to be empiricists, we have nowhere to turn for the locus of possibility other than to thought and language. In other words, an empiricist position must entail that the philosophical explication of modality, even of its occurrence in science, is to be part of the theory of meaning.

Scientific models may, without detriment to their function, contain much structure which corresponds to no elements of reality at all. The part of the model which represents reality includes the representation of actual observable phenomena, and *perhaps* something more. But it is explicitly allowed to be only a proper part of the whole model.

This gives us I think the required leeway for a programme in the theory of meaning. If the link between language and reality is mediated by models, it may be a very incomplete link—without depriving the language of a complete semantic structure. The idea is that the interpretation of language is not simply an association of real denotata with grammatical expressions. Instead the interpretation proceeds in two steps. First, certain expressions are assigned values in the family of models and their logical relations derive from relations among these values. Next, reference or denotation is gained indirectly because certain parts of the model *may* correspond to elements of reality. The exploration of modal discourse may then draw largely on structure in the models which outstrips their representation of reality.

A graphic, if somewhat inaccurate way to put this would be: causal and modal discourse describes features of our models, not features of the world. The view of language presented here—that discourse is guided by models or pictures, and that the logic of discourse is constituted by this guidance—I recommend as a general empiricist approach for a theory of meaning without metaphysics. I have tried to support this approach to modality elsewhere, and shall not discuss it further here.[35]

PART III
Symmetry as Guide to Theory

INTRODUCTION

In this Part and the next we shall address issues in general philosophy of science. I mean by this issues which arise when we are not inquiring into the foundations of one of the special sciences, but into the content and structure of scientific theories in general. My aim here is to contribute to part four of the programme described in the preceding chapter, partly by elucidating it further, and partly by realizing it in concrete fashion.

The approach I take is that of the semantic view. That does not mean, for me, a systematic attempt to put everything into some standard form. The semantic approach gives a view of what theories are, and orients us toward models rather than language. It need, as such, do no more, but it does give us the task to explore and elaborate concepts which may be useful when one turns to a theory in this way. Once we have some such concepts, we take them along to whatever science catches our interest, and see if they help us to gain some insight there. Such tools of the trade had better be used lightly; it is not much good to hammer in a screw, even if that hammer is your favourite tool. Nor is it appropriate to refer to any of these concepts as property of the semantic approach—that approach only leads us to appreciate them in a certain way.

My favourite concept relating to models is symmetry. I take it to be the primary clue to the theoretically constructed world.

9.
Introduction to the Semantic Approach

THE semantic approach, briefly explained in the preceding chapter, focuses on models rather than on the linguistic formulation of theories (it is 'semantic' rather than 'syntactic'). Its difference from other approaches is largely one of attitude, orientation, and tactics rather than in doctrines or theses. The conviction involved is that concepts relating to models will be the more fruitful in the philosophical analysis of science.

1. PHENOMENA, DATA, AND THEORIES

When I was at Yale, philosophers met for lunch almost every day in each other's colleges, and discussed innumerable philosophical puzzles and paradoxes. One of the subjects was the concept of shadow, and some remarks were eventually published by two of us.[1] The word 'shadow' has two senses, one as count-noun (as in 'there were more soldiers than shadows, for some soldiers were inside'), and one as mass-noun (as in 'there was more shadow than water in the square' or 'how much shadow is there in the photograph?'). The following is a theory of shadow, meaning that term in its mass-noun sense. The letter *X* stands for any physical object.

I. If *X* casts any shadow, then some light is falling directly on *X*.
II. *X* cannot cast shadow through an opaque object.
III. All shadow is shadow of something.

Each principle appears to be accepted as universally correct, at least on first sight. Yet this theory is not adequate. To see that, imagine a barn on a sunny day, and a robin which flies through the shadow cast by the barn. Drawing straight lines from the sun, touching the robin at a certain instant, we mark out a small region on the ground. There is certainly shadow there (assuming the sun is the only illumination). But by I, it is not shadow of the robin and by II it is not shadow of the barn. There is certainly nothing else of which it could be shadow. But III requires that all shadow be shadow of something.

This little theory and its demise are certainly not of scientific interest (except perhaps to the linguist) but it serves to introduce some important distinctions. The theory is not inconsistent. Taken by itself, it is logically impeccable. But there are phenomena that do not fit this theory—and our little thought experiment points to a large class of these. So we have here two distinct concepts of inadequacy: *inconsistency*, and what I propose to call *empirical inadequacy*.

If a theory is not empirically adequate—so some actual phenomenon does not fit this theory—then it is certainly not true. However, it could be empirically adequate and still not true. That means that all phenomena fit the theory, but the theory also says that there are certain things or kinds of things which in fact there are not. Hence *falsity* is a third type of inadequacy. However it may be disputed whether it matters if a theory is false, as long as it is consistent and empirically adequate.

The example that refuted this little theory was a situation described in the following way: certain physical objects were listed, and the shadow of the barn was mentioned. The relations between sun, barn, robin, and shadow were for the most part tacitly understood. This tacit understanding could be made explicit, and part of it is that there was nothing unusual which was left unmentioned (such as a large hole in the barn through which light falls on the robin). When we don't leave such a lot up to tacit understanding and imagination, we say that we have constructed a *model*.

In this case, it would be a model of that situation or phenomenon, but not a model of that theory. A model is called a model of a theory exactly if the theory is entirely true if considered with respect to this model alone. (Figuratively: the theory would be true if this model was the whole world.) An inconsistent theory has no models at all. A consistent but empirically inadequate theory does have models, but none of them can accommodate all actual phenomena.

Here is an example of a little theory and one of its models. The theory has three principles, which I've taken (essentially) from Hilbert's axioms for Euclidean geometry.

*A*1. For any two lines, at most one point lies on both.
*A*2. For any two points, exactly one line lies on both.
*A*3. On every line there are at least two points.

Let us call this Theory *G*. One model is just a single line with two points on it. A more interesting but still finite model is what is called the Seven Point Space (see Fig. 9.1). The seven points are clearly marked. To satisfy *A*2 we need a line that lies on D and F—that is the circle, whose points are D, F, B. The only points in this model are those marked and labelled, the rest is just there for visual presentation. Each line contains exactly three points.

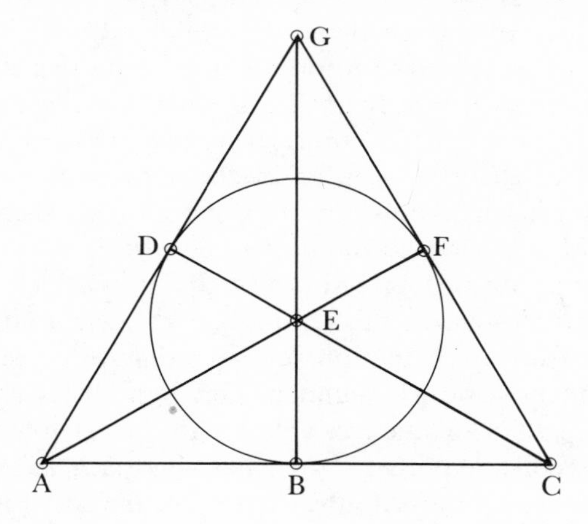

FIG. 9.1. The Seven Point Space

This model is finite, you could construct it from wires and buttons. Also you can draw it in its entirety on a blackboard. When a model is infinite you can't draw it, but only describe it. So it is first constructed in the imagination; and whether or not a person constructs it, it always exists as an abstract object (or as we also say, a mathematical object).

The Seven Point Space is not a model of Euclidean geometry, which has only infinite models. But the fact that I could draw it establishes an interesting relationship. That little structure can be *embedded* in Euclidean space. That means it is isomorphic to a part of Euclidean space.

This relation is important because it is also the exact relation a phenomenon bears to some model or theory, if that theory is empirically adequate. The shadow theory was inadequate because for each of its models there is some actual phenomenon which is not isomorphic to some part of that model.[2]

There is one more issue we must take up here about theories and models. This is the difference between theoretical and empirical equivalence of theories. The first means that they really say the same thing; each can be deduced from the other (or, more accurately, they have the same set of models). The second only means that if some phenomena fit a model of one theory then they also fit a model of the other.

In the history of science there are celebrated cases of putative equivalence. In some cases several theories really turned out to be the same, in others it was claimed only that one theory would do just as well as the other for empirical science. (Comte claimed this about the rival theories of heat in the nineteenth century, for instance.) But such claims are treacherous for there may be phenomena of a sort we have not conceived yet.

The first explicitly recognized use of equivalence concerned two hypotheses concerning the sun's motion. The ancient Greeks noted that the sun's apparent motion in the zodiac appeared to vary, fastest in winter, slowest in summer. One hypothesis was that the sun's annual path is a circle concentric with the earth, with varying speed. The second hypothesis was that this path is an eccentric circle (i.e. not concentric with the earth), but the speed is constant. Apollonius proved in addition that motion along an eccentric circle can be perfectly duplicated by motion on an epicycle of a concentric circle. If all relevant phenomena are observations of the sky from the earth, there can be no decision between such hypotheses on empirical grounds. And Apollonius envisaged no other sorts of observable phenomena.

2. FROM THE AXIOMATIC TO THE SEMANTIC APPROACH

In what is now called the 'received view' (developed by the logical positivists and their immediate heirs), a theory was conceived of as an axiomatic theory. That means, as a set of sentences, defined as the class of logical consequences of a smaller set, the axioms of that theory. A distinction was drawn: since the class of axioms was normally taken to be effectively presentable, and hence syntactically describable, the theory could be thought of as in itself uninterpreted. The distinction is then that scientific theories have an associated interpretation, which links their terms with their intended domain.

The story of misery and pitfalls that followed is well known. Only two varieties of this view of scientific theories as a special sort of interpreted theories emerged as anywhere near tenable. The first variety insists on the formal character of the theory as such, and links it to the world by a partial interpretation. Of this variety the most appealing to me is still Reichenbach's which said that the theoretical relations have *physical correlates*. Their partial character stands out when we look at the paradigm example: light rays provide the physical correlate for straight lines. It will be immediately clear that not every line is the path of an actual light ray, so the language-world link is partial. The second variety, which came to maturity in Hempel's later writings, hinges for its success on treating the axioms as already stated in natural language. The interpretative principles have evolved into axioms among axioms. This means that the class of axioms may be divided into those which are purely theoretical, in which all non-logical terms are ones specially introduced to write the theory, and those which are mixed, in which non-theoretical terms also appear.

It will be readily appreciated that in both these developments, despite lip-service to the contrary, the so-called problem of interpretation was left behind. We do not have the option of interpreting theoretical terms—we only have the choice of regarding them as either (*a*) terms we do not fully understand but know how to use in our reasoning, without detriment to the success of science, or (*b*) terms which are now part of natural language, and no less well understood than its other parts. The choice, the correct view about the meaning and understanding of newly introduced terms, makes no practical difference to philosophy of science, as far as one can tell. It is a good problem to pose to philosophers of language, and to leave them to it.

In any tragedy,[3] we suspect that some crucial mistake was made at the very beginning. The mistake, I think, was to confuse a theory with the formulation of a theory in a particular language. The first to turn the tide was Patrick Suppes with his well-known slogan: the correct tool for philosophy of science is mathematics, *not* metamathematics. This happened in the 1950s—bewitched by the wonders of logic and the theory of meaning, few wanted to listen. Suppes's idea was simple: *to present a theory, we define the class of its models directly*, without paying any attention to questions of axiomatizability, in any special language, however relevant or simple or logically interesting that might be. And if the theory as such, is to be identified with anything at all—if theories are to be reified—then a theory should be identified with its class of models.[4]

This procedure is in any case common in modern mathematics, where Suppes had found his inspiration. In a modern presentation of geometry we find not the axioms of Euclidean geometry, but the definition of the class of Euclidean spaces. Similarly Suppes and his collaborators sought to reformulate the foundations of Newtonian mechanics, by replacing Newton's axioms with the definition of a Newtonian mechanical system. This gives us, by example, a *format* for scientific theories.

3. THEORY STRUCTURE: MODELS AND THEIR LOGICAL SPACE

The semantic view of theories makes language largely irrelevant to the subject. Of course, to present a theory, we must present it in and by language. That is a trivial point. Any effective communication proceeds by language, except in those rare cases in which information can be conveyed by the immediate display of an object or happening. In addition, both because of our own history—the history of philosophy of science which became intensely language-oriented during the first half of this century—and because of its intrinsic importance, we cannot ignore the language of science. But in discussion of the structure of theories it can largely be ignored.

In Ronald Giere's recent encapsulation of the semantic approach, a theory consists of (*a*) the *theoretical definition*, which defines a certain class of systems; (*b*) a *theoretical hypothesis*, which asserts that certain (sorts of) real systems are among (or related in some way to) members of that class.[5]

This is a step forward in the direction of less shallow analysis of the structure of a scientific theory. The first level of analysis addresses the notion of theory *überhaupt*. We can go a bit further by making a division between relativistic and non-relativistic theories. In the latter, the systems are physical entities developing in time. They have accordingly a space of possible states, which they take on and change during this development. This introduces the idea of a cluster of models united by a common *state-space*; each has in addition a domain of objects plus a 'history function' which assigns to each object a history, i.e. a trajectory in that space. A real theory will have many such clusters of models, each with its state-space. So the presentation of the theory must proceed by describing a class of *state-space types*.[6]

For such a non-relativistic theory, we can see at once that there must be basic equations of two types, which correspond to the traditional ideas of *laws of co-existence* and *laws of succession*. The former type, of which Boyle's law $PV = rT$ is a typical example, restricts positions in the state-space. Selection and super-selection rules are the quantum theoretical version of such restrictions. The other type has Newton's laws of motion and Schroedinger's equation as typical examples: they restrict trajectories in (through) the state-space. Symmetries of the model—of which Galileo's relativity is the classical example *par excellence*—are 'deeper' because they tell us something beforehand about what the laws of coexistence and succession can look like. It is in the twentieth century's quantum theory that symmetry, coexistence, and succession became most elegantly joined and most intricately connected.

What I have just described are laws of the model: important features by which models may be described and classified. The distinction between these features and others that characterize the model equally well is in the eye of the theoretician; it does not, to my mind, correspond to any division in nature. As I have indicated too, these sort of laws pertain to non-relativistic models only.

One reason why the distinction between the laws of the model and other features should not be taken as corresponding to a division in nature, is that structural revisions in this respect need not affect adequacy. If we replace one model by another, whose state-space is just what remains in the first after the laws of coexistence are used to rule out some set of states, we have not diminished what can be modelled. Similarly, if we include in the state some functions of time, laws of succession may be replaced by laws of coexistence. One way to read Newton's laws is this: the second derivative with respect to time, of position conceived as a function of time, is one parameter represented in the state, just as well as the total force impressed and the mass. The second law of motion, $F = ma$, is then a law of coexistence, rather than of succession: it rules out certain states as impossible. This sort of rewriting has been strikingly exhibited in certain philosophical articles about Newtonian mechanics.

In the case of relativistic theories, early formulations can be described roughly as relativistically invariant descriptions of objects developing in time—say in their proper time, or in the universal time of a special cosmological model (e.g., Robertson–Walker models). A more general approach, developed by Glymour and Michael Friedman, takes space-times themselves as the systems. Presentation of a space-time theory T may then proceed as follows: a (*T*-)*space-time* is a four-dimensional differentiable manifold M, with certain geometrical objects (defined on M) required to satisfy the *field equations* (of T), and a special class of curves (the possible trajectories of a certain class of physical particles) singled out by the *equations of motion* (of T).

Clearly we can further differentiate both sorts of theories in other general ways, for example with respect to the stochastic or deterministic character of imposed laws. (It must be noted however that except in such special cases as the flat space-time of special relativity—its curvature is independent of the matter-energy distribution—there are serious conceptual obstacles to the introduction of indeterminism into the space-time picture.)

I must leave aside details of foundational research in the sciences. But I want to insist that the point of view which I have been outlining—the *semantic view* as opposed to the received view—is much closer to practice there. The scientific literature on a theory makes it relatively easy to identify and isolate classes of structures to be included in the class of theoretical models. It is on the contrary usually quite hard to find laws which could be used as axioms for the theory as a whole. Apparent laws which frequently appear are often partial descriptions of special subclasses of models, their generalization being left vague and often shading off into logical vacuity.

Let me give two examples. The first is from quantum mechanics: *Schroedinger's equation*. This is perhaps its best known and most pervasively employed law—but it cannot very well be an axiom of the theory since it holds only for conservative systems. If we look into the general case, we find that we can prove the equation to hold, for some constant Hamiltonian, under certain conditions—

but this is a mathematical fact, hence empirically vacuous. The second is the *Hardy-Weinberg* law in population genetics. Again, it appears in any foundational discussion of the subject. But it could hardly be an axiom of the theory, since it holds only under certain special conditions. If we look into the general case, we find again a logical fact: that certain assumptions imply that the equation describes an equilibrium which can be reached in a single generation, and maintained. The assumptions are very special, and more complex variants of the law can be deduced for more realistic assumptions—in an open and indefinite sequence of sophistications.

What we have found, in the semantic approach, is how to describe relevant structures in ways that are also directly relevant, and seen to be relevant, to our subject matter. The scholastic logistical distinctions that the logical positivist tradition produced—observational and theoretical vocabulary, Craig reductions, Ramsey sentences, first-order axiomatizable theories, and also projectible predicates, reduction sentences, disposition terms, and all the unholy rest of it—had moved us *mille milles de toute habitation scientifique*, isolated in our own abstract dreams. Since Suppes's call to return to a non-linguistic orientation, now about thirty years ago, we have slowly regained contact.

4. WHAT IS INTERPRETATION? THE THREE-TIERED THEORY

When Pierre Duhem wrote *The Aim and Structure of Physical Theory* he included an Appendix which presented his own interpretation of physical science, tinged with the very metaphysics he had insistently banished from science. Is it really possible to separate the categories neatly? I think not.

Leaving realist and anti-realist disagreements about aim aside, we can all go on to a co-operative venture: to analyse the theories science gives us. We must analyse (*a*) the theory's relation to the phenomena reported in our data, (*b*) the theory's relation to its users, who explain and predict, (*c*) the theory all by itself, its *structure* and *content*. In this section let us focus on task (*c*), which it would be most felicitous to complete first. The semantic view has a simple schema for delineating structure and content, in principle.

Above I mentioned Giere's elegant capsule formulation of the semantic view: a theory is presented by giving the definition of a certain kind (or kinds) of systems plus one or more hypotheses about the relation of certain real (kinds of) systems to the defined class(es). We speak then of the *theoretical definition* and the *theoretical hypothesis* which together constitute the given formulation of the theory. A 'little' theory might for example define the class of Newtonian mechanical systems and assert that our solar system belongs to this class.

Truth and falsity offer no *special* perplexities in this context. The theory is true if those real systems in the world really do belong to the indicated defined classes. From a logical, or more generally semantic, point of view we may consider as implicitly given, models of the world as a whole, which are as the theoretical hypothesis says it is. In a very large class of models of the world as a whole, our solar system is a Newtonian mechanical system. In one such model, nothing except this solar system exists at all; in another the fixed stars also exist, and in a third, the solar system exists and dolphins are its only rational inhabitants. Now the world must be one way or another; so the theory is true if the real world itself is (or is isomorphic to) one of these models. This is equivalent to either of two familiar sorts of formulations of the same point: the theory is true exactly if (*a*) one of the possible worlds allowed by the theory is the real world; or (*b*) all real things are the way the theory says they are.

The analysis of structure provided therefore leads also to a schematic description of content. For what a theory says, that is its content, and we know what it says if we know under what conditions it is true. But in practice we do not find the analysis going so smoothly. The real answers found in the literature, to questions about what a theory says the world is like, are at the same time sketchy and very far-reaching. *Any question about content is, in actuality, met with an interpretation.* Newton himself placed the reality of time and space at the centre of his answer; Laplace insisted that this same theory describes systems evolving with perfect determinism in this space and time. Mach, harking back to Leibniz and Berkeley, argued that Newton had gone beyond the logical content of his own theory in postulating absolute space. John Earman has recently shown how much extrapolation and assumption was present in Laplace's understanding of the theory.

Of course, in the limiting case, a perfectly faithful interpretation would not go beyond the theory at all. But we tend to have many additional questions that, for us, cry out for an answer. Since the theory is not formalized, takes time to grow into a definitive form, is subject to disputes, and was in any case developed with attention focused on certain phenomena in the experimental limelight, we are also not altogether clear on where exact content ends and extrapolation begins. The demarcation is vague, and comes to light mainly when rival interpretations appear. We must however insist on one crucial criterion. A putative interpretation is no longer an interpretation, but a new scientific theory, if its empirical predictions either go beyond or contradict those of the original.

This criterion is surely incontrovertible. Despite the fact that it too must rule an objectively vague domain, the distinction it marks is workable and indispensable. We discern three tiers: the theory's representation of the phenomena, the theory in itself, the interpretations of the theory. Because of that criterion, however, the demarcation between first and second tier is needed also, however, to delineate the third. We have come therefore again to this relation to the phenomena—the theory's *empirical adequacy* as opposed to its *truth* as such—as central concern.

In *The Scientific Image* I proposed a new explication of empirical adequacy. The logical positivist tradition had given us a formulation of a concept of empirical adequacy which was not only woefully inadequate but had created a whole cluster of 'artifact problems' (by this I mean, problems which are artifacts of the philosophical approach, and not inherent in its subject). In rough terms, the empirical content of a theory was identified with a set of sentences, the consequences of that theory in a certain 'observational' vocabulary. In my own studies, I first came across formulations of more adequate concepts in the work of certain Polish writers (Przelewski, Wojcicki), of Dalla Chiara and Toraldo di Francia, and finally of course in Patrick Suppes's own writings on what he calls empirical algebras and data models.[8] While some of these formulations were still more language-oriented than I liked, the similarity in their approach was clear: certain parts of the models were to be identified as *empirical substructures*, and these were the candidates for representation of the observable phenomena which science can confront within our experience.

At this point it seemed that the relationship thus explicated corresponds exactly to the one Reichenbach attempted to identify through this concept of coordinative definitions, once we abstract from the linguistic element. Thus in a space–time the geodesics are the candidates for the paths of light rays and particles in free fall. More generally, the identified spatio-temporal relations provide candidates for the relational structures constituted by actual genidentity and signal connections. These actual physical structures are to be embeddable in certain substructures of space–time, which allows however for many different possibilities, of which the actual is, so to say, some arbitrary fragment.

Thus we see that the empirical structures in the world are the parts which are at once *actual* and *observable*; and empirical adequacy consists in the embeddability of all these parts in some single model of the world allowed by the theory.

Patrick Suppes carefully investigated the construction of data models, and the empirical constraint they place on theoretical models. When thought of as concerned with exactly this topic, much apparently 'a prioristic' theorizing on the foundations of physics takes on a new intelligibility. A reflection on the possible forms of structures definable from joint experimental outcomes yields constraints on the general form of the models of the theories 'from below'; that class of models can then be narrowed down by the imposition of postulated general laws, symmetry constraints, and the like, 'from above'.

5. THEORIZING: DATA MODELS AND THEORETICAL MODELS

New theories are constructed under the pressure of new phenomena, whether actually encountered or imagined. By 'new' I mean here that there is no room for these phenomena in the models provided by the accepted theory. There is no room for a mutable quantity with a discrete set of possible values in the models of a theory which says that all change is continuous. In such a case the old theory does not allow for the phenomenon's description, let alone its prediction.

I take it also that the response to such pressure has two stages, logically if not chronologically distinguishable. First the existing theoretical framework is widened so as to allow the possibility of those newly envisaged phenomena. And then it is narrowed again, to exclude a large class of the thereby admitted possibilities. The first move is meant to ensure empirical adequacy, to provide room for all actual phenomena, the rock-bottom necessary condition of success. The second move is meant to regain empirical import, informativeness, predictive power.

It need hardly be added that the moves are not made under logical compulsion. When a new phenomenon, say *X*, is described it is no doubt possible to react with the assertion: if it looked as if *X* occurred, one would only conclude that familiar fact or event *Y* had occurred. A discrete quantity can be approximated by a continuous one, and an underlying continuous change can be postulated. From a purely logical point of view, it will always be up to the scientists to take a newly described phenomenon seriously or to dismiss it. Logic knows no bounds to *ad hoc* postulation. This also brings out the fact of creativity in the process that brings us the phenomena to be saved. Ian Hacking put this to me in graphic terms when he described the quark hunters as seeking to create new phenomena. It also makes the point long emphasized by Patrick Suppes that theory is not confronted with raw data but with models of the data, and that the construction of these data models is a sophisticated and creative process. To these models of data, the dress in which the débutante phenomena make their début, I shall return shortly.

In any case, the process of new theory construction starts when described (actual or imagined) phenomena are taken seriously as described. At that point there certainly is logical compulsion, dimly felt and, usually much later, demonstrated. Today Bell's Inequality argument makes the point that some quantum mechanical phenomena cannot be accommodated by theories which begin with certain classical assumptions. This vindicates, a half-century after the fact, the physicists' intuition that a radical departure was needed in physical theory.

Of the two aspects of theorizing, the widening of the theoretical framework, and its narrowing to restore predictive power, I wish here to discuss the former only. There we see first of all a procedure so general and common that we recognize it readily as a primary problem-solving method in the mathematical and social as well as the natural sciences—any place where theories are constructed, including such diverse areas familiar to philosophers as logic and semantics. This method may be described in two ways: as *introducing hidden structure*, or 'dually' as *embedding*. Here is one example.

Cartesian mechanics hoped to restrict its basic quantities to ones definable from the notions of space and time alone, the so-called kinematic quantities. Success of the mechanics required that later values of the basic quantities depend functionally on the earlier values. There exists no such function. Functionality in the description of nature was regained by Newton, who introduced the additional quantities of mass and force. Behold the introduction of hidden parameters.

The word 'hidden' in 'hidden parameters' does not refer to lack of experimental access. It signifies that we see parameters in the solution which do not appear in the statement of the problem.

We can 'dually' describe the solution as follows: the kinematic relational structures are embedded in structures which are much larger—larger in the sense that there are additional parameters (whether relations or quantities or entities). *The phenomena are small but chaotic; they are treated as fragments of a 'whole' that is much larger but orderly and simple.* This point could, I believe, be illustrated by examples from every stage of the history of science.[9] When a point has such generality one assumes that it must be banal, and carry little insight. In such a general inquiry as ours, however, perspective is all; and we need general clues to find a general perspective.

6. EXPERIMENTATION: AS TEST AND AS MEANS OF INQUIRY

Theory construction I have described as being ideally divisible into two stages: the construction of sufficiently rich models to allow for the possibility of described phenomena, and the narrowing down of the family of models so as to give the theory greater empirical content. There must be a constant interplay between the theoretician's desk and the experimenter's laboratory. Here too I wish to distinguish two aspects, which must be sharply distinguished.

The best known function of experimentation is hypothesis testing. The experimenter reads over the theoretician's shoulder, and designs experiments to test whether the narrowing down has not gone too far and made the theory empirically inadequate. This characterization is simple and appealing; it is unfortunately over-simple. It overlooks first of all the fact that hypothesis testing is in general comparative, and ends not with support or refutation of a single hypothesis but with support of one hypothesis against another. It also overlooks a second feature: *that testability has to do as well with informativeness.* A theory may be empirically adequate—or at least adequate with respect to a certain class of phenomena—but not sufficiently informative about them to allow the design of a test which could bring it support. Hence when we try to evaluate theories on the basis of their record even in the most assiduous testing, the ranking will reflect not only adequacy or truth, but also informativeness.

Certainly truth and informativeness are both virtues, and both are epistemic values to be pursued (as has been cogently argued by Isaac Levi), but this means that it is a mistake to think of testing in terms of pure confirmation. To give an analogy: the number of votes a candidate receives is a measure of voter support for his platform but it is also a function of media exposure, and so it is not a pure measure of voter support.[10]

There is another function of experimentation which is less often discussed and in the present context more interesting.

This second function is the one we describe in the language of discovery. Chadwick discovered the neutron, Millikan the charge of the electron, and Livingstone the Zambesi river. Millikan's case is a good illustration. He observed oil droplets drifting down in the air between two plates, which he could connect and disconnect with a battery. By friction with the air, the droplets could acquire an electrostatic charge; and Millikan observed their drifting behaviour, calculating their apparent charges from their motion. Thus he found the largest number of which all apparent charges were integral multiples, and concluded that number to be the charge of a single electron. That number was *discovered.* The number of such charges per Faraday equals Avogadro's number! No one could have predicted that! You would think that empiricists would be especially bothered by such a scientific press release. For it sounds as if by carefully designed experiment we can discover facts about the unobservable entities behind the phenomena.

Let me outline the alternative interpretation of what experiment does. Note first that the division of experiments into means of testing and means of discovery is a division neither by experimental procedure nor by experimental apparatus. The set-up and operations performed would have been just the same if Millikan had made bold conjectures beforehand about that number, and had set out to test these conjectures. The division is rather by function *vis-à-vis* the ongoing process of theory construction.

The theory is written, so to say, step by step. At some point, the principles laid down so far imply that the electron has a negative charge. A blank is left for the magnitude of the charge. If we wish to continue now, we can go two ways. We could certainly proceed by trial and error, hypothesizing a value, testing it, offering a second guess, testing again. But alternatively, we can let the experimental apparatus write a number in the blank. What I mean is: in this case the experiment shows that unless a certain number (or a number not outside a certain interval) is written in the blank, the theory will become empirically inadequate. For the experiment has shown by actual example, that no other number will do; that is the sense in which it has filled in the blank. So regarded, *experimentation is the continuation of theory construction by other means*.

Recalling the similar saying about war and diplomacy, I should like to call this view the 'Clausewitz doctrine of experimentation'. It makes the language of construction, rather than of discovery, appropriate for experimentation as much as for theorizing.

We have now seen a considerable number of topics in philosophy of science, as they look in the semantic view. It is clear that in each case we had only a sketch, and the discussion was to some extent programmatic. We need to develop and increase the conceptual tools used in our approach. We have to carry out the programme proposed. We have to get on with the analyses promised. I have already cited a number of contributions to that work, and offer the following chapters to this end.

10.

Symmetry Arguments in Science and Metaphysics

Socrates: 'Then, if we are not able to hunt the goose with one idea, with three we may take our prey; Beauty, Symmetry, Truth are the three . . .

Plato, *Philebus* 65 a.

THE philosophical study of science as an inquiry into the laws of nature had a presupposition. It assumed that the structure of science is to be understood as a reflection of the structure of nature. I propose that we embark on a study of the structure of science—its theories and models—in itself. The clue, I shall suggest, is this: at the most basic level of theorizing, *sive* model construction, lies the pursuit of symmetry.

Symmetry has many uses. The most impressive are in 'symmetry arguments' which look entirely a priori and have astonishingly far-reaching conclusions. Yet nothing contingent and about the world can be deduced by logic alone. The a priori appearance is therefore deceptive—but tells us that we have here an argumentative technique of unusual power and elegance.

There are two forms of argument which reach their conclusion 'on the basis of considerations of symmetry'. One, the symmetry argument proper, relies on a meta-principle: that structurally similar problems must receive correspondingly similar solutions. A solution must 'respect the symmetries' of the problem. The second form, rather less important, assumes a symmetry in its subject, or assumes that an asymmetry can only come from a preceding asymmetry. Both exert a strong and immediate appeal, that may hide substantial tacit assumptions.

1. MIRROR IMAGES: SYMMETRY AS PROOF TECHNIQUE

The paradigm of symmetry is the mirror image. I and my image are a symmetric pair. To a lesser extent, the left and right parts of my body are each other's mirror image—that is the extent to which my body is most obviously symmetric. Other kinds of symmetry exist, but this is the one to begin with. It is called *bilateral symmetry*.

To make this precise, consider the case of figures in a plane. Figure F_1 and F_2 may be related as follows: there is a straight line that separates them (*the line of reflection*) and each point in F_1 can be connected to a corresponding point in F_2 by means of a line perpendicular to m. The correspondence connects all points in F_1 and F_2 and is such that corresponding points are equidistant from the line of reflection. In space, the definition is similar, but with a *plane of reflection* (the surface of the mirror, so to say).

Here is a symmetry argument (Fig. 10.1).[1] Farmer Able walks each morning to the creek, fills a pail with water, and takes it to the hen-house. Where should he reach the creek, in order to walk the shortest distance? There is an easy solution. Recall that the classic answer to 'what is the shortest path?' is 'a straight line'. Unfortunately there is no straight line from Able's house via the creek to the hens. But imagine he has a twin brother, Baker, living in bilateral symmetry to him, who has a similar problem with geese. The shortest path from Baker to Able's hen-house is a straight line! But Baker can also see that his journey to Able's hens is exactly as long as that to his geese, if he reaches the creek at the same point.

If Able and Baker meet at arbitrary point P' on the creek, they will walk equally far, whether they go together to the same fowl-house, or separately to different fowl. So the four possible problems, of finding the point of the creek that yields the shortest path from house A or house B, to fowl-house H or G, are *all equivalent*. One of these four has a very simple solution: if Baker goes from B to H, he should follow a straight line, through P. So P is the answer for all four problems—hence also for Able's original problem.

There are two questions we should carefully investigate. What was the method followed to reach this solution? Is it a *general* method? And secondly, what presuppositions did it have—what was tacitly taken for granted?

Let us start with the second question. If we discover a hill or pond on the farm, lying on the line from A to P for example, that defeats the solution. Indeed, we took Able's practical problem and conceived of it as a problem of geometry in the Euclidean plane.

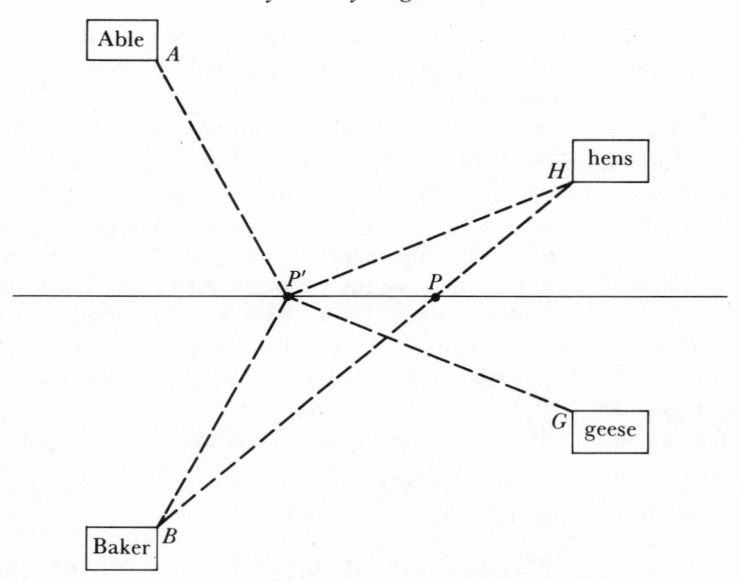

FIG. 10.1. A symmetry argument for shortest paths

That was our model. Then we thought about symmetry in this model. That is very important. The symmetry argument begins after the choice of some model or theoretical context—which may be quite general or vague, but its assumptions were strong enough to allow the argument to take off.

The first question was: what method did we use? Well, we had a relatively difficult problem (did you think of using calculus? Shortest path problems are typical calculus exercises) and *transformed* it into another problem that was easy to prove. That is, we found another *equivalent* problem, one that was 'essentially the same'. Thus our guiding idea was: *essentially similar problems have essentially similar solutions*. That the four problems, Able's and Baker's, were essentially the same was guaranteed by the symmetry of the diagram. And that of course must be a typical way in which problems are going to be essentially the same: that they pertain to symmetrically related situations.[2]

What is 'essentially the same'? It means that the *essential aspects* remained the same, that the new problem was like the old one in all *relevant respects*. These two ways of putting it, however, are not the same. The word 'relevant' is context-dependent—relevant to what?—while 'essential' purports to be objective. We'll have to be careful not to let anyone get philosophical mileage out of this purported objectivity.

Let me describe the method in yet another way. We have understood a problem 'in its full generality' only when we know exactly what counts as essentially the same problem. That means: when we know exactly which transformations do and which do not change the situation in relevant respects. Now, to state a problem in its full generality is to achieve the proper degree of abstraction: to abstract the problem itself from the concrete guise of its appearance. Generality, abstraction, transformation, equivalence of problems—some very old philosophical ideas are here mobilized in new logical form.

We can summarize these reflections to some extent in a slogan:

> *Symmetry Requirement*: Problems which are essentially the same must receive essentially the same solution.

This slogan may look, as slogans do, rather banal. It will, I hope, also be clearer by the end of this chapter, why it is appropriate to call this a (or *the*) symmetry requirement. By introducing related notions one by one through examples, I hope to reach the point where it will be natural to say that this requirement *is* the requirement to 'respect the symmetries' of the problem. And in subsequent chapters, where the examples will be much more concrete, the slogan will become descriptive of a natural and fruitful methodology.

Proofs and illustrations

In a very successful symmetry argument, the solution is shown to be uniquely determined by the Symmetry Requirement. More often, the requirement only simplifies the problem, by showing that the solution must take a certain form. I will end this section by giving an elementary example of this simplifying role. The problem is suggested by the famous one of the Seven Bridges of Königsberg, which has no solution. Let us consider the six bridges variant (see Fig. 10.2). What path can be followed which crosses each bridge exactly once? This problem has a solution, but it is not unique. The bridges are numbered; *A* and *B* are islands, while *X* and *Y* are the opposite shores. In the abstract, a path is a sequence of six distinct bridges, in some order; for example 123456. But not all of these (there are approximately a thousand of them) can be followed consecutively by a pedestrian.

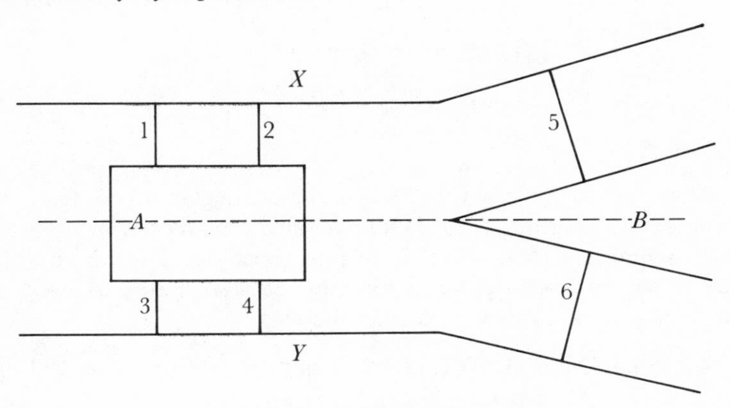

FIG. 10.2. The six bridges

Let us classify them by where they start and where they end. For example, 134256 begins on *X* (because it starts with 13) and ends on *Y*. Now we see at once that for any path that begins on *X*, there must equally be one that begins on *Y*. This is because the diagram is symmetric about the horizontal line *m*. So in our classification of cases into 16, we only need to look seriously at the top 12. The same symmetry tells us that if there is a path beginning on *A* or *B* and ending on *Y*, there is equally one ending on *X*. Moreover, there is a sort of temporal symmetry: if a path begins in one place and ends in another, there is also a path (the exact reverse) which ends in the former and begins in the latter. Let us cross off all the cases we accordingly need not inquire into separately (see Fig. 10.3). At this point we can use another consideration to rule out a path from *B* to *A*: since *B* has only two bridges, if you begin on *B* you must end there. This new fact is indicated by the circle in the *BA* box, eliminating that one too. But equally if you begin on either *A* or *B* you will have a bad problem with *X*: it has three bridges, so if you do not start *X* then you will cross in, then out, then in again eventually—and not be able to leave. So if you start with *A* or *B*, you'll end up on *X*—but the diagram already shows that beginning on *A* or *B* you must end on *A* or *B*. Hence we have eliminated from consideration all paths, except those beginning on *X*.

	X	*B*	*A*	*Y*
X				
B	×		○	×
A	×	×		×
Y	×	×	×	×

FIG. 10.3

Now let us classify those paths which begin on *X*. They can begin with bridges 1, 2, or 5. But we have a further symmetry: any path remains a path if we uniformly substitute 1 for 2 and/or 3 for 4. So, by ignoring the distinctions between paths which differ in such permutations we see the following:

Course 1: *X*, *B* as 1st and 2nd places require *Y*, *A* as 3rd and 4th places

Course 2: *X*, *A* as 1st and 2nd places give choice of *X* or *Y* as 3rd. The former requires *B*, *Y*, *A*, *Y* as 4th, 5th, 6th, 7th.

So now we have found one completely determinate path unique (up to the described permutations) from *X* to *Y*. And we know that any remaining ones must either begin as *X*, *B*, *Y*, *A* or as *X*, *A*, *Y*.

At this point the solution can be completed by inspection, and we arrive at the following possible sequences of places:

1. *X*, *A*, *X*, *B*, *Y*, *A*, *Y*
2. *X*, *A*, *Y*, *B*, *X*, *A*, *Y*
3. *X*, *A*, *Y*, *A*, *X*, *B*, *Y*
4. *X*, *B*, *Y*, *A*, *X*, *A*, *Y*

Because of the spatial symmetry, each of these has another possible sequence associated with it: the one that results from interchanging *X* and *Y*. The temporal symmetry of possible path following also allows us to construct a new one by inversion. However, as it happens, the results are the same as those which interchange *X* and *Y*. (For 1 and 2 the two changes have the same result; for 3 and 4, the inverse of each is the *X*/*Y* permutation of the other.) Therefore we have in all eight possible sequences of places.

For each of these sequences of places we can choose a representative sequence of bridges; e.g. 125634 for *XAXBYAY*. But this sequence of bridges remains a possible path if we choose to permute 1 and 2; and also if we permute 3 and 4. Thus we have in total $8 \times 2 \times 2 = 32$ possible sequences of six bridges crossed consecutively. By exploiting the spatial, temporal, and permutation symmetries, we have arrived at a complete and systematic classification of solutions.

2. THE SYMMETRY INSTINCT: THIRST FOR HIDDEN VARIABLES

When we hear that some result has been reached on the basis of symmetry considerations, the method may have been the pure one of the preceding section. A model may have had its symmetries exploited logically in some elegant fashion. But it is also possible that something else is meant: that the result was reached on the basis of explicit assumptions about symmetry in nature.

The deepest such assumption is not that any one particular subject—space, time, matter—is symmetric in some particular respect. Rather it is the general conviction that *an asymmetry must always come from an asymmetry*.

Buridan's ass is the classic illustration of this conviction at work. This donkey is hungry, and confronts two bales of hay, equidistant to its left and its right. It has no sufficient reason to turn to one or to the other. Does it eat anyway? Then we believe there was some asymmetry in the situation after all: a small difference that made one bale more attractive to the donkey, though not even apparent to us; or myopia in its left eye; or a difference between the left and right hemispheres of its brain. To believe this, on no evidence other than that the animal eats, is to honour that conviction or instinct in question. It is the instinct which says that, if the donkey turned to the left, then *really* there was no symmetry of left and right in this situation beforehand after all.

Leibniz's God was Buridan's ass magnified. This God makes no choice without sufficient reason; when it comes to creating a world his only reasons can have to do with how good the world is. Thus the fact that we exist shows that there was a unique world which was better than all other possible worlds—ours. For if there had been two or more equally good which were better than the rest, or if for each good one there had been another still better, God could have had no sufficient reason to choose among them, and he would not have created any. Hence our world is the best of all possible worlds.

If we look for a moment again at the donkey we see that this sort of conviction pretty well rules out indeterminism from the beginning. Suppose the initial situation was perfectly symmetric— and then spontaneously an event occurs, the donkey eats the left. By hypothesis, the preceding situation did not contain a causal factor that favoured the left, nor was subject to a constraint that says: if no other considerations apply, choose the left. For such a constraint would *be* an asymmetry. Thus the indeterministic story which makes the donkey's choice a chance event, is in violation of the conviction. An asymmetry would have been created *ex nihilo* and would not have come from a preceding asymmetry. The conviction is one which, carried to its logical extreme, requires determinism.

Ernst Mach gave us a modern illustration. As a boy, he witnessed what was to all appearances a miracle. A wire runs parallel to a suspended magnetized needle, in the same vertical plane. The wire is electrified and the needle turns out of the plane—let us say to the north-east if it was pointing to the north before. But how could it choose between east and west? The situation—geometrically two straight line segments parallel in a plane, one directed south-north—is symmetric between east and west.

Immediately we want to tell young Mach about the direction of an electric current. Maybe he speculated at once that electrification— the closing of the switch—is a process which, despite the appearance to his eye, introduces an asymmetry. Or perhaps he speculated that the relation to some third entity (such as the earth) entailed a relevant asymmetry. The point is that as long as there appears to be an asymmetry born from a symmetric process, it looks like a mystery or miracle. At once we want to introduce a hidden asymmetry, a hidden parameter.

What is to be said of this fundamental, profound principle that an asymmetry can only come from an asymmetry? The first reply is that *qua* general principle it is most likely false and certainly untenable. That I made clear by showing how an indeterministic account of Buridan's ass violates the principle. Of course we expect symmetries in chance events: if the ass does not choose the left about half the time in such situations, we'll feel the same conviction return. But even if no statistical asymmetry appears, the individual event still violates that principle. On the positive side we must say that the conviction is a good guide for humans looking for theories. Their speculation about hidden asymmetries often pays off. It is only important not to raise a tactic to the status of strategy.

We all have the same reaction to Mach's needle, and this reaction is partly a shrewd idea of how to construct working hypotheses, and partly an inclination to metaphysics. It is important therefore to think also of real examples in which the supposed principle is not accepted. Quantum mechanics gives us examples of indeterminism for individual events, similar to my indeterministic donkey. Therefore it allows for the spontaneous introduction of new factual asymmetries where there weren't any before. In addition it gives us strange statistics. The strangest is in the correlation experiments which violate Bell's Inequalities, foreshadowed in the famous Einstein–Podolsky–Rosen paradox and lately carried out most beautifully by Aspect and his colleagues. We can use it as an illustration of Mach's sort of miracle.

We have a source that sends out pairs of photons, toward two vertical polarization filters. It turns out that at each filter, half the time the photon goes through, and half the time not. But there is a perfect correlation between the two: if the left photon goes through its filter, the other does not, and vice versa.

Immediately we experience Mach's instinct and suspect that the photon pairs are pre-programmed. They are like twins, with each twin-set having one male and one female. But now the filters are rotated, remaining parallel, and the perfect correlation persists. It is not possible to prepare a set of photons of which half is vertically polarized, half horizontally polarized, and also half polarized in an intermediate orientation! Next we suspect interactions or messages: how else does the photon on the left know which direction of polarization to choose, so as to be opposite to its twin? But there can be no such messages or interactions. So now we have, at the left-hand filter, an asymmetry—the arriving photon 'chooses' to pass it or not—but the only asymmetry it could come from (the behaviour or 'choice' of the twin photon) is a factor it could *not* come from because that cannot have any influence.

Like the young Mach, we are faced with an apparent miracle. But in our case unlike in his, there is no possible way to satisfy the thirst for hidden variables behind the scenes. My little sketch here is not enough to prove this, but it is well documented in current literature on quantum mechanics and Bell's inequalities.

In conclusion then it is very important to distinguish symmetry arguments *proper*—logical exploitations of the symmetries of a problem as studied—from arguments based on substantive assumptions about symmetry in the world. How could this distinction fail or be fudged? Leibniz had a wonderful programme in metaphysics that would have nullified this distinction. For he thought that one could deduce a priori the existence of the God he described, whom we know a priori to make no choice without sufficient reason. Then it would follow that a solution which respects all the symmetries of a problem must fit what actually happens, because that is how God would act. Later philosophy has not been kind to a priori existence proofs—existence becomes manifest in experience or else is unknown, we tell the rationalists. With this conviction about the limits of logic and reason, we *must* respect the distinction between the two sorts of arguments outlined above.

Proofs and illustrations

The conviction that an asymmetry can only come from an asymmetry, appears in many other contexts. One is the early Kant's 'proof' of the reality of absolute space, on the basis of enantiomorphs (such as the left and right hand), which are congruent but cannot be brought into coincidence by any continuous rigid motion.[3] A related one is Peirce's 'disproof' of the Corpuscular Philosophy, on the basis of the existence of the optically active substances.[4]

> Now I maintain that the original segregation of levo-molecules, or molecules with a left-handed twist, from dextro-molecules, or molecules with a right-handed twist, is absolutely incapable of mechanical explanation . . .. The three laws of motion draw no dynamical distinction between right-handed and left-handed screws, and a mechanical explanation is an explanation founded on the three laws of motion. There then is a physical phenomenon absolutely inexplicable by mechanical action. This single instance suffices to overthrow the Corpuscular Philosophy.

But the most important further illustration is undoubtedly the long debates over the 'direction of time' and the related 'paradoxes' of statistical thermodynamics. Is it possible to have a pervasive temporal asymmetry on the macro-level, while all processes on the micro-level are reversible? After all, logs burn up and are not reconstituted, waves spread outward and we never observe 'inverse waves' contracting to a point, coffee and milk mix and do not unmix. To put the question a different way: if the basic equations describing the behaviour of the individual particles are invariant under time reversal, can the theory possibly fit aggregate phenomena which are not?[5]

3. SYMMETRY AND INVARIANCE

At this point we must make our concepts precise. I shall do this here quite abstractly and formally. The *Proofs and illustrations* however will make the same ideas clear with the concrete example of temperature scales.

Symmetries are transformations (technically one-to-one functions which map onto their codomain) that leave all relevant structure intact—the result is always exactly like the original, in all *relevant* respects. What the relevant respects are will differ from context to context. So settle on some respect you like: colour or height or cardinality or charm or some combination thereof. You have now partitioned your domain of discourse into *equivalence classes* (see Fig. 10.4). The little square in which individual x is located represents its equivalence class $S(x)$: the class of those individuals which are exactly like x in all the respects you designated as relevant. These equivalence classes form a *partition*: that is, their sum is the whole domain, and they are disjoint one from another.

That is easy to see. With S standing for the relation of complete relevant alikeness, it is obvious that each individual falls into one of those little squares. For it must lie at least in its own equivalence class. And it is equally obvious that it cannot be in two of them: for if x is in $S(y)$, then y is also in $S(x)$. But then everything that bears relation S to y—i.e. is relevantly just like y—also bears it to x. So all of $S(y)$ is part of $S(x)$ then; and conversely too; so $S(x)$ and $S(y)$ are the same. To sum up, S is like this:

1. (*a*) everything bears S to itself (S is *reflexive*)
   (*b*) if a bears S to b, then so does b to a (S is *symmetric*)
   (*c*) if a bears S to b, and b bears S to c, then also a bears S to c (S is *transitive*)

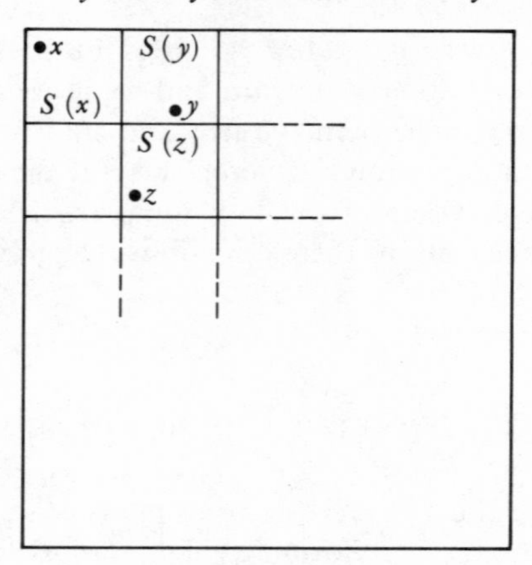

FIG. 10.4

This terminology is that of logic, and in logic any relation which satisfies (*a*)–(*c*) is called an *equivalence relation*.

We have connected two important ideas: *equivalence* and *partition*. Now I want to spell out exactly the same subject in terms of *transformation* and *invariance*. The interesting transformations in this context are those which leave each individual the same in all relevant respects. It does not matter whether the transformation is an actual physical change, or only a function that associates each individual with another one. Suppose that the respect you designated was *sex*, the property of being male or female. One transformation of the domain of humans is this: replace each man with his father, and each woman with her mother. Obviously the result of the transformation has the same sex as the original—the result is *equivalent* to the original with respect to sex, or as we also say, this transformation leaves the sex of the individuals *invariant*.

So to say that a transformation leaves your relevant respect invariant, means that the original will bear *S* to the result. Let us look at a large family *G* of such transformations. *G* is a set of transformations of which the result bears *S* to the original. What is *G* like? Or better yet, what could it be like, and still be quite simple to describe?

Because *S* is reflexive, we can have the identity transformation *I* in *G*. And because *S* is symmetric, we could have *inverses* in *G*: for every transformation, another one that gets you back to the original. Finally, because *S* is transitive, if one transformation is followed by another, and each leaves the relevant respect intact, then the final result still bears S to the original. So we can have all the *products* in G—where the product of transformations f and g is gf which consists in first applying f and then applying g to the result. To sum up:

2. I. G contains the *identity* transformation I: $Ix = x$ for all x
   II. G is closed with respect to *product*: if f and g are in G, then so is another transformation gf such that $gf(x) = g(f(x))$ for all x
   III. G has *inverses*: if f is in G, so is another transformation f^{-1} such that $ff^{-1} = f^{-1}f = \mathrm{I}$

If a family G has property II it is called a *semi-group* and if it has both I and II, a *monoid*. A family that has all three properties is called a *group*.

Note that the transformations in a group must all be one-to-one. That means that if $f(x) = y$ and also $f(z) = y$ then $x = z$. The reason is property III: $f^{-1}(y)$ has to be whatever y 'came from' by f, and this cannot be at the same time the object x and the object z, if those are distinct. Obviously we chose very nice properties for this family G, and now we need to justify this choice. This we can do by showing that the idea of a group of transformations does indeed correspond exactly to the initial idea of an equivalence relation.

3. Define xSy to be the case exactly if $y = f(x)$ for some member f of G. Then, if G is a group, S is an equivalence relation.

The proof is easy. Obviously xSx because the identity I is in G. Also if xSy, let $y = f(x)$. But then $x = f^{-1}(y)$ so also ySx.

Finally if xSy and ySz, let $y = f(x)$ and $z = g(y)$. But then $z = g(f(x)) = gf(x)$ and hence xSz. The converse theorem is more important, but not much more difficult.

4. Suppose S is an equivalence relation. Then there is a group G such that S can be defined by the relationship: xSy if and only if $y = f(x)$ for some f in G.

As we saw at the outset, the equivalence relation S partitions the domain into disjoint sets $S(x)$. So the above result 4. follows at once from the next:

5. Let domain D have the partition X_1, X_2, ... Then there is a group G such that each class X_r is an exhaustive class of

individuals that can be transformed into each other by members of G.

We begin with this partition: call these sets X_1, X_2, ... its *cells*. Then it is easy to see what must go into the family G. We let it be the family of all those one-to-one functions which never take anything out of its cell. Is G a group? Obviously the identity function has this property: $Ix = x$ therefore Ix is in Sx. A one-to-one function always has an inverse, so property II also holds. And finally, suppose f and g are in G. Then $gf(x) = g(f(x))$ which lies in $Sf(x)$. But $f(x)$ lies in Sx because f is in G, and $f(x)$ is also in $Sf(x)$ by reflexivity of S. Thus these two compartments overlap, and must therefore be the same. Hence $gf(x)$ also lies in Sx, and so G is closed with respect to products. Finally suppose that some part X of D is an *invariant set* of G, that is, that no member of G takes any member of X out of X. In that case, X must be a sum of cells. For suppose that y is in X and some other member z of the same cell as y is not in X. Then define the function:

$$g(z) = y \qquad g(y) = z$$
$$g(x) = x \text{ whenever } y \neq x \neq z$$

Obviously g is in G, but it takes y which is in X, into z which is outside X, so then X is not an invariant set of G. This ends the proof.

We have now in effect seen that the three concepts: *equivalence relation*, *partition*, and *group of transformations*, amount really to the same concept. For we have demonstrated: (*a*) if S is an equivalence relation, its equivalence classes form a partition; (*b*) for any partition there is a group whose invariant sets are essentially the cells in that partition; and (*c*) each group defines an equivalence relation: namely, the relation of being transformable into each other by members of that group. In the *Proofs and Illustrations* we will discuss further how the group is related to what it leaves invariant.

Proofs and illustrations

Temperature scales furnish a neat illustration. That a body has a temperature of zero degrees Celsius (or centigrade) is just the same fact as that its temperature is 32 degrees Fahrenheit. The scale of Celsius is defined by noting that it sets 0 and 100 at the freezing and boiling points of water respectively.[6] Anglo-Saxons prefer the scale of Fahrenheit, who placed 32 and 212 at those points. Due to Napoleon's forceful example, Celsius' scale is used on the Continent, and due to its logical superiority, its use will eventually be universal. But meanwhile we all learn to convert as we travel. This conversion uses transformations which correspond to simple, linear calculations with numbers. For simplicity, I shall ignore the fact that since Lord Kelvin we conceive of temperature as having an absolute zero.

Let us denote the usual scales as C and F and the transformations accordingly:

$$S_{CF}(x) = 32 + x(9/5)$$
$$S_{FC}(x) = (x - 32)\ (5/9) = -32\ (5/9) + x\ (5/9)$$

It is easy to check that these are each other's inverse:

$$S_{CF}S_{FC}\ (x) = S_{FC}S_{CF}(x) = x.$$

But Reaumur also devised a scale, and so did Kelvin. Once Fahrenheit is allowed, many more scales can be produced in the same fashion. For scale R to be fixed it will suffice to define the transformation of C into R:

$$S_{CR}(x) = k + rx$$

with real numbers k and r, where r must be positive. We may call k the zero-point: a reading of k on scale R is the same as a reading of *zero* on scale C. Let us call any function of this sort a *scale transformation*. Scales linked by such transformations I'll call *equivalent*. The family of *standard temperature scales* can be defined as the set of scales equivalent to C (or, to be fair, equally well as the set of those equivalent to F).

It is easy to visualize which transformations are admissible. Represent one of the scales—say, Celsius' centigrade scale again—by the X-axis in a diagram, and every other scale by a curve (see Fig. 10.5). The line labelled F represents Fahrenheit, hence has 32 where $C = 0$, 122 where $C = 50$, and 212 where $C = 100$. This shows at once what is invariant: when C has two intervals equal, so does F (for $100 - 50 = 50 - 0$ and also $212 - 122 = 122 - 32$). And more generally, if C has one interval N times another, so does F (e.g. $100 - 0 = 5 \times (20 - 0)$ and correspondingly $212 - 32 = 5 \times (68 - 32)$).

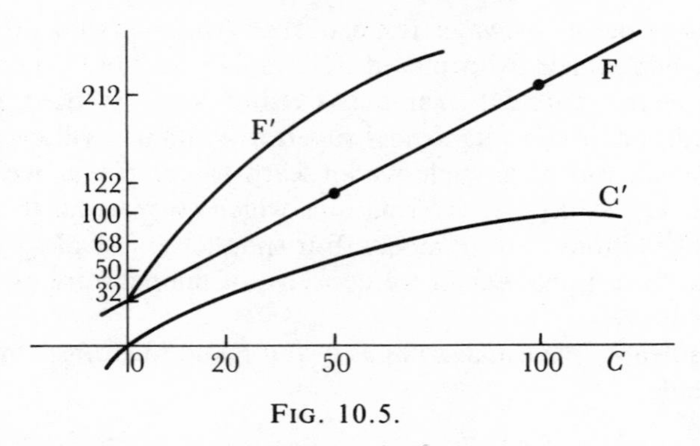

FIG. 10.5.

Ratios of intervals are invariant

Nothing hinged on the choice of straight lines here. Suppose we used curve C' to represent centigrade degrees. Then the X-axis does not represent a temperature scale at all. Every temperature scale is then represented by some curve—like the one labelled F'—which has the 'same' shape as C'. The reasoning about intervals would be the same.

When I defined 'standard temperature scale', I assumed that our transformations form a group, otherwise it would have made no sense to use the term 'equivalent'. This we need to check. Obviously the identity I is there, as the transformation $S_{CC} = S_{FF}$ etc. ($k = 0$, $r = 1$). Next let us check that products have the requisite form. Let

$$s(x) = a + bx$$
$$u(x) = c + dx$$

Then

$$us(x) = u(a + bx) = c + d(a + bx), \text{ i.e.}$$
$$us(x) = (c + da) + dbx$$

which has the required form. Finally we need the inverses to be there. The inverse of s must be a scale transformation u such that $su = us = \text{I}$. But we know what us looks like now. The requirement is therefore that $(c + da) = 0$ and $db = 1$. That means $d = 1/b$ and $c = -da = -(a/b)$ which is just fine. We have deduced the inverse of s to be

$$s^{-1}(x) = -(a/b) + x(1/b) = (x - a)/b$$

happily just the form of what we wrote down above for the transformation of Fahrenheit into Celsius.

What exactly is invariant under these transformations? Very little. The numerical order, of greater and less, is preserved:

if k is positive then $s(x + k) > s(x)$, because
$s(x + k) = a + b(x + k) = a + bk + bx > a + bx = s(x)$

because the coefficient b that multiplies x is supposed to be positive too. Secondly, as we saw in the diagram, equality of intervals is also preserved, and even the relative magnitude of such temperature differences:

if $x - y = k(z - w)$ then $s(x) - s(y) = k(s(z) - s(w))$.

To prove this we note that $s(x) - s(y) = a + bx - a - by = b(x - y)$. Similarly $s(z) - s(w) = b(z - w)$ accordingly, and the conclusion follows at once. This applies *a fortiori* if we take the simplest illustration

bodies A and B have temperature 10 and 80 on the Celsius scale. Therefore, B is 8 times as hot as A on any scale with the same zero point as Celsius

but allows us moreover to see an identical structure also with other scales:

With A and B as above, the difference in temperatures between B and freezing water is 8 times the difference in temperatures between A and freezing water, on any standard scale.

Thus *ratios of intervals* and *order* are invariant under the scale transformations. These invariants represent the objective content of our temperature discourse. All else is just true relative to one scale or another; but ratios of intervals are the same for all scales.

These invariants also characterize our transformation group. For suppose we ask: what transformations leave the features of a scale invariant? A different choice of zero-point won't matter; so we can always add a constant. If we have the zero-point fixed, what can we do? We must keep the order the same, so it is now determined what becomes positive and what negative. Moreover, if $y = bx$ then $y - 0 = b(x - 0)$ so we must have $s(y) = bs(x)$, i.e. $s(bx) = bs(x)$. This means that, with zero fixed, s is entirely determined by the choice of $s(1)$; for $s(y) = ys(1)$. But $s(1)$ is just another (positive) constant. So we can only have transformations which combine these two choices of constants, which we can represent as follows:

Every transformation s is the product of a translation and a dilation: $s = uv$, where $u(x) = x + a$
$v(x) = bx$ with b positive

The notion of product is the following: that $s = uv$ means that $s(x) = u[v(x)]$.

Thus we find the invariant structure of the temperature scales by looking at their admissible transformations—but equally the group of these transformations is determined by what that invariant structure is.

We should not generalize this nice point too far: it is not in general true that groups with the same invariant sets, are the same. Suppose we call a *finite permutation* of the natural numbers, any function which permutes the first n numbers, and maps the rest into themselves. The product of any two of these is still a finite permutation, and so is the inverse of any one. The whole set N of natural numbers is an invariant set, of course, but is any smaller non-empty set? Suppose X has m in it, but not n, and let k be bigger than both. Then there is certainly a permutation of the first k numbers, which takes m into n, so X is not an invariant set. Now consider the one-to-one function $f(m) = m + 1$. It is not a finite permutation. If we add it, and generate a bigger group, we certainly can't make new sets invariant; and of course, N is still invariant. So here we have two different groups with the same invariant sets. However, we can still distinguish the two groups in terms of invariant structure in the following way: in the first case, and not the second, is it true to say that each finite subgroup leaves some finite set of numbers invariant. Without going into further details, we can say in a rough but not trivial sense that for symmetry groups, what is significant is exactly what their invariants are.

4. SYMMETRIES OF TIME: WHAT IS DETERMINISM?

Indeterminism was a heresy to modern science, looked down upon by all scientifically enlightened people. The idea was known of course from Lucretius: in his universe, indivisible atoms moved continuously and predictably, except for the occasional spontaneous swerve:

> In this connection there is an other fact I want you to grasp. *When the atoms are travelling straight down through empty space by their own weight, at quite indeterminate times and places they swerve ever so little from their course*, just so much that you can call it a change of direction. If it were not for this swerve, everything would fall downwards like rain-drops through the abyss of space. No collision would take place and no impact of atom on atom would be created. Thus nature would never have created anything.[7]

A universe of that sort is however another example of Buridan's ass's predicament. For how does the moving atom 'choose' the time, direction, and magnitude of the swerve? It is possible to imagine a fictional follower of Lucretius who insisted that any asymmetry must come from an asymmetry. Suppose a Lucretian atom is travelling along a straight line, and that the universe has rotational symmetry, with that line of travel being the axis of rotation. Then the atom cannot swerve in any direction away from that line without introducing a new asymmetry. Hence this fictional Lucretius follower would have to say that the occurrence of such swerves is possible only if the situation already lacked that rotational symmetry. But if he pursues this line of reasoning, he will find that any direction, or combination of directions, of swerve, will be incompatible with many previous symmetries. The swerve will keep losing more and more of its apparent spontaneity, and finally be determined—perhaps entirely determined—by the structure of the universe.

Lucretius had no such followers, nor did he have sophisticated detractors: the history of modern science, characterized by a deep conviction of determinism, is also marked by nearly total indifference to any exploration of the concepts of determinism and indeterminism as such.

Let us pick up the thread very late: Bertrand Russell's attempt to define determinism in his essay on causation.[8] Intuitively, according to Russell, a system is deterministic exactly if its previous states determine its later states in the exact sense in which the arguments of a function determine its values.

To make this precise, let us say that a given system S has a family of possible states, and that it has one such state at each instant t during its existence. Thus we write: $s(t) =$ the state of system S at time t. The little letter s thus denotes a function, whose arguments are times (instants) and whose values are possible states of S. Any such function describes a *trajectory* of the system in the space of its possible states. But s is the function that describes the trajectory which S *actually* follows.

Now Russell suggested that to be deterministic, the state at time $t + b$ must be determined uniquely, given the state at previous time t, plus the time t, and the length b of the time interval. Hence, S is deterministic exactly if:

(1) There is a function f such that, for all times t and positive numbers b,
$s(t + b) = f(s(t), t, b)$

But this is trivially satisfied by every system. The reason is that $s(t + b)$ is uniquely determined already given the numbers t and b alone—because s itself is a function. Russell noticed this, but did not manage to suggest a wholly satisfactory improvement.

There are two ways to approach this problem. The first is to say that Russell was on the right track, and that a system is deterministic exactly if its *actual* history has a certain character. It seems Russell would have insisted on that, for (as he makes quite clear in that article) he wants to explain what is possible in terms of what happens *sometimes*. To follow this approach we would have to improve (1) so as to make it non-vacuous. One of Russell's remarks here can be interpreted as the suggestion of:

(2) There is a function f such that for all times t and positive numbers b,
$s(t + b) = f(s(t), b)$

This describes a certain *periodicity* to the actual history, for it means the same as

(2*A*) For any times t, t', if $s(t) = s(t')$ then $s(t + b) = s(t' + b)$

It means also of course that the time as such makes no difference to how the system changes. Periodicity is a certain kind of symmetry in time. It is not mirror-image ('bilateral') symmetry, but the symmetry of identical repetition, as we also see in wallpaper or ornamental tiling. Before looking more closely at this, let us consider the second approach.

The second response we could make is that Russell was wrong to concentrate on the actual history. To say that the past determines the future means not that the actual changes fall into a certain pattern, but that only certain possibilities are open to the system. If we modify (1) minimally with this idea in mind we arrive at

(3) There is a function f such that for all times t, all positive numbers b, and all possible trajectories s',
$s'(t + b) = f(s'(t), t, b)$

Another way to put this, which shows that it is not vacuous, is this:

(3*A*) For all times t, and possible trajectories u and v, if $u(t) = v(t)$, then $u(t') = v(t')$ for all times t' after t.

We lack the periodicity of (2) in this condition, for (3) allows that the system could have the same state twice over, but evolve differently from it. This would mean that the time as such appears to make a difference. But if we look at two different histories, we see that if ever they are the same at a given time, they remain the same thereafter.

The lack of periodicity is certainly a shortcoming. For what better reason could we give for a system being indeterministic, than to point out that, with the same state realized twice, it evolved nevertheless in two different ways? If time itself has a causal influence, the system is not isolated but is being interfered with. Either this interference ought to be described as an aspect of the state, or we should regard the system *taken in itself* as not having its future determined by its past.

We see here an ambiguity in the notion of determinism. If the state of a system does not determine its subsequent states, that may be because of an element of pure chance in its evolution, or else because some other factor is involved. Within the context in which the system is being described, the two cases are the same, and so we should say for both that the system, *taken as characterized*, is not a deterministic one.

A second difficulty with the lack of periodicity in (3) concerns the relation of possible histories via a single time. Suppose I describe a history that is not the actual one: which part of it is simultaneous with the actual events of today? That question appears to make no sense. There is no criterion of simultaneity for merely possible events happening in different possible worlds. I could describe the alternative history using my own calendar, but then a redescription which resets time $t = 0$ must be equally admissible. In the description of histories, actual or possible, we should allow clocks and calendars to be reset however we like. That means

(4) If s is a possible trajectory, and s' is definable by the equation $s'(t) = s(t + b)$, where b is any real number, then s' is a possible trajectory also.

Of course we can then call s and s' equivalent descriptions of what happens, in terms of a different clock.

But this addition gives (3) the periodicity which it lacked taken by itself. For now, if $u(t) = u(t + x)$ we look at the possible history v defined by: $v(t') = u(t' + x)$ for all times t'. By (4) v is also a possible history and by definition, $v(t) = u(t + x) = u(t)$. So by (3A) we deduce that v and u agree from the time t on. Let y be any positive number; then $u(t + y) = v(t + y)$ by this result; but then by definition $v(t + y) = u(t + x + y)$. So we see that the history described by u from t on is the same as that history from $t + x$ on. The same state, twice realized, is followed by the same changes. To sum up, (3) and (4) together gives us

(4A) If u and v are possible histories, and $u(t) = v(t')$ then for all positive numbers b, $u(t + b) = v(t' + b)$

which in effect combines (2) and (3) into a single condition.

But why did we not just stick with the simple amendment of Russell's (1) to our (2)? Why not stay with what is actual, and eschew this talk of possibilities? Richard Montague gave the simple counterexample that shows that (2) does not capture our idea of determinism.[9] Suppose a system evolves, and every state is different from every other one. Should it follow that this system is deterministic?

A good example is one of Lucretius' atoms, if it never goes back on its route. Suppose it moves, from all eternity to all eternity, swerving a little from time to time, but always forward in roughly the same direction, approximately along the line that connects our Sun and the North Star. Describe its state for a moment solely in terms of the spatial coordinate: $s(t)$ is the atom's position coordinate along that line. Then if $t \neq t'$, it follows that $s(t) \neq s(t')$. As a result, vacuously, it fulfils condition (2). Just because it never has the same state twice, it follows vacuously that when it does have the same state (i.e. never), it remains the same thereafter. But any definition that makes such a Lucretian atom deterministic is definitely defective. It appears then that to define determinism, we really need to pay the price of taking possibility seriously. Before looking into how high the price is, however, we should tie all this in with our continuing story about symmetry.

At this point, (4*A*) is the condition that defines what is meant by saying that the system *S* is deterministic. Define the operations U_b on the possible states of our system, by means of the equation:

(5) $U_b v(t) = v(t + b)$.

This defines U_b uniquely, for (4*A*) tells us that if we choose any other time t'' and any other possible trajectory *w*, and $w(t') = v(t)$, then $w(t' + b) = v(t + b)$. So these operators are well defined. We prove:

(6) The family of operators $\{U_b: b \geqslant 0\}$ is a monoid.

A monoid is a semi-group with identity; and this monoid is also called the *dynamic semi-group* of the system. Obviously U_0 is the identity, and the product of two operators is given by

(7) $U_b U_c = U_{b+c}$

We have now seen that if (4*A*) holds, then the system has a dynamic semi-group of operators. The converse is also true, as is obvious: if $u(t) = v(t')$ then $u(t + b) = U_b u(t) = U_b v(t') = v(t' + b)$. Hence (4*A*) is equivalent to

(8*A*) The system *S* has a dynamic semi-group, namely a family $\{U_b : b \geqslant 0\}$ such that (5) and (7) hold

which we can equivalently take as the definition of determinism.

There is also a stronger notion in the literature, *bi-determinism*, which means that the past must also have been the same if the present is. This is defined equivalently by the conditions

(4*B*) For any real number *b*, any times *t*, *t*′, any possible trajectories *u* and *v*, if $u(t) = v(t')$ then $u(t + b) = v(t' + b)$.

(8*B*) The system has a *dynamic group*, namely a family $\{U_b$: *b* a real number$\}$ such that (5) and (7) hold and $U_b^{-1} = U_{-b}$.

The operators U_b are called *evolution operators*. One way of looking at mechanics is to say that the laws of motion (or Schroedinger's equation, in quantum mechanics) describe the dynamic group.

I promised to return to the philosophically touchy issue of possibility. To begin, the assertion of determinism may be quite ambiguous in certain cases. Am I a deterministic system? Well, what goes into the description of my state? Suppose we choose the parameters treated in mechanics, and we include all mechanically relevant information about my environment, and finally, assume classical mechanics is true. Then the answer may be *Yes*, but what is the question? The question has become: whether *qua* mechanical system, my evolution in time is deterministic, and that is answered *Yes*. But it leaves open that there are other significant parameters as well. The answer is compatible with the assertion that *qua* thinking being I am not a deterministic system. If you toss me out of the window, I will fall; but who is to say what I will think or feel if you do?

The distinction can be made less fancifully. Consider a pendulum; describe its state first of all solely in terms of the position of the bob, in the plane that contains its trajectory. If the bob ever has a certain location, say with coordinates (1, 1), then it does so again and again. So we have, for example, $s(15) = (1, 1)$ and $s(25) = (1, 1)$. But does it follow that $s(16) = s(26)$? No, it does not because we don't know whether the bob is travelling in the same direction at those two times. In other words, *as characterized* the system is indeterministic. If we go on to a different characterization of its set of states, we may find that *so characterized* it is indeed deterministic.

Now this means that the attribution of determinism pertains in the first instance to a *kind of system*. Real individual systems we can point to, but kinds of systems we define. And indeed, we can define a particular kind of system by listing its set of possible states and its set of possible histories (trajectories). If that *defines* the kind of system—for example Newtonian mechanical system—the question of determinism for this kind is logically settled. The contingent or empirical question is then only whether a given real individual system belongs to this kind, or whether its observable behaviour conforms to what is logically possible for this kind.

With this conception, the apparently factual question about determinism in an individual case, has been factored into a logical question plus a question of classification. So conceived, the price of taking possibility and necessity seriously is not very high. This means that no noxious metaphysics has been allowed to creep in by the back door after all.

Proofs and illustrations

John Earman's recent book, *A Primer on Determinism*, is full of challenge to naïve ideas about determinism, as well as many other subjects we have touched on.[10] One especially nice example he gives shows that if we ignore the global conservation laws then classical physics allows for the possibility of unpredictable phenomena.

Here is the example: a UFO (perhaps a comet) approaches the centre of our solar system. Its speed diminishes rapidly. Indeed, in the past half year it travelled two miles towards us, in the preceding quarter four miles, . . ., in the preceding $(\frac{1}{2})^N$ year, it travelled 2^N miles towards us, for each $N = 1, 2, 3, \ldots$. So for quite a while now it has been going rather slowly, but a year ago we had no inkling of its imminent arrival. We can calculate how far away it still was from us for each moment in the preceding year. Assume that *now* is noon, 1 January 1987, Table 10.1 its approximate distances from us at certain dates and times.

TABLE 10.1. *Distances of UFO from centre of our solar system*

Date	Time	Distance from us (miles)
14 Feb. 1986		14
23 Jan. 1986		30
7 Jan. 1986		126
2 Jan. 1986	5:11 a.m.	1,022
1 Jan. 1986	2:08 p.m.	8,190
	0.32 p.m.	32,776
	0:08 p.m.	131,070
	0:02 p.m.	524,286
	0:01 p.m.	1,048,574

In that first minute after noon a year ago, however, it was clearly decelerating from astronomical speeds. Where was it at noon exactly? Or before that? Calculation has come to an end; it was at no distance from us at all. It did not yet exist in our universe. Yet there was never an instant of creation, no first instant of the comet's existence.

Today we would rule out this example on the basis that nothing can travel faster than light. The classical imagination was not stopped by any such rule. Yet we can see that, if a comet arrives thus, then even Laplace's demon—who knew the total state of the universe two years before, and had unlimited ability to predict its evolution by means of physics—could not have predicted the arrival of this mysterious comet. Yet its behaviour is so regular and unexceptional, during its existence, that we are not at once ready to see a failure of determinism. If there is a failure, it does not lie in what the object *does* at all.[11]

The arrival of this UFO would violate certain conservation laws: of mass, of momentum, of energy.[12] These are meant to apply to isolated systems. But however 'isolation' and 'universe' are understood, the universe must be an isolated system. The example shows that there is something inalienably global about these conservation laws. Their apparent deduction, in mechanics, from Newton's laws of motion rests therefore on premisses that remain non-trivial even for the universe as a whole, and are not guaranteed by those laws alone. The universe is by definition isolated but *not* by definition conservative! In the twentieth century we have learned to say that every symmetry yields a conservation law. But as this example illustrates, whatever the conservation law comes from, it must have a global character. Merely local theories about how individuals develop and interact cannot entail conservation, or more generally determinism, for a system as a whole.

5. TRUE GENERALITY: SYMMETRY ARGUMENTS IN PERSPECTIVE

We have now had a number of examples of symmetry, and symmetry arguments. Some were examples of symmetry arguments properly so-called; and some were of the other sort, deductions based on substantial assumptions about symmetry. There will be many more examples in the next three chapters. In this section I want to gather up the concepts introduced so far, and make as clear as possible the distinction between these two types of arguments.

Symmetry, transformation, invariance: this is the crucial triad of concepts utilized in all varieties of symmetry argument. In a symmetry argument one proceeds as follows: a problem is stated, and we first endeavour to isolate the essential features of the problem situation—that is, the features relevant to the solution. This does not necessarily call for great insight: the precise statement of the problem generally specifies what is intended. Thus if I ask how to boil an egg, I am asking for a method that does not depend on peculiarities of the present circumstances ('step one: hand the egg to your mother . . .'). But what counts as peculiarity may be vague (should the method work at any elevation above sea level, or is elevation to be kept fixed, as part of the relevant structure?) and insight, or decision, may be needed to eliminate this vagueness. Sometimes the achievement consists in reconceiving the problem in greater generality, as prolegomenon to a good solution. Sometimes also the problem as stated has no solution: Descartes, for example, apparently conceived of the problem of mechanics as that of finding an adequate, deterministic account which relies solely on kinematic parameters. There is none.

Isolating the essential or relevant structure is equivalent to defining the set of transformations that leave the problem essentially the same. These transformations are the *symmetries* of the problem. Only with these at least implicitly specified can we insist: *problems which are essentially the same must have essentially the same solution.* This is the great Symmetry Requirement, the principle of methodology that generates symmetry arguments. To put it somewhat differently: once the relevant parameters are isolated, the solution must consist in a rule (i.e. function) that depends on those parameters alone.

What does it mean for a rule or function to depend or not depend on some parameter? Its name does not give the right clues. What about

$$h(x) = x + 1; \qquad g(x, y) = x + y$$

Since $h(x) = g(x, 1)$, does it depend on 1? or only on x?

This is where the relevant set of transformations comes in useful. Instead of asking: does the rule depend solely on the *relevant* parameters, we can ask instead: if the problem is transformed by one of the admissible transformations, i.e. by a symmetry, and the rule is applied to the result, then does the rule give the same answer as when applied to the original?

The Symmetry Requirement can be illustrated by a diagram which may be familiar from other contexts (see Fig. 10.6). Suppose

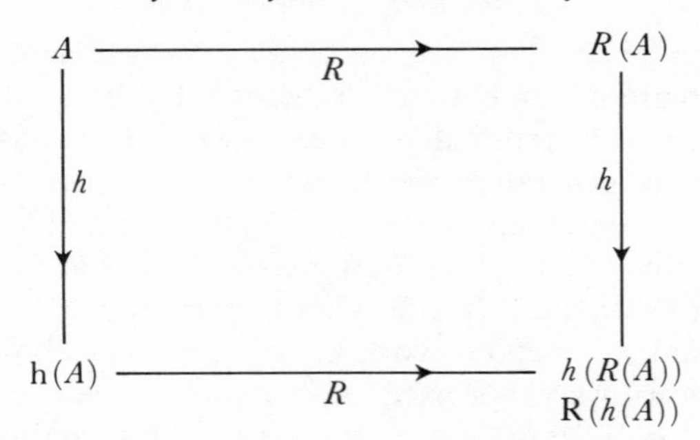

FIG. 10.6. The symmetry requirement

we have a rule or function R for solving a class of problems. In each case the problem has certain data—the *input* for the rule. The solution which the rule prescribes for it is the corresponding *output*. But we also have a class of transformations h which leave the problem essentially the same. If a given problem with input A is transformed by transformation h, it becomes the problem with input $h(A)$. Notice that at the bottom right we have two entries at the vertex: the transformed original output, and the output of the transformed problem. But if the two problems are essentially the same, then the two solutions must be too!

Symmetry Requirement: $h(R(A)) = R(h(A))$

The jargon for this is that the diagram must commute (in the function product notation, $hR = Rh$). This is our new, and more sophisticated formulation of the principle introduced in our first section. In subsequent chapters we will see its subtle power.

It is most important to recognize not only the power of this method of argument, so amply illustrated in modern science, but also its limits. As I pointed out in the first section, the elegant solution to farmer Able's problem carried a presupposition. This presupposition lay in our conceiving his problem as one in Euclidean geometry. That is the more striking because farms are typically riddled with hills, ponds, ditches, and pigpens. We could exploit symmetries only after deliberate choice of a model—and then the symmetry carried us swiftly to the end—but that initial choice has no a priori guarantee of adequacy.

Symmetry arguments have that lovely air of the a priori, flattering what William James called the sentiment of rationality. And they are a priori, and powerful; but they carry us forward from an initial position of empirical risk, to a final point with still exactly the same risk. The degree of empirical fallibility remains invariant.

11.
Symmetries Guiding Modern Science[1]

THE symmetries of time, space, and motion determine the structure of modern science to a surprisingly large extent. We began this story in the section on determinism. Here we will see that relativity is but another aspect of symmetry. Via the idea of relativity, the connection between symmetry and generality will emerge even more strikingly. And we will see how stringently symmetry dictates theory. Symmetry takes the theoretician's hand and runs away with him, at great speed and very far, propelled solely by what seemed like his most elementary, even trivial, assumptions.

I. SYMMETRIES OF SPACE

If I put a cardboard capital letter F on the table, I cannot produce the figure ꟻ just by moving that cardboard letter around on the table-top. I could do it by picking the letter up and turning it upside-down. A reflection in the plane can be duplicated by a rotation in three-dimensional space. But a reflection through a mirror (three-dimensional reflection) cannot be produced by rotations, because we have no fourth dimension to go through or into.

Reflections and rotations are transformations which leave all Euclidean geometric relations the same. In modern geometry it is shown that Euclidean geometry is especially simple, because everything can be defined in terms of *distance*. Thus for a transformation to leave *all* geometric structure the same, it suffices (in this geometry) that all distances remain the same.

The general notion of symmetry is this: *a symmetry is a transformation that leaves all relevant structure the same*. It follows then that a symmetry of Euclidean space is a transformation that leaves all distances the same. Such transformations are called *isometries* ('iso' means 'same'; an isometry leaves all metric structure, i.e. all distances, the same).

The simplest sort of isometry is neither reflection nor rotation, but translation. A translation just moves everything over, a fixed distance, in a fixed direction (see Fig. 11.1). Looking from the left, the first F is moved over so as to coincide with the third, by the translation which just moves everything 8 units in the positive direction along the X-axis. But now note that the second F is a reflection of the first, and the third a reflection of the second. The X-axis was the first line of reflection, and the line m (at $X = 4$) was the second line.

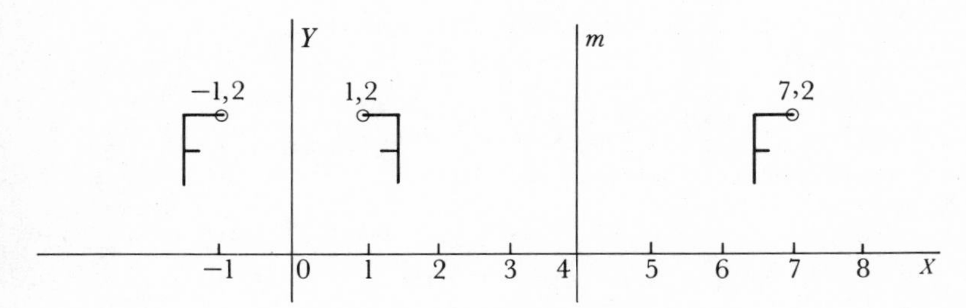

FIG. 11.1 Reflection and translation

I have now illustrated what a translation is, but at the same time have implicitly proved a little theorem: *every translation can be defined as the result of reflections* (with parallel lines of reflection). We can prove the same fact for rotations. Fig. 11.2 shows a typical rotation. The second F is formed by rotating the first one through an angle of 40 degrees, in clockwise direction. But now look at the result of two successive reflections (Fig. 11.3). Starting from the left, the first F is reflected over line m, and the result is the second F. But then the second F is reflected about line n (*which intersects line* m!) to form the third. Obviously the third is a rotated image of the first. But it is too far over. We can translate it back along line k, until its bottom vertex coincides with that of the first F. That we know we can do with two reflections. So here is another little theorem: *any rotation can be defined as the result of reflections.*

These two little theorems are parts of a real theorem:

Theorem: Each symmetry of a Euclidean space (i.e. every isometry) is the result of a finite number of successive reflections.

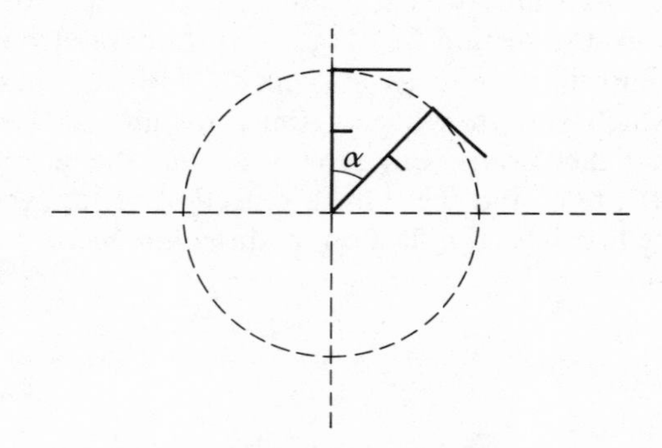

FIG. 11.2. Rotation

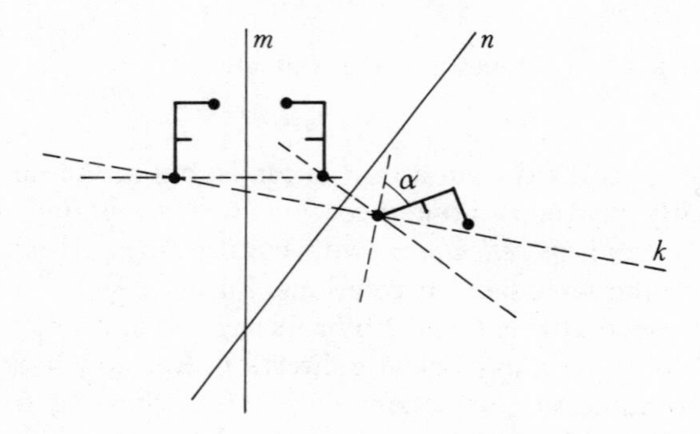

FIG. 11.3. Rotation by reflections

We count here as a special case the trivial or *identity* transformation, which leaves everything in its place. Obviously it is the result of doing one and the same reflection twice, because that puts everything back where it was.

Every isometry has an *inverse*, i.e. another isometry which puts everything back. A reflection is its own inverse. If you do two reflections successively, then you can invert the action by doing them again in opposite order. We write this as follows:

If s, t are reflections, then $s^{-1} = s$ and $(st)^{-1} = ts$. In general, if s and t are any two isometries, then $(st)^{-1} = t^{-1} s^{-1}$. We call s^{-1} the *inverse* of s.

It is easy to see that $ss^{-1} = s^{-1}s =$ the *identity*, which I will always denote I. The composite st we call the *product* of s and t; note that the order matters, and that st is not generally the same as ts.

Before we look at this in a more general setting still, we should get a better grasp on exactly what the isometries are. Of course we know now one simple, comprehensive answer: they are the products of series of reflections. But there is a curious and deep difference between reflections on the one hand, and rotations or translations on the other. A left hand is an exact copy of a right hand, yet they can't just be moved around or rotated so as to become orientated in the same way.

When I proved those little theorems by illustration, I exhibited a translation as a product of two reflections, and a rotation as a product of four. In general we call every *even* product of reflections a *proper motion* and every *odd* product an *improper motion*. Sometimes we use the term 'rigid motion'; this will mean proper motion. Now I want to explain why this terminology is important.

Imagine that a certain figure, say the letter F, moves across the plane. To fix your imagination suppose the figure is cut out of cardboard and we slide it over a table top, without ever lifting it from the surface. We start at time zero, and continue till time t. Now look at where it was at two intervening times, t and $t + b$. At time t, for example we may have the figure with tip at $(-2, 2)$ and at $t + b$ at $(3, 1)$. Obviously the second figure must be the image of the first by some isometry.

The important fact is this. Suppose each point on the figure traces out a continuous path. Suppose in addition, for any time t, and any positive number b (however small), that the figure at $t + b$ is an isometric image of the one at time t. Then none of these isometries that connect the figures at successive times can be improper motions. Single reflections, for example, can never be used in this process. For if you reflected the figure around some line, at a definite time t, then at least one of the tips of the figure would 'make a jump' at that time. It would not trace a continuous path at all. Therefore the proper motions are properly so-called— for they represent real motion of real objects. The proper motions of the plane represent real motion on flat surfaces, and the proper motions of space represent real motions in three dimensions.

A product of proper motions is always again a proper motion. But a product of improper motions need not be improper. For example, a product of two reflections can be a translation, which is proper. So the proper motions by themselves form a natural family—in fact, a *group*—while the improper motions do not.

Let us now utilize the abstract motions of group and invariance, to describe geometry more precisely.

> An *isometry* is a one-one function which preserves distance between x and y (for all pairs of points x, y). The isometries form a group. The corresponding equivalence relation is *congruence*.

The relevant respect of equivalence here is of course the structure definable in terms of *distance*; and this is all there is to Euclidean geometric structure. Klein's famous Erlanger programme generalized the subject of geometry by taking various groups—not just the isometries—as topic of inquiry. Table 11.1 contains a list of

TABLE 11.1. *Geometric transformations and their invariants*

Group	Invariant structure
1. All transformations (= one-one, onto functions)	
2. Collineations	preserve the property of being a line
3. Dilations	collineations which preserve parallelism between lines
4. Similarities	multiply all distances by a constant factor
5. Isometries	preserve all distances
6. Proper motions	preserve distances and parity
7. Translations	preserve distances and parity; and each acts as identity on a family of parallel lines which together contain all points
8. The trivial group	has the identity as sole element

geometric transformation groups, defined by the structure they preserve; we begin with the largest, which leaves no peculiarly geometric relationships intact at all. Each group contains the one below it. In the case of Euclidean geometry, the collineations and dilations are exactly the same, because there two lines can be defined to be parallel by the property of having no points in common. These transformations (collineations, hence dilations, of Euclidean space) are also called *affine*, and form the subject matter of affine geometry.

Note that the improper motions do not appear as an item in the list. A reflection can be defined as an isometry which acts as the identity on a single plane (or line, in two-dimensional)—the plane of reflection. These reflections *generate* the group of isometries, meaning that each isometry is a finite product of reflections. But this generating family generates the proper as well as the improper motions, as we know.

Besides symmetries of the space as a whole we can also study symmetries of specific figures. As examples I will take some letters (see Fig. 11.4). A symmetry of a specific figure is an isometry which leaves this figure invariant, i.e. acts like the identity on this figure. The letter R has no symmetries except the identity itself. The letter A has the identity plus reflection in the line *m*. The letter Q has identity plus reflection in the line *n*. Of course, O has a large family: all rotations around the centre, and all reflections in lines that pass through its centre, and all products of series of these rotations and reflections. The letter S has as symmetry the rotation through 180 degrees, that turns it upside down, but no reflections. None of them is brought into a coincidence with itself by a translation, of course.

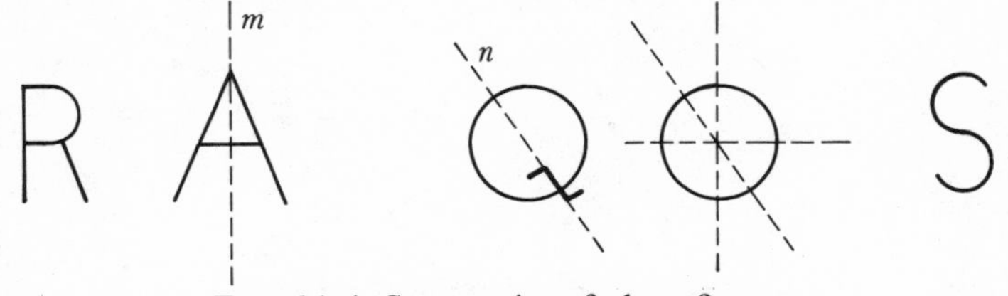

FIG. 11.4. Symmetries of plane figures

For each figure there is therefore a specific family of symmetries. This family is a group. That follows at once because we define the family, in effect, in terms of an equivalence relation. The original is the space full of figures, the result is another space full of figures. Call the results *equivalent* to the first if the designated figure appears in exactly the same place in the same way, and if all distances are everywhere the same. The family of symmetries of the figure is the family of transformations corresponding to this equivalence relationship, so we know it is a group. But you could also prove this directly by checking properties I–III that define a group.[2]

Proofs and illustrations

The well-known parallelogram law for the composition of forces is dictated almost entirely by considerations of spatial symmetry and continuity. Its deduction became successively clearer in the work of Stevinus (1605); Lami, Varignon, and Newton (all in 1687). Daniel Bernoulli argued that the principle is a purely geometric truth, independent of experience.[3] In our century, George David Birkhoff gave an exact demonstration, which isolates all operative assumptions.[4] I will here reconstruct its central part, which shows clearly how far symmetry dictates theory here (see Fig. 11.5). We have two forces *OA* and *OB* acting at a point *O*; what is the resultant force on *O*? Everybody knows that the mathematical representation of forces leads to answer *OC*. But can't we imagine a possible world in which this representation is just the wrong one to use? A world in which forces 'sum' together in some different way? Well, let us put the question in very precise form:

> find a function f which, given forces OA and OB yields a force $OX = f(OA, OB)$

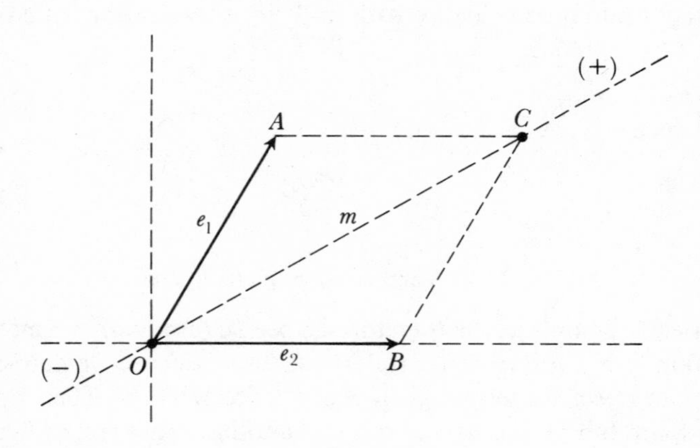

FIG. 11.5. The force parallelogram

There are many such functions. But if we now add that different, but equivalent ways of stating the problem must receive equivalent answers, *we require* in addition that this function 'looks the same' in all spatial frames of reference. This spatial symmetry requirement rules out all but a very narrow class of solutions. I may add that continuity requirements impose another drastic limitation. So we see that our imagined possible worlds are really very limited in this regard—there really is an enormous distance between merely saying 'let us imagine' and genuinely imagining.

The spatial symmetry requirement can be restated in the terms we have just introduced. *Any symmetry of the figure consisting of* OA *and* OB, *must also be a symmetry of* OX. Since reflection of the plane, with OC as line of reflection, is obviously such a symmetry, it follows at once that X must lie on the line OC. That is the core of the demonstration of the parallelogram law. Let us just follow the argument a little further.

Here is Birkhoff's argument, reconstructed for the simple case of two forces e_1 and e_2 of equal magnitude. We shall moreover be concerned with the direction of the resultant only and not its magnitude. It is assumed the forces are represented by vectors in the usual way. We make some further assumptions about the function f, which I shall state very precisely here:

Premiss 1. $f(e_1, e_2) = f(e_2, e_1)$
Premiss 2. $f(e, e) = 2e$
Premiss 3. If T is a Euclidean transformation (isometry) then $f(Te_1, Te_2) = Tf(e_1, e_2)$

The third premiss is the symmetry principle that will be invoked; in words it says that f remains invariant under Euclidean transformations. Let m be the line that lies in the e_1, e_2 plane and bisects the angle between them. Choose T to be the rotation around axis m such that $Te_1 = e_2$ and $Te_2 = e_2$. Then by Premiss 3, $Tf(e_1, e_2) = f(Te_1, Te_2) = f(e_2, e_1)$ which equals $f(e_1, e_2)$ by Premiss 1. So $f(e_1, e_2)$ remains unaffected by T. But any vector in this Euclidean space which does not lie along line m, is changed by T. So the resultant $f(e_1, e_2)$ lies along the line m; we only don't know whether it has the $(+)$ or $(-)$ direction. To establish this we need a further premiss which introduces a continuity requirement. Suppose we fix e_1; the identity of e_2 we pick out by two factors, namely, the plane e_1, e_2 and the angle θ between e_1 and e_2 (draw a circle in the e_1, e_2 plane with e_1 and e_2 as radii, and let θ be the smaller of the two angles between them).

Premiss 4. $f(e_1, e_2) = g(e_1, e_1e_2, \theta)$ for a function g which varies continuously with θ

Use e_1 as picking out the X and Y axes in the e_1, e_2 plane as I show in the diagram, keeping e_1 fixed. If now $e_1 = e_2$ then $f(e_1, e_2) = 2e_1$ by Premiss 2, so it lies along m and has the $(+)$ direction. If θ now increases toward $\pi/2$, vectors along m must lie in either the upper right or lower left quadrant. So since discontinuous jumps have been ruled out, the resultants for $\theta \geqslant \pi/2$ *all* have the $(+)$ direction. The argument can easily be completed for all values of θ. This ends the proof.

2. RELATIVITY AS SYMMETRY

Aristotle's *Physics* shows a good deal of concern with symmetry, in the discussions of the structure of the world, the motions of the spheres, and the possibility of a vacuum. But one sort of symmetry dominates the differences between the Aristotelian tradition and the new sciences of the Renaissance and since. Moving frames of reference were not equivalent, according to the older tradition. Once the challenge was raised, experiments were devised by Aristotelians and often enough claimed to have been carried out and to have confirmed the non-equivalence. One example is the experiment of dropping a weight from the crow's nest at the top of the mast, in a moving ship. The weight takes a little time to reach the deck. Will it land directly below where the crow's nest was, or at the bottom of the mast, a spot which has moved forward during the time of fall? Even today, our intuitions are none too steadfast when confronted with thought experiments of this sort. We know that the correct answer—and this is *Galilean relativity*—implies that an observer in constant motion cannot detect this motion by such experiments, but only by observing the shore or some other point of reference. So the weight falls to the bottom of the mast. But should we, in medieval terms, conclude that the hand imparted some forward motion to the weight? What if God created an identical weight just beside my hand, at the moment of release, to fall side by side with mine? Would the two weights land at different places on the deck? What of the strange hypothesis that God instantaneously destroys the weight that I release, and replaces it by another identical weight—is this hypothesis testable because if it were true, the weight would land somewhere behind the mast? Having been taught Galileo's way already in our school days, we resist the temptation even to feel puzzled—but we can still perceive that we are recalling here a veritable conceptual revolution.

Another example uses the earth's motion rather than a ship's: if cannon-balls of equal shape and mass are shot from cannons with equal powder charges, in the directions east, west, and north, do they travel equally far as measured on the surface of the earth? As thought experiments these are correctly conceived and telling; in technological reality they would be too crude to settle the question. It was the success of a larger dynamics, covering both celestial and sublunar motions—culminating in Newton's mechanics and system of the heavens—which as a whole led to the acceptance of this new principle of equivalence: the *classical* or *Galilean* principle of relativity.

Abstractly conceived, we have here a new symmetry for space–time, the mathematical space with three spatial and one temporal dimension. We know now that this treatment would also need to be corrected. The theory of light and electromagnetism developed in the nineteenth century violated Galilean relativity, and yet appeared to save the phenomena. The reaction was not a return to the older conviction that uniformly moving frames of reference are not equivalent after all, but the acceptance of a more radical principle of relativity. (This was of course Einstein's, which made it possible that even a constant speed, the speed of light, could be the same for all observers.) But let us stay with the simple subject of classical relativity.

Consider two observers in constant motion with respect to each other. Imagine that to begin their frames coincide exactly, but with time, the X-axis of the one slides along that of the other with velocity v. In Fig. 11.6 Q and R are the spots marked (0, 0, 0) in the second frame F' *at* $t = 0$ and $t = k$ but marked respectively (0, 0, 0) and $(vk, 0, 0)$ in the first frame, at all times. Question: at time $t = k$, what are the F' coordinates of P, which has (7, 7, 0) in F? Obviously it has its same Y and Z coordinates 7 and 0, but its X-coordinate has changed from 7 to $7 - vk$. In fact, a moment's reflection shows that the F'-coordinates x', y', z' are related to the F-coordinates x, y, z as follows:

$$x' = x - vt$$
$$y' = y$$
$$z' = z$$

for all times t. This is not a Euclidean but a *Galilean transformation.*

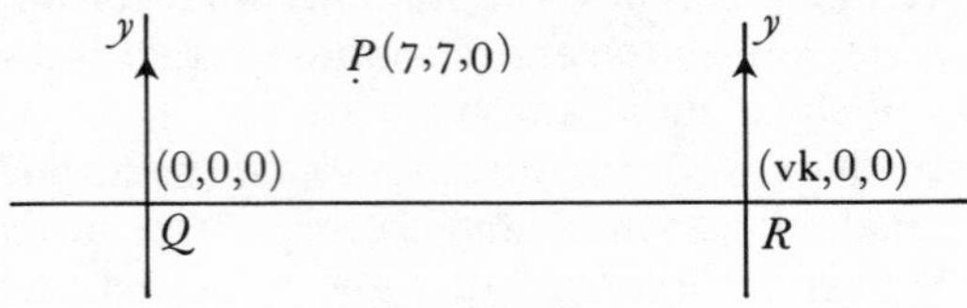

FIG. 11.6

The above form defines the Galilean transformations in the narrow sense. In the broader sense, this term refers to the entire group generated by the Euclidean transformations plus the Galilean transformations in the narrow sense. Thus we can have moving frames which are at an angle with respect to each other, or reflected. All uniform motions (motions at constant velocity) are comprised.

Invariant under Galilean transformations are all *distances*, and the magnitudes of *relative velocity*, and *acceleration*. Not invariant are location or velocity *per se*. Let us just check this for acceleration. Imagine I am at Q on the station platform, you are at R on a moving train. Our situation is like that in Figure 11.6, with the velocity v equalling, say, 60 miles per hour. Thus if QR equals 60 miles, you reached R one hour after the moment we coincided. Our friend Timothy walks on one of your flat carriages, at a speed of 10 m.p.h. forward in your frame. Then he is walking at speed $60 + 10 = 70$ m.p.h. in mine. Now he speeds up. Over an interval of 2 hours, he increases his speed from 10 to 14 m.p.h., as measured in your frame. His acceleration according to you equals 2 m.p.h./h. In my frame he speeds up to 74 m.p.h. over 2 hours, so his acceleration according to me equals $(74 - 70)/2 =$ 2 m.p.h./h, just like for you.

Galilean relativity has very direct consequences for natural science. Let us reflect first on the law of inertia. Suppose we have the following principle: *under conditions C, an object which is at rest, remains at rest*. And suppose that a given object actually exists in conditions C, and that it is at rest in frame F. Then its velocity in another frame can be different, but its relative velocity to, say, the centre of F—i.e. *zero*—must be the same in all frames. Since the centre of F can only be in constant motion with respect to other frames, we see at once that the principle entails the apparently more general principle: *under conditions C, an object retains its velocity*, whatever it is.

A more interesting example is found in the theory of collisions. In 1669, Christiaan Huygens sent the Royal Society of London his theory, which had the now familiar form, deducing the velocities after impact via conservation of momentum.[5] In his posthumous treatise *De Motu Corporum ex Percussione* (1703) it was found that his conclusions were almost entirely derived by means of true symmetry arguments.

We must distinguish collision of elastic bodies, which rebound, from that of inelastic bodies which (in the extreme case) stick together. For the former, Huygens's illustration shows two men, one on the shore, and one on a boat moving past the shore. Each man has his arms outstretched, and at the moment of the experiment, their postures coincide. From their outstretched hands, two solid balls hang suspended on ropes. These balls swing together, and hit in the middle with equal and opposite velocity, as reckoned by the man on the shore. They rebound with the same speed, each having changed only its direction of motion. But suppose this speed, call it v, is exactly the speed with which the boat moves past the shore. Then from the boatman's point of view, the initial velocities of the balls were $2v$ and 0 respectively, and their final velocities are 0 and $2v$. More generally, if the boatman was moving at velocity w, the velocities are:

initial	final
$v + w$	$v - w$
$v - w$	$v + w$

So here is the principle: in a collision of perfectly elastic bodies, with equal masses, the velocities are exchanged. This is the general solution demanded by Galilean relativity.

To make the details of such an argument perfectly clear, it is more perspicuous to focus on the case of bodies which, when they collide, stick together and thereafter move as one (inelastic collisions). This is a situation approximated by cars colliding on a highway; not by a rubber ball hitting a wall. I shall also restrict myself to the case in which the two bodies are initially moving along a simple straight line. We make two initial assumptions. The first sets the stage for the symmetry argument: it is the assumption that the masses and velocities alone determine the final motion. The second concerns a special case:

Collision I: If the two bodies have equal but opposite momentum (= mass times velocity) then they come to rest when they meet.

We now consider two bodies A, B of arbitrary masses m_A, m_B and velocities $v_A > v_B$ in a given frame F, which are approaching each other. So it is possible to change to some frame F' such that the speed of A becomes zero and that of B becomes $v = v_B - v_A$ which is negative. We expect that in *this* frame F', the bodies will move together to the left. If we have heard of the law of conservation of momentum (definitely not assumed at this point) we also expect that the final velocity w will be such that $w(m_A + m_B) = vm_B$. But how can we deduce this?

Well we transform F' too, into a frame F'' moving to the left at speed b with respect to F'. This motion must be such that in F'', the two momenta are equal but opposite, so that Collision I will be applicable. Hence we choose b such that

$$bm_A = (v - b)\ m_B$$

which is solved for b by

$$b = vm_B/(m_A + m_B)$$

Changing back to frame F' we see that the final motion of the two collided bodies—which are now at rest in F''—is to the left at speed b.

Now F' came from F by setting the speed of A equal to zero, so F' was moving relative to F with speed v_A. Hence in F the two collided bodies move with speed $b + v_A$. Thus we have deduced the rule that determines the final velocity for two arbitrary colliding bodies in an arbitrarily given frame (recalling v is in magnitude equal to $v_B - v_A$):

$$v_A + (v_B - v_A)m_B/(m_A + m_B)$$

If we now multiply this by the total mass we get

$$v_A(m_A + m_B) + (v_B - v_A)\ m_B = v_A m_A + v_B m_B$$

which is indeed equal to the initial total momentum. Thus we have deduced:

Collision II: If two bodies moving along a single straight line collide and remain together, they will move along that same line with final constant momentum equal to their initial total momentum.

Thus the complete 'law', i.e. rule to determine the momentum of the composite body, has been deduced (via Galilean relativity) from the weak principle Collision I which governed a special case.

The conservation of momentum, as is well known, applies much more widely, to all classical mechanical phenomena. We will consider this later, after our general concepts have been made more precise.

The third and final example I'll give concerns light, and has an incidental lesson for philosophers. In attempts to explicate the concept of a law of nature, much attention was paid to *generality* (or *universality*). True generality cannot lie in the form of words used to express a proposition—so what does it consist in? Let me put it this way: could *all robins are red* or *sodium always burns yellow* be a law or couldn't it? In physical theory, the generality sought is *invariance* under all symmetries. And so the answer must be: no, it couldn't. For colour is not invariant under Galilean transformations, just as velocity is not. That all robins are red can no more be a law than that all force-free bodies are at rest.

This is an example in which the physical theory implies an answer to the question whether the proposition has significant generality. Another theory could imply the opposite answer. Hence the answer cannot be a matter of logic alone.

The colour of an object is measured by the frequency of the light it reflects or emits. Light propagates as a wave, spreading spherically from its origin. Being located at a certain point successive crests or wave fronts reach me, and I register such a crest, say, every T minutes. Obviously, the same must happen at the locations of the preceding and succeeding crests. So T minutes is also the time that it takes a crest to move that distance. We have therefore two 'speeds': the *frequency* f which is the rate of crests per unit time registered at a given point, and the *speed of advance* v, the speed at which the crest appears to travel one *wavelength* λ (distance from one crest to the next). What we just saw was that $f = 1/T$ and $\lambda = vT$, so frequency and wavelength are related as

$$\lambda = \frac{v}{f} \tag{1}$$

So far we have described the situation in terms of the rest frame, the frame of reference in which the light source S and my location are at rest. Shift now to a moving frame, that of an observer moving at a constant velocity k.

What is invariant here? All distances—the distance λ between two successive crests (*however their locations be labelled*) must be the same. But velocities are not invariant: the velocity v in the rest frame is only $(v - k)$ in the frame moving to the left at speed k. So as measured in that frame, one crest gets from its position at given time t to where the preceding one was at t, in $\lambda/(v - k) = T'$ minutes—a longer interval than T. The frequency of oscillation, as viewed in that frame is accordingly $f' = 1/T'$.

We see therefore that the equation which connects wavelength and frequency *has the same form* for each frame, but only the number λ is an invariant magnitude:

$$\lambda = \frac{v}{f}; \quad \text{also} \quad \lambda = \frac{v-k}{f'}$$

$$\text{Hence} \quad f' = \frac{(v-k)\,f}{v} < f$$

Thus the source S as seen in the moving frame emits light of lower frequency—more reddish. This is the *red shift* of light from a receding source; a special case of the general phenomenon of the Doppler shift.

Obviously the physical constitution of the light is in no way affected by being observed by a moving observer. The shift is a perspectival effect; colour is not an invariant, it is merely relative.

Proofs and illustrations

As we know from the preceding section, the Euclidean transformations are generated by the set of reflections. Once we use coordinates, it is more perspicuous, however, to think of them as generated by three simple sorts of transformations of frames of reference. The first is *translation*, which consists simply in adding a constant to each coordinate.

$$\text{translation:} \quad \begin{aligned} x' &= x + a \\ y' &= y + b \end{aligned}$$

The second is reflection, around one of the coordinate axes. It is a simple change of sign.

reflection: $x' = -x$ or $y' = -y$

In both cases, the coordinate axes remain parallel to their originals. The third rotates them through an angle θ, and the new coordinates are found by trigonometry, as the following diagram shows.

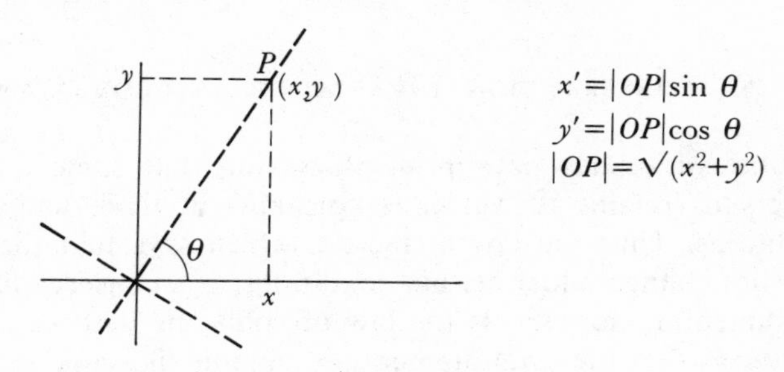

FIG. 11.7

The group of all Galilean transformations, generated by these Euclidean ones plus the moving frame transformation

$$x'(t) = x - vt$$
$$y'(t) = y$$
$$z'(t) = z$$

together comprise all uniform motions. This must be properly understood: the group does not include frames fixed to rotating disks. It can include, for example, the transformation t_1t_2 where:

t_1 = move forward at constant velocity v
t_2 = reflect through the $Y - Z$ plane

In the description of t_2, what is meant is the $Y - Z$ plane of the frame to which it is applied. So t_2 is defined by the equation:

$t_2\ (a, b, c) = (-a, b, c)$

Now it is easy to see that t_1t_2 produces a moving frame, that has its origin at R at time $t = k$, by a constant motion along PR. But the X-coordinates have their sign changed. So at $t = k$, the point Q has in the moving reflected frame not the X-coordinate $-vk$ but $+vk$. And the point P has X-coordinate $vk - 7$ rather than $7 - vk$. This is correct no matter what time k is; it is *not* as if the frame is reflected at time k and reflected *again* at $k + 1$. There is no uniform motion that consists in a series of flips (or rotations).

What if we had reflected the frame first? Then we could have produced the same motion only by moving it *backward* rather than *forward* along the X-axis. So $t_2t_1 \neq t_1t_2$ but $t_2t_1 \neq t_1^* \ t_2$ where $t_1^* \ (a, b, c) = (a + vt, b, c)$. Of course we note at once that t_1^* is the *inverse* t_1^{-1} of t_1.

3. CONSERVATION, INVARIANCE, AND COVARIANCE

A conservation law is a principle saying that some quantity is conserved (retains its value, is constant) in time, under certain conditions. Thus the law of inertia, which says that the velocity does not change under certain conditions, is a conservation law of a rudimentary sort. So is the law of collisions that we discussed, for it says that the total momentum remains the same in that sort of phenomenon. Newton's general law of the conservation of momentum is a better example; laws of conservation of mass, energy, charge are of course famous in the history of physics. As we shall see below, these laws are often intimately connected with symmetries.

We must at this point give more precision to some of the concepts involved.[6] First, a (physical) quantity is (represented by) a function of a special sort: its arguments are the 'bearers' of that quantity, which are in general not mathematical entities at all. Thus, a temperature is the temperature of a body, or of a body at a certain point on that body. The values generally are mathematical entities: a temperature is a real number, a position or velocity is a vector (which is a finite sequence of numbers). As is clear from these examples, the value is generally relative to a scale or frame of reference.

This means also that the value of a quantity is just the sort of thing that is a point in a scale or in a frame of reference. Thus points on temperature scales are numbers—temperatures are numbers. Points in a spatial frame of reference are three-dimensional vectors (x, y, z)—so are a body's position, velocity, acceleration, momentum. Transformations which turn one frame of reference into another can therefore also be applied to such quantities. In the case of temperature and position this is obvious, and indeed gives the correct result at once, *ipso facto*. But in the case of, for example, velocity, though the transformation is applicable, it does not always give the right result.

Consider frame F' moving at speed v along the X-axis of F, in the positive direction. Then the X component of position of a body is transformed by

$$s: x' = x - vt$$

What about its velocity? Let us suppose that the body moves from the origin to point (a, b) in F in two hours. Then the X-component of its (average) velocity in the X-direction equals a/z, with respect to this frame. In the same time it moved, in F', from the origin to $(a - 2v, b)$ and so the X'-component of its velocity was $(a - 2v)/2$. Now could we have arrived at that answer by applying transformation s to the X-component? For time $t = 2$,

$$s(a/2) = a/2 - 2v$$

which is clearly different from $(a - 2v)/2 = a/2 - v$. Obviously the velocity is not invariant, and does transform *systematically* along with the position, but *not in the same way*. This is also obvious, perhaps even more so, for the case of a Euclidean translation linking two frames of reference at rest with each other. The origins are different but the X-axis the same, let us say; now velocities along the X-axis are equal in the two frames, but positions on that axis of course are not. Hence the transformations that correctly transform the latter must give the wrong result if applied to the former.

Yet a quantity may vary in the same way as the coordinates, as far as some transformations are concerned. This possibility is given the name covariance. Here is the definition given by Herman Weyl[7] summarized and with the notation slightly changed:

> Let G be the group of all linear transformations between the normal coordinate systems in space or space–time (i.e. the group of Euclidean *rotations* or the group of Lorentz transformations). A quantity Q which has n components $Q_F = (a_1, ..., a_n)$ relative to any coordinate system F is called *covariant with respect to G* provided $Q_{F'} = sQ_F$ when F' results from F by transformation s, for any s in group G.

Notice that in the case of space he mentions the group of rotations;

it is quite possible for a quantity to be covariant with respect to the rotations but not with respect to the translations or reflections. The crucial ingredient in this passage is the equation

$$Q_{F'} = sQ_F \text{ for all } s \text{ in } G \tag{1}$$

which can be used to extend the meaning of 'covariant with respect to G" to the case of any group at all. For example, if the quantity Q_F is the velocity at time t of a certain body, relative to frame F, then it is represented by a vector, just like its position is, and so the same transformation s can be applied to it. Indeed, this will give the correct result if the transformation is a rotation, for the direction alone of the velocity is then affected, and it is rotated through the same angle as the coordinate system. (For example, if the rotation is around the z-axis, 180°, so that the x-coordinate of each point is mutiplied by -1, then the x-component of velocity v_x is transformed in the same way, into $-v_x$.) So velocity is covariant with respect to the group of rotations, though not with respect to the entire group of Galilean transformations.

We have now seen what conservation, invariance, and covariance mean for quantities. The notion of covariance is however also applied to theories, laws, principles, statements, equations. I think that in this area we tend to see equivocation between two notions: the one applies to the linguistic or notational forms in which a proposition can be expressed and the other to propositions themselves. Let us look at the latter first.

Here we can quite easily extrapolate from quantities. We may put it in general form as follows: if certain transformations form a group, and we call frames of reference related by transformations of this group *equivalent*, then:

(2) Equation $Q_F(x) = f(Q^1{}_F(x), \ldots, Q^n{}_F(x))$ is covariant exactly if its truth value is the same in equivalent frames.

Another way to say this, more perspicuous but less precise, relies on the intuitive notion of a sequence of numbers *satisfying* our equation. For example, -3 and $+3$ both satisfy $x^2 = 9$, $<4,3>$ satisfies $x^2 + y = 19$, and $<1,2,3>$ satisfies $x + y + z = 6x$. Then:

an equation EQ is covariant with respect to group G exactly if the following is the case: for all transformations g in G, if $<x_1, \ldots, x_n>$ satisfies EQ, so does $<g(x_1), \ldots, g(x_n)>$.

When all frames of reference can be produced from any one frame by the group (as is the case for all inertial frames of reference and the Galileo group) then all frames are equivalent to each other, and (2) becomes:

(3) The equation is *covariant* exactly if it is either true for all frames of reference or for none.

Thus covariance as applied to an equation is a type of *generality*. It is the sort of generality we want for all putative laws of nature—or, to give up that metaphor, for all basic principles of science.

At the beginning of this discussion I said that there appeared to be two notions of covariance applied to propositions. The one we have just arrived at has nothing to do with the notation or linguistic form in which the equation is expressed. But there is also another idea of covariant form, often presented in the context of a methodological requirement that all theories should be stated in covariant form. This idea, most prominently associated with Einstein, is well presented by Max Jammer.[8] A formula, as he explains, is covariant or form-invariant under a given family of (syntactic) transformations if its logical form is not affected by these transformations. Thus the formula $x \neq y$ is not covariant under substitution, for the substitution of x for y turns it into $x \neq x$, which does not have the same form as the original. But $y \neq y$ is covariant under substitution. Now this must somehow be connected with transformations of coordinate systems, which are not syntactic transformations. This is possible, if not easy to do with logical precision; the usual rather half-hearted attempt results in such hybrids as 'A law is stated in covariant form exactly if its restatement for any coordinate system takes the same form.' An obvious way to achieve this is to find coordinate-free formulations, and that is generally what is at issue. But I offer this conjecture about methodology: the *important* notion of covariance, and the virtue really sought, is the one explicated by (3) above.[9]

4. CASE-STUDY: CONSERVATION OF MOMENTUM

As concrete illustration of these various notions concerning (in)variance I shall now discuss conservation of momentum. Newton's system of the heavens can be briefly described thus: there is a privileged frame of reference, in which the centre of our solar system and the fixed stars are at rest; the laws of mechanics hold in this frame; there is exactly one sort of force which accounts for all celestial phenomena, namely the force described by the law of gravity.

His laws of mechanics and gravity hold not only in this frame; if they hold there they hold in all its Galilean transforms as well. For these propositions are all covariant with respect to that group of transformations. The class of inertial frames of reference—the frames for which Newton's theory was written—can therefore be taken as the class of frames connected to the privileged one by the group of Galilean transformations. I should add that the success of his theory was such that within a century, for example, Euler was willing to define the class of inertial frames by the fact that Newton's laws hold in them, even if the real frame which Newton pointed to, fulfilled that condition only approximately.

The laws of mechanics are three. The first is the law of inertia: the velocity of a body to which no force, or total force of magnitude zero, is applied, remains constant. The second states that the force equals the mass times the acceleration. The third has as slogan formulation: action equals reaction. They should be prefaced by the statement that every acceleration (change of velocity) has a cause, namely the application of a force. Indeed, Newton's forces may probably be defined, as far as his thinking was concerned, as the causes of acceleration. But of course the theory was streamlined so that it became usable equally without the acceptance of any idea of causation: taken in textbook form it only says that for each acceleration there exists a force incident on the body, equal to the acceleration times the mass.

To go on we must state the theory in modern form.[10] Position, velocity, acceleration, and force are all represented by vectors. That is, relative to each frame of reference, each has, at each instant, three components associated with the axes of the frame. The velocity is the time derivative of the position (rate of change of position) and the acceleration the time derivative of the velocity. A vector can be decomposed into (i.e. written as a sum of) other vectors; these notions I will now assume to be familiar. Thus the laws are, in obvious notation:

I. if $F = 0$ then $a = 0$

II. $F = ma$

III. given a set S of bodies and the forces F_α on member α of S, these forces can be decomposed into two sets:
internal: $F_{\alpha\beta}$, directed along the line joining α and β;
external: F^e_α whose component along any such line is zero;
such that $F_\alpha = F^e_\alpha + \sum_\beta F_{\alpha\beta}$
and $F_{\beta\alpha} = -F_{\alpha\beta}$ for all α, β in S

This is cryptic; obviously each of these notions applies to a body at an instant (or even more precisely, to a point at an instant) and the relativization to time is implicit. A system is called isolated if the external forces are all *zero*. The law of gravity describes the internal forces that account for celestial phenomena: the magnitude of $F_{\beta\alpha}$ equals the product of their masses divided by the square of their distance, and the direction is from α to β (attraction to β).

The quantity of position q is just the set of coordinates in the frames of reference. The magnitude of acceleration is invariant under Galilean transformations, and given law II, we are guaranteed that this is also true of the force. We shall take a closer look at invariance and covariance now, in the special case of the law of conservation of momentum.

That theorem, that total momentum is conserved in time, for an isolated system, is quite easy to prove. Consider a system of two bodies, α and β, and let $u(t)$, $v(t)$ be the functions which represent their velocities. Because of law III we know that in such an isolated system, the incident forces are exactly two, $F_{\alpha\beta} = -F_{\beta\alpha}$. By law II these bodies have masses m_α and m_β such that

$$F_{\beta\alpha} = m_\beta \, \mathrm{d}v/\mathrm{d}t \qquad F_{\alpha\beta} = m_\alpha \, \mathrm{d}u/\mathrm{d}t$$

But because the forces are equal and opposite, their sum equals zero:

$$m_\beta \, \mathrm{d}v/\mathrm{d}t + \mathrm{m}_\alpha \, \mathrm{d}u/\mathrm{d}t = 0$$

By the linearity of the derivative operation, that means

$$\frac{\mathrm{d}}{\mathrm{dt}}(m_\beta \, v + m_\alpha u) = 0$$

which by integration entails that this total momentum, $m_\beta v + m_\alpha u$, is constant. The theorem has a nice corollary: the composite system has a common centre of mass which is at rest or in uniform motion (as Newton expressed it). For the total momentum stays constant, and the velocity of the system as a whole is the total momentum divided by the total mass.

Let us now look more closely at the equation that states this law. Let our frame of reference be F, and our bodies $\alpha(1), ..., \alpha(n)$, and let the functions which represent their masses and velocities in this frame be $m(1), ..., m(n)$, $v_1(t), ..., v_n(t)$. We consider an alternative frame F' which results from F by the Galilean transformation of coordinates $q' = (q + vt)$. Of course each body $\alpha(i)$ has a position function $q_i(t)$ and velocity $v_i = \frac{d}{dt}q_i$. So the total momentum, relative to frames F and F' equals:

$$M_F(t) = \sum_{i=1} m(i)v_i(t)$$

$$M_{F'}(t) = \sum_{i=1} m(i)v'_i(t) \qquad \text{where } v'_i = \frac{d}{dt}q'_i$$

therefore

$$\begin{aligned} M_{F'} &= \sum m(i)\frac{d}{dt}(q_i + vt) \\ &= \sum m(i)(v_i + v) \\ &= v\sum m(i) + \sum m(i)v_i = v\sum m(i) + M_F \end{aligned}$$

I have suppressed the time variable: M_F is short for $M_F(t)$ and v_i for $v_i(t)$. Now the quantity v itself is constant—Galilean transformations proper relate frames in constant relative motion with respect to each other. Therefore we see at once that if M_F is constant, so is $M_{F'}$.

In other words, the statement that the total momentum is conserved, has that special generality of covariance: if it is true in one frame, then it is true in all. This quantity itself is not at all invariant under Galilean transformations and in general it need not be constant at all. But its logical status guarantees that a statement to the effect that it is constant in time, will be true relative to all frames if it is for one. Before knowing that it is a law, before appealing to Newton's laws of motion, we already know that its logical status makes it fit for the status of a basic principle of mechanics.

Proofs and illustrations

Let us continue this discussion of the momentum conservation laws in a higher key, where the symmetries expressed in Galilean relativity play a more central role.[11]

We deduced earlier that in a Newtonian system, the total momentum with respect to F is constant in time, either for all frames or else for none. The equation $dp/dt = 0$ which expresses this concisely, is therefore covariant in the sense of Galilean determinate (to coin a phrase which relativizes Carnap's 'logically determinate').

But the law of momentum conservation says that in all *closed, conservative* Newtonian systems, the first alternative holds: the momentum with respect to F is constant for all F. What is meant by this? A system is closed if it is not subject to interference from or interaction with outside—there are no external forces at work. And it is conservative if the internal forces depend solely on the distances between the particles, and not on their other mutable attributes such as velocity. Thus gravity, which depends on the distances and the constant masses, is conservative. A rocket which spews out fuel is not *by itself* a closed system; the rocket plus fuel is, but the masses here are not constant unless we analyse the objects into constituent particles. Obviously we have here certain restrictions on the Newtonian scheme. The elegant formal restriction following this line of thought is the sub-theory of *Hamiltonian mechanics*.

A Hamiltonian system is a Newtonian system for which there exists a *Hamiltonian* (*function*) H such that

$$dq_i/dt = \delta H/\delta p_i \tag{1}$$
$$dp_i/dt = -\,\delta H/\delta q_i \tag{2}$$

for the position coordinates and p_i ($i = n, n+1, n+2$) of the nth particle in the system. Note that p_i and q_i are themselves functions of time. Hence H, obviously a function of these position and momentum coordinates, is also a function of time. But we call the system a *conservative Hamiltonian system* exactly if H is in fact conserved, i.e. constant in time.

To fill all the logical gaps, one should like to see a proof here that all and only the closed, conservative Newtonian systems are conservative Hamiltonian systems. I doubt that this is a logical fact. Treatises on mechanics tend to introduce sub-theories, such as Hamiltonian mechanics, and then focus on them without attempting to fill all logical gaps. But the equation $dH/dt = 0$

which holds exactly if the system is Hamiltonian conservative, is covariant (Galilean determinate). The symmetry argument for the laws of conservation of momentum starts with it.

> *Theorem*: In a conservative Hamiltonian system, whose Hamiltonian function is invariant under space translations, the total momentum is constant.

The coordinates q_i and p_i are taken, of course, with respect to a single, arbitrary, inertial frame of reference F. The total momentum is the vector sum Σp_i and the assertion is

$$d(\Sigma p_i)/dt = \Sigma dp_i/dt = 0 \tag{3}$$

which we know, by (2) to follow from

$$\Sigma \delta H/\delta q_i = 0 \tag{4}$$

To evaluate (4), we use a bit of old-fashioned calculus legerdemain. Let Δx stand for an infinitesimal increment in the variable x. The dependence of H on the coordinates, and its invariance under space translation, we express then as

$$H(q_1 \ldots, q_i, \ldots, q_{3n}) = H(q_1 + \Delta\, q, \ldots, q_i + \Delta q, \ldots, q_{3n} + \Delta q) \tag{5}$$

By expansion into a Taylor series, and discarding[12] 'negligible' terms, the right-hand side is equal to

$$H(\ldots q_i, \ldots) + \Sigma(\delta H/\delta q_i)\Delta q \tag{6}$$

Putting (5) and (6) together we deduce

$$\Sigma(\delta H/\delta q_i)\Delta q = 0 \tag{7}$$

and since the infinitesimal quantity Δq is not zero, we have, by (7) and (2)

$$\Sigma dp_i/dt = 0 \tag{8}$$

which is the conservation of momentum equation.

Essentially similar deductions prove that:

(9) if H is invariant under time translation, then H is constant in time, and *a fortiori* the total energy is conserved

(10) if H is invariant under spatial rotation, the total angular momentum is conserved.

These last three facts, (8)–(10) are the principal classical conservation laws. This subject was definitively characterized by Emmy Noether, who proved the theorem we know under the slogan: *for every symmetry a conservation law*.

5. TRUE GENERALITY: WEYL ON POSSIBLE WORLDS

In philosophical thought, laws of nature were characterized especially by generality (universality) and necessity. Now, in our look at classical physics, we have seen quite a number of deep principles traditionally called laws, and we have also seen intimate connections between symmetry and generality. Is it possible, perhaps, that we have found the wherewithal to vindicate that tradition of laws and nature?

That there may be a bridge to the philosophical idea of law, appears to be expressed, though cryptically, early in Herman Weyl's book *Symmetry*. There is a striking passage in which he discusses Leibniz's views and the principle of sufficient reason:

> If nature were all lawfulness then every phenomenon would share the full symmetry of the universal laws of nature. . . . The mere fact that this is not so proves that *contingency* is an essential feature of the world. . . . The truth as we see it today is this: The laws of nature do not determine uniquely the one world that actually exists, not even if one concedes that two worlds arising from one another by . . . a transformation which preserves the universal laws of nature, are to be considered the same world.[13]

To explicate this simultaneously clear and mysterious prose, let us recapitulate what we have just learned about generality. In section 2 we found that significant generality is not a matter of linguistic form. A statement cannot wear generality on its sleeve, so to say. The form of 'all robins are red', 'all blue-jays are blue' and 'sodium always burns yellow' is general enough. But they are still automatically disqualified from the status of true generality, once we realize that what is red for one observer, is not red for another. To be truly general, a statement must be covariant, it must have this logical status: *it is either true in all frames of reference or true in none*. Equivalently: its truth value must remain invariant under all admissible transformations. But 'admissible' is theory-relative.

What is thus truly general to the classical scientist, is not always so for contemporary science.

The nature of this generality, and how it relates to the structure of models, was well illustrated by the conservation law for momentum. In the case of an isolated system, with constant masses, the total momentum is not an invariant quantity. Its value varies from one frame of reference to another. But the statement that this total momentum is constant in time is covariant: it is true for this system either in all frames of reference or in none.

Could the class of true covariant statements be a candidate (even if we ignore the theoretical context dependence) for the traditional notion of law? No, for the admissible transformations leave much too much invariant. A law was supposed to tell us not merely what is and does happen, but what must be and must happen. There are exactly six planets, Kepler thought, and no theory of the heavens is adequate unless it entail this fact. So he thought that this fact is not a mere accident, but something that had to be so. Newton disagreed: he had no more reason than Kepler to think that there were more planets, but for him the completeness of a celestial mechanics was not linked to this question. His theory was meant to apply regardless and to leave this question open, because as far as he could see it was not a matter of law at all.

This is what Weyl talked about in the passage I have just quoted. Let us discuss it in the concrete context of Newton's ideas for a moment, to illustrate it. Newton *did* offer hypotheses sometimes; and such was the hypothesis that the centre of mass of our solar system is at absolute rest. A Galilean transformation can change rest to motion, so this was a hypothesis not of a law, but of an accidental feature of the universe. No Galilean transformation, however, changes number; thus the number of planets is invariant. *Yet it is not a matter of law either*. This is sufficient to establish Weyl's point that: 'The laws of nature do not determine uniquely the one world that actually exists, not even if one concedes that two worlds arising from one another by . . . a transformation which preserves the universal laws of nature, are to be considered the same world.' Thus even if one disregards the distinction between absolute rest and motion, Newton's laws still leave us with many possible worlds, differing for example in the number of planets. That is what Weyl meant, and it withholds rather than gives content for the metaphor of law.

We have indeed found a significant notion of true generality, but not one of necessity.[14] And that significant generality pertains to our description of the structure of models, not the structure of nature. The conceptual triad of symmetry, transformations, and invariance does not explicate or vindicate the old notion of law—it plays the counterpoint melody on the side of representation.

PART IV

Symmetry and the Illusion of Logical Probability

INTRODUCTION

As will have been clear in Part I, and even more in Part II, probability has become a rich field of study for philosophy. Diverging views about probability also play an ever more important role in philosophical controversy—in general epistemology as well as in philosophy of science proper.

But probability is also a prime area for applications of symmetry arguments. I think these two points are not unconnected. Several times in Part I we came across the idea of a unique probability singled out on purely logical grounds—*logical probability*. In each case I asserted that this does not exist, that it is a philosophical will-o'-the-wisp. Here I return to this point, with a good deal of historical argument to draw on, for it concerns the first use of symmetries in probability theory. It is true that the historical controversy extended into our century, but I regard it as clearly settled now that probability is not uniquely assignable on the basis of a Principle of Indifference, or any other logical grounds.

Similarly in Part II, a crucial role was played by assertions about change of probabilities by Conditionalization. Paradoxically, it appears that this rule has the status of logic, and also that it need not be obeyed. This will be investigated in the last chapter, and central to it is a symmetry argument which fixes the form of admissible rules in probability kinematics.

In the choice of these topics I have been specially concerned to bring us back full circle to earlier parts of this book. The more technical points investigated here—exploiting symmetry arguments—can substantiate or destroy positions taken in general philosophy of science and epistemology. As a result I have ignored much of intrinsic interest—for example, the most famous symmetry result of all, De Finetti's representation theorem for exchangeable (i.e. permutation invariant) probability functions. I have also ignored here to some extent my aim of elucidating the role of symmetry in theory and model construction in physical science. But such omissions can be made good especially fittingly within philosophical discussions of quantum mechanics.

12.
Indifference: The Symmetries of Probability

> On estime la probabilité d'un événement par le nombre des cas favourables divisé par le nombre des cas possibles. La difficulté ne consiste que dans l'énumération des cas.
>
> Lagrange, quoted as epigraph to ch. 1 of J. Bertrand, *Calcul des probabilités*.

SINCE its inception in the seventeenth century, probability theory has often been guided by the conviction that symmetry can dictate probability. The conviction is expressed in such slogan formulations as that equipossibility implies equal probability, and honoured by such terms as indifference and sufficient reason. As in science generally we can find here symmetry arguments proper that are truly a priori, as well as arguments that simply assume contingent symmetries, and 'arguments' that reflect the thirst for a hidden, determining reality. The great failure of symmetry thinking was found here, when indifference disintegrated into paradox; and great success as well, sometimes real, sometimes apparent. The story is especially important for philosophy, since it shows the impossibility of the ideal of logical probability.

1. INTUITIVE PROBABILITY

A traveller approaches a river spanned by bridges that connect its shores and islands. There has been a great storm the night before, and each bridge was as likely as not to be washed away. How probable is it that the traveller can still cross? This puzzle, devised by Marcus Moore, clearly depends on the pattern of bridges represented in Figure 12.1.

It also depends on whether the survival of a bridge affects the survival of another. The traveller believes not. Thus for him each bridge had an independent 50 per cent probability of washing away.

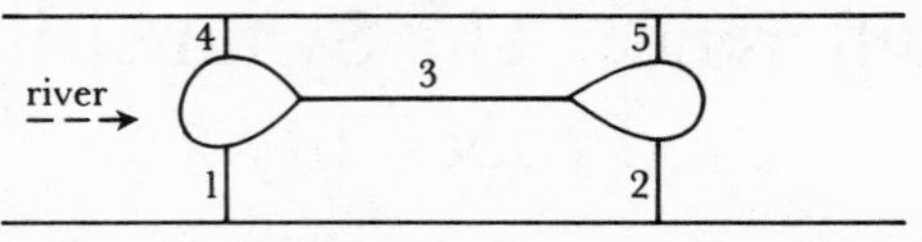

FIG. 12.1. A symmetry argument for probabilities

There is a simple but plodding solution (see *Proofs and illustrations*). But there is a symmetry argument too. Imagine that besides the traveller, there is also a boat moving downstream. The boatman's problem is to get through, which is possible if sufficiently many bridges have been washed away. What is the probability he can get through? Our first observation is that he faces a problem with the same abstract structure. For the traveller, the *entries* are bridges 1 and 2, while for the boatman they are 1 and 4. The *exits* are 4 and 5 for traveller, and are 2 and 5 for boatman. For both there is a *connector*, namely bridge 3. So each sees lying before him the 'maze'

entry   exit
            connector
entry   exit

Good and bad are reversed for traveller and boatman; but suppose that for each, the good state of a bridge has the same independent probability of 50 per cent. Now, by the great Symmetry Requirement, essentially similar problems must have the same solution. Hence:

1. Probability (traveller crosses) = Probability (boat gets through)

But the problems are not only similar; they are also related. For if the traveller has some unbroken path across, the boat cannot get through; and vice versa. Therefore:

2. Probability (boat gets through) = Probability (traveller does not cross)
3. [from 1 and 2] Probability (traveller crosses) = Probability (traveller does not cross)

So it is exactly as likely as not that the traveller will cross—the probability is 50 per cent.

This is a remarkable example, not only as a pure instance of a symmetry argument, but because it introduces all the basic ingredients in the three centuries of controversy over the relation between symmetry and probability. In this problem, the *initial probabilities* are given: 50 per cent for any bridge that it will wash away. We are also given the crucial probability datum about how these eventualities are related: they are *independent*. That means that the collapse of one bridge is neither more nor less probable, on the supposition that some other bridge is washed away. (We are here distinguishing simple probability from *conditional* probability, marked by such terms as 'on the supposition that' or 'given that'.) Then, purely a priori reasoning gives us the probabilities for the events of interest.

The great question for classical probability theory was: can the initial probabilities themselves be deduced too, on the basis of symmetry considerations? If we knew absolutely nothing about storms and bridges, except that one can wash away the other, would rationality not have required us to regard both possible outcomes as equally likely? Once the answer seemed to be obviously *Yes*, and now it seems self-evidently to be *No*, to many of us. But our century also saw the most sophisticated defences of the *yes* answer. And the history of the controversy spun off important and lasting insights.

Proofs and illustrations

In our example, the symmetry transformation used mapped bridge 2 into 4, and vice versa, leaving the others fixed. The *entry-connector-exit* structure is invariant, as is the probability of 'good' (i.e. *whole* for traveller and *broken* for boatman). The reader is invited to consider similar patterns with 1, 3, 4, 5 islands, and to generalize.

The single probability calculus principle that was utilized was—writing '*P*' for 'Probability':

$$P(A) = 1 - P(A)$$

which itself is an immediate corollary to the two axioms

I. $0 = P(\text{contradiction}) \leqslant P(A) \leqslant P(\text{tautology}) = 1$
II. $P(A) + P(B) = P(A \text{ or } B) + P(A \text{ and } B)$

which together exhaust the entire finitary probability theory. For our present purposes, it is not necessary to focus on this calculus (which will be explored further in the next chapter), but the following notions will be relevant (and will be employed intuitively in this chapter):

> The *conditional probability* $P(A|B)$ of A *given* that B equals $P(A$ and $B)/P(B)$
> A and B are (*stochastically* or *statistically*) *independent* exactly if $P(A|B) = P(A)$

That conditional probability $P(A|B)$ is defined only if the *antecedent* B has probability $P(B) \neq 0$. The independence condition is equivalent to

> $P(B|A) = P(B)$
> $P(A \text{ and } B) = P(A)P(B)$

always provided the conditional probabilities are defined. The last equation shows clearly, of course, that the condition is symmetric in A and B.

2. CELESTIAL PRIOR PROBABILITIES

The modern history of probability began with the Pascal–Fermat correspondence of 1654. The problems they discussed concerned gambling, games of chance. If someone wanted to draw practical advantage from these studies, he would learn from them how to calculate probabilities of winning (or expectation of gain) from initial probabilities in the gambling set-up. But of course he would have to know those initial probabilities already. While we cannot attribute much sophistication here to the gambler, we may plausibly believe that he takes a hard-nosed empirical stance on this. He believes that the dice are fair exactly if all possible numerical combinations come up equally often—and that this assertion is readily testable even in a small number of tosses. Daggers and rapiers will be drawn if a challenged and tested die comes up even three sixes in a row. We know of course from the play *Rosencrantz and Guildenstern Are Dead* how inconclusive such tests must be on a more sophisticated understanding of probability. But the crucial role and status of initial probability hypotheses appears much more clearly in a different sort of problem.

The Academy of Sciences in Paris proposed a prize subject for 1732 and 1734: the configuration of planetary orbits in our solar system. This configuration may be described as follows: each planet orbits in a plane inclined no more than 7.5° to the sun's equator, and the orbits all have the same direction.[2]

The prize was divided between John Bernoulli and his son Daniel. The latter included three arguments that this configuration cannot be attributed to mere chance. Of these the third argument is a typical eighteenth-century 'calculation' of initial probabilities: 7.5° is $\frac{1}{12}$ of 90° (possible maximum inclination of orbit to equator if we ignore direction); there are six (known) planets, so the probability of this configuration happening 'by chance' is $(\frac{1}{12})^6$, which is negligibly small (*circa* 3 in 10 million).

Daniel Bernoulli has here made two assumptions: of a certain *uniformity* (the probability of at most $\frac{1}{12}$ of the maximum, equals $\frac{1}{12}$) and of *independence* (the joint probability of the six statements is the product of their individual probabilities). Before scrutinizing these assumptions, let us look at two more examples.

Buffon, in his *Historie naturelle* gives an argument similar to Daniel Bernoulli's.[3] Buffon says that the mutual inclination of any two planetary orbits is at most 7.5° Taking direction into account, the maximum is 180°, so the chance of this equals $\frac{1}{24}$. Taking now one planet as fixed, we have five others. The joint probability of all five orbits to be inclined no more than 7.5° is therefore $(\frac{1}{24})^5$. This probability (*circa* 1 in 10 million) is approximately three times smaller than the one noted by Bernoulli. Independently Buffon notes that the probability that all six planets should move in the same west to east direction for us, equals $(\frac{1}{2})^6$. It is clear that he is calculating initial probabilities by the same assumptions as Daniel Bernoulli.

In Laplace's writings on celestial mechanics we find another such example.[4] Bernoulli and Buffon argued for a common origin of the planets, that is, a common cause, on the basis of the improbability of mere chance or coincidence. Laplace argues conversely that a certain fact is not initially improbable, and therefore needs no common-cause explanation. The fact in question was that among the many observed comets, not a single hyperbolic trajectory has been reported.[5] Laplace demonstrates that the probability of a comet with hyperbolic orbit is exceedingly low. The demonstration is based on a uniform distribution of probability over the possible directions of motion of comets entering the sun's gravitational field at some large given distance from the sun.

3. INDIFFERENCE AND SUFFICIENT REASON

It is clear that each of these authors is entertaining what we may call a chance hypothesis: that the phenomenon in question arises 'by mere chance', that is, without the presence of causal or other factors constraining the outcome. There is an ambiguity here: are the probabilities assigned the correct ones (*a*) given *no* hypotheses or assumptions about the physical situation, or (*b*) given a substantial, contingent hypothesis about the absence of certain physical features?

If the former is the case, we have here typical symmetry thinking: the fact that certain information is absent in the statement of the problem, is used as a constraint on the solution. If the latter, we are in the presence of a metaphysical assumption, which may have empirical import: that nature, when certain physical constraints are absent, is equally likely to produce any of the unconstrained possibilities, and therefore tends to produce each equally often.

Ian Hacking locates the first theoretical discussion of this topic in Leibniz's memorandum '*De incerti aestimatione*' (1678).[6] In this note Leibniz equates probability with gradations of possibility ('*probabilitas est gradus possibilitas*'). He states the Principle of Indifference, that equipossible cases have the same probability, and asserts that such a principle can be 'proved by metaphysics'.

We can only speculate what metaphysical proof Leibniz envisaged, but it must surely be based on his Principle of Sufficient Reason. Leibniz's programme set out in the *Discourse of Metaphysics* was to deduce the structure of reality from the nature of God. As a first step, this nature entails that God does, or creates, nothing without sufficient reason. In this marriage of metaphysics with divine epistemology, the difference between points (*a*) and (*b*) above vanishes. For Leibniz's God solves the problem of what nature shall do without contributing factors of his own to destroy the symmetries of the problem-as-stated.

This is how Leibniz must have derived symmetry principles governing nature—determining what the real, objective probabilities shall be in a physical situation. We cannot be sure on the basis of this brief note, but he must have given the principle of sufficient reason also this form: that a rational being should assign equal probabilities to distinct possibilities unless there be explicit reason to differentiate them. Since Leibniz clearly appreciated the great value of such an equation for metaphysics, he must have appreciated that strictly speaking, his new beginning for metaphysics effects a collapse of two logically distinct problems.

It was certainly in the terminology of sufficient reasons—perhaps always with a equivocation between (*we have reason*) and (*there is reason*)—that principles of indifference were formulated. There were two; we have seen both at work in the arguments of Bernoulli, Buffon, and Laplace.

The first is the *Principle of Uniform Distribution*. Suppose I shoot bullets at a target and am such a poor marksman that it makes no difference at which point of the target I aim. Then any two equal areas on the target are equally likely to be hit. We call this a uniform distribution. The first indifference principle for assigning probabilities is *to assume a uniform distribution in the absence of reasons to the contrary*.

The second is the *Principle of Stochastic Independence*. I explained independence above; let me illustrate it here. Suppose we are told that 40 per cent of the population smokes and 10 per cent has lung cancer. This gives me the probability that a randomly chosen person is a smoker, or has lung cancer, but does not tell me the joint probability of these two characteristics. There are three cases (see Fig. 12.2). Each of the three lines p, q, r has 10 per cent of the area below it. In the case of the horizontal line q, the joint probability of lung cancer *and* smoking is 10 per cent of 40 per cent, namely 4 per cent. For p it is larger and for r it is smaller. The second indifference principle is *to assume statistical independence, in the absence of reasons to the contrary*.

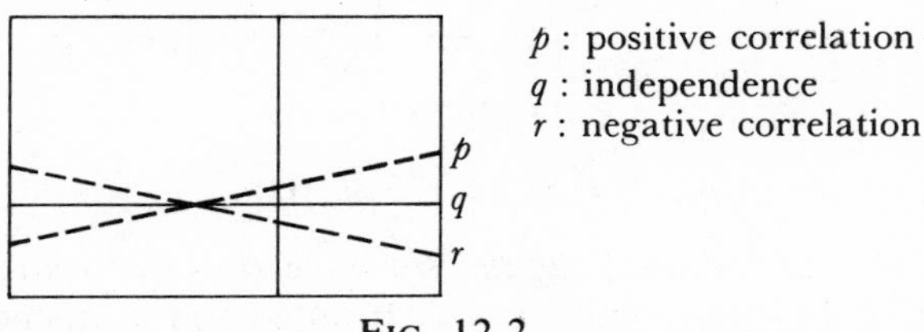

FIG. 12.2

Are these two principles consistent with each other? The joint probability of two events is the same as the ordinary probability of a single complex event. It seems possible therefore that the two principles could be made to apply to the same example, and offer contradictory advice. In the *Proofs and illustrations* we will see that this is not so; the two are consistent with each other.

Proofs and illustrations

Let us consider two variables, say *height* h and *weight* w. Suppose height varies from zero to 10 and weight from zero to 100. Given no other information (hence no reasons to diverge from uniformity or independence), assign probabilities to all possibilities.

The first procedure is to choose uniform distributions for each:

1. $P(0 \leqslant h \leqslant a) = a/10 \qquad P(0 \leqslant w \leqslant b) = b/100$

Then calculate the joint probability by assuming independence:

2. $P(0 \leqslant h \leqslant a \text{ and } 0 \leqslant w \leqslant b) = (a/10)(b/100)$

The other procedure is to look at the complex variable hw which has pairs of numbers as values. A person with height 6 and weight 60 has hw equal to $<6, 60>$. The big rectangle in Fig. 12.3 encompasses all possibilities ($0 \leqslant h \leqslant 10$ *and* $0 \leqslant w \leqslant 100$) while the smaller one describes the possibility of having hw fall between $<0, 0>$ and $<a, b>$ in the proper sense of 'between'. Uniformity alone applies now and demands a probability proportional to the area:

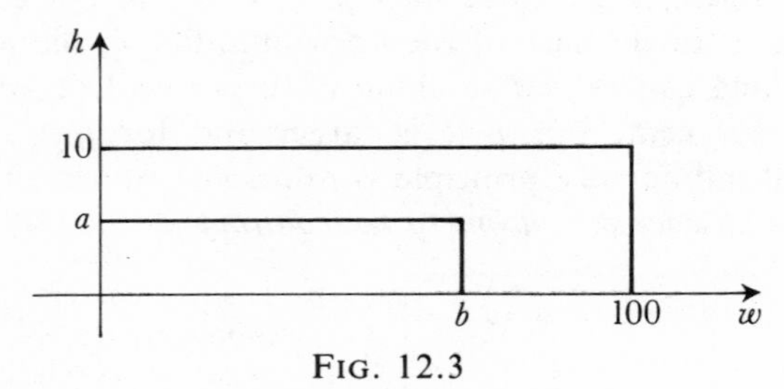

FIG. 12.3

3. $P(<0,0> \leqslant hw \leqslant <a, b>) = ab/1000$

But as we see, 2 and 3 agree. We have proved in effect that if variables h and w are uniformly distributed and independent, then the complex variable hw is uniformly distributed. Hence the two principles are mutually consistent and together constitute the great symmetry principle of classical probability theory—the *Principle of Indifference*.

4. BUFFON'S NEEDLE: EMPIRICAL IMPORT OF INDIFFERENCE

If we must assign initial probabilities, in the absence of relevant information, reason bids us be like Buridan's ass. Do not choose between $P(A) > P(-A)$ and $P(A) < P(-A)$, but set them equal. Similarly in such a case, do not choose between $P(A \text{ and } B) > P(A) \,.\, P(B)$ and $P(A \text{ and } B) < P(A) \,.\, P(B)$, but set those equal as well. Very well; but will nature oblige us with frequencies to which these initial probabilities have a good fit? Is this dictate of reason one that will let reason unlock the mysteries of nature?

An empiricist will ask these questions with a distinct tinge of mockery to his voice. But here we should report a marvellous example in which calculation by the Principle of Indifference led to beautifully confirmed empirical results. This is Buffon's needle problem. It is much more probative than planetary orbit and comet examples, where one only finds *explanation*—that beautiful but airy creature of the fecund imagination—and not *prediction*.

Buffon's needle problem[7]

Given: a large number of parallel lines are drawn on the floor, and a needle is dropped. What is the probability that the needle cuts one of the lines?

To simplify the problem without loss of essential generality, let the lines be exactly two needle lengths apart. Touching will count as cutting, but clearly at most one line is cut. We may even speak sensibly of the line nearest the needle's point (choose either if the point is exactly halfway between). Then our question is equivalent to: what is the probability that the needle cuts this nearest line? In Fig. 12.4 the needle point is a distance $0 \leqslant d \leqslant 1$ away from line L, and its inclination to L is the angle θ. Thus we have:

favourable cases: the needle cuts L exactly if $d \leqslant y = \sin \theta$

This θ varies from zero to 2π ($= 360$ degrees), and so we can diagram the situation with an area of 1 (needle length) by 2π (radians) as in Fig. 12.5. To distinguish the favourable cases from the unfavourable ones, we draw in the sine curve and shade the area where $y \geqslant d$. Assuming independence and uniform distribution, the probability of the favourable cases must be proportional to the

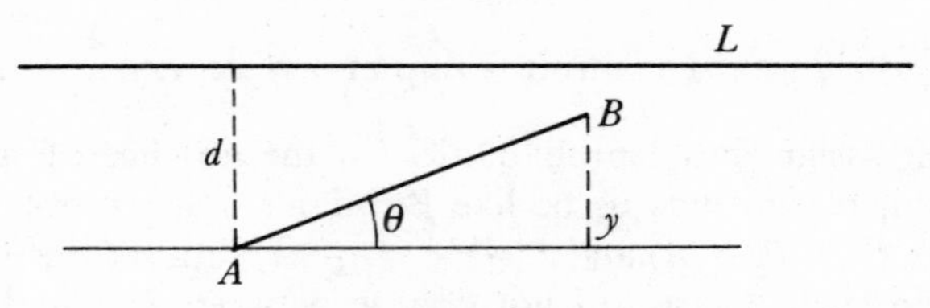

FIG. 12.4. Buffon's needle

shaded area. Since a little calculus quickly demonstrates that this area equals 2, we arrive at the number $2/2\pi$:

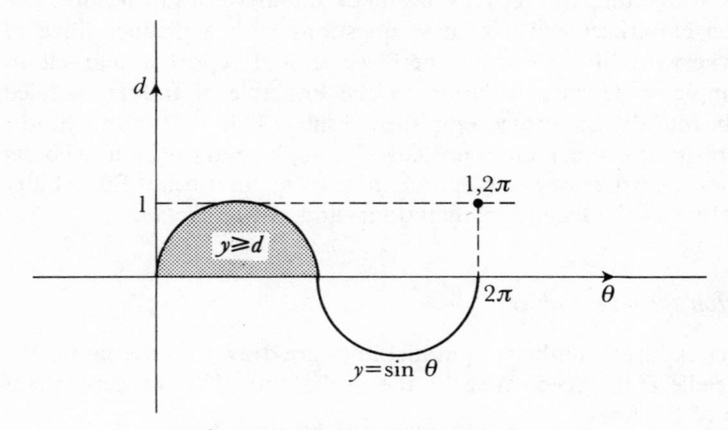

FIG. 12.5. Buffon's probability calculation

The probability of a favourable case equals $1/\pi$, the solution Buffon himself found for his problem.

Since the experiment can be carried out, this is an empirical prediction. It has been carried out a number of times and the outcomes have been in excellent agreement with Buffon's prediction.[8] Now is this not marvellous and a result to make the rationalist metaphysician squeal with delight? For the assumption of symmetry in the probabilities of equipossible cases has here led to a true prediction made a priori.

5. THE CHALLENGE: BERTRAND'S PARADOXES

What I have so far recounted has been very favourable to the Principle of Indifference. Many readers, knowing of its later rejection, but perhaps less familiar with attempts to refine and save it, may already be a little impatient. I will argue for the rejection of its uncritical versions—the empirical phenomena cannot be predicted a priori—but this will be a rejection of naïve symmetry arguments in favour of deeper symmetries, with due respect for the insights that were gained along the way.

We have seen that the Principle has two parts, which are indeed consistent with each other. We have also seen the significant successes of explanations and predictions arrived at in the eighteenth century by means of this Principle. But the challenge to this attempt to calculate initial probabilities on the basis of physical symmetry came exactly from the fundamental principle of symmetry arguments. If two problems are essentially the same, they must receive essentially the same solution. So *a fortiori* if a situation can be equally described in terms of different parameters, we should arrive at the same probabilities if we apply the Principle of Indifference to these other parameters. There will be a logical difficulty—indeed, straightforward inconsistency—if different descriptions of the problem lead via Indifference to distinct solutions.

This logical difficulty with the idea was expounded systematically in a series of paradoxes by Joseph Bertrand at the end of the nineteenth century.[9] Leaving his rather complex geometric examples for *Proofs and Illustrations*, let us turn immediately to a paradigmatic but simple example: the perfect cube factory.[10]

> A precision tool factory produces iron cubes with edge length $\leqslant 2$ cm. What is the probability that a cube has length $\leqslant 1$ cm, given that it was produced by that factory?

A naïve application of the Principle of Indifference consists in choosing length l as parameter and assuming a uniform distribution. The answer is then $\frac{1}{2}$. But the problem could have been stated in different words, but logically equivalent form:

Possible cases	Favourable
edge length $\leqslant 2$	length $\leqslant 1$
area of side $\leqslant 4$	area $\leqslant 1$
volume $\leqslant 8$	volume $\leqslant 1$

Treating each statement of the problem naïvely we arrive at answers $\frac{1}{2}, \frac{1}{4}, \frac{1}{8}$. These contradict each other.

The correspondence $l^m \leftrightarrow l^n$, for a parameter l with range $(0, k)$ is one to one, but does not preserve equality of intervals.

Hence uniform distribution on l^m entails non-uniform distribution on l^n. Now sometimes the problem is indeed constrained by symmetries. The cubes example illustrates how these constraints may be so minimal as to leave the set of possible solutions unreduced. More information about the factory could improve the situation. But the Indifference Principle is supposed to fill the gap left by missing information!

Even taken by itself, the example is devastating. But since we shall discuss various attempts to salvage Indifference, it is important to assess two more examples, with somewhat different logical features.

Von Kries posed a problem which is like that of the perfect cube factory, in that several parameters are related by a simple logical transformation. Consider volume and density of a liquid. If mass is set equal to 1, then these parameters are related by:

density = 1/volume; volume = 1/density.

But a uniform distribution on parameter x is automatically non-uniform on $y = (1/x)$. For example,

x is between 1 and 2 exactly if y is between $\frac{1}{2}$ and 1
x is between 2 and 3 exactly if y is between $\frac{1}{3}$ and $\frac{1}{2}$.

Here the two intervals for x are equal in length, but the corresponding ones for y are not. Thus Indifference appears to give us two conflicting probability assignments again.

Von Mises's example of a Bertrand-type paradox concerned a mixture of two liquids, wine and water. We have a glass container, with a mixture of water and wine. To remove division by zero from every inversion, let the following be data:

the glass contains 10 cc of liquid, of which at least 1 cc is water and at least 1 cc is wine.

What is the probability that at least 5 cc is water? Let the parameters be:

a = proportion of wine to total: $(1/10) \leqslant a \leqslant (9/10)$
b = proportion of water to total: $(1/10) \leqslant b \leqslant (9/10)$
x = proportion of wine to water: $(1/9) \leqslant x \leqslant 9$
y = proportion of water to wine: $(1/9) \leqslant y \leqslant 9$

Obviously $b = (1 - a)$, $x = (a/b)$, $y = 1/x$, and $a = x/(1 + x)$, so descriptions of the situation by means of any parameter can be completely translated into any other parameter. It is easy to see that the same problem recurs. Here are two equal intervals for the proportion of wine to total:

a = Proportion of wine to total	x = Proportion of wine to water
4/10	4/6 = 2/3
5/10	5/5 = 1
6/10	6/4 = 3/2

Since $1 - (2/3)$ is not equal to $(3/2) - 1$, it is clear that a uniform distribution on the proportion a entails a non-uniform proportion on proportion x.

In each case the Principle of Uniformity is applied to one perfectly adequate description of the problem. The statements of the problem, both as to sets of possible cases and set of favourable cases, differ only verbally. But the great underlying principle of symmetry thinking is that essentially similar problems must receive the same solution. Thus the attempt to assign uniform distribution on the basis of symmetries in these *statements* of the problem, is drastically misguided—it violates symmetry in a deeper sense.

Most writers commenting on Bertrand have described the problems set by his paradoxical examples as not well posed. In such a case, the problem as initially stated is really not one problem but many. To solve it we must be told *what* is random; which means, *which* events are equiprobable; which means, *which* parameter should be assumed to be uniformly distributed.

But that response asserts that in the absence of further information we have no way to determine the initial probabilities. In other words, this response rejects the Principle of Indifference altogether. After all, if we were told as part of the problem which parameter should receive a uniform distribution, no such Principle would be needed. It was exactly the function of the Principle to turn an incompletely described physical problem into a definite problem in the probability calculus.

There have been different reactions. We have to list Henri Poincaré, E. T. Jaynes, and Rudolph Carnap among the writers who believed that the Principle of Indifference could be refined and sophisticated, and thus saved from paradox.

Proofs and illustrations

The famous chord problem asks for the problem that a stick, tossed randomly on a circle, will mark out a chord of given length. For a definite standard of comparison we inscribe an equilateral triangle ABC in the circle (see Fig. 12.6). However we draw the triangle, it is clear that the separated arcs, like arc AEB, must each be $\frac{1}{3}$ of the circumference. Thus the length of the side of any such triangle is the same. In fact it is $r\sqrt{3}$, where r is the radius, and the point D is exactly halfway along the radius OE.

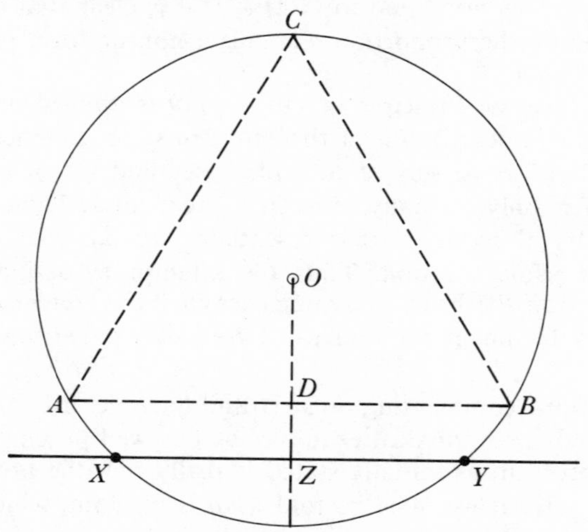

FIG. 12.6. Bertrand's chord problem

What is the probability that chord XY is greater than side AB? If we try to answer this question on the basis of the Principle of Indifference, we actually find three variables which might be asserted to have uniform distribution:

$XY > AB$ exactly if any of the following holds:

(*a*) $OZ < r/2$

(*b*) Y is located between $\frac{1}{3}$ and $\frac{2}{3}$ of the circumference away from X, as measured along the circumference

(*c*) the point Z falls within the 'inner' circle with centre 0 and radius $r/2$.

This gives us three possible applications of the Principle of Uniformity.

Using description (*a*) we reason: *OZ* can be anything from 0 to *r*; the interval [0, *r*/*z*] of favourable cases has length $\frac{1}{2}$ of the interval [0, *r*] of possible cases; hence the probability equals $\frac{1}{2}$ (*Solution A*).

Using (*b*) we reason: each point of contact *X*, *Y* can be any point on the circle. So given the point *X*, we can find point *Y* at any fraction between 0 and 1 of the circumference, measuring counter-clockwise. Of these possible locations, $\frac{1}{3}$ fall in the favourable interval $[\frac{1}{3}, \frac{2}{3}]$; hence the probability equals $\frac{1}{3}$ (*Solution B*).

Using (*c*) we note that the centre *Z* of the stick can fall anywhere in the whole circle. In the favourable cases it falls in the 'inner' circle with radius *r*/2—which has an area $\frac{1}{4}$ that of the big circle. Hence the probability equals $\frac{1}{4}$ (*Solution C*).

6. SYMMETRIES TO THE RESCUE?

Henri Poincaré and E. T. Jaynes both argued that if we pay attention to the geometric symmetries in Bertrand's problem, we do arrive at a unique solution.[11] Their general idea applies to all apparent ambiguities in the Principle of Indifference: a careful consideration of the exact symmetries of the problem will remove the inconsistency, provided we focus on the symmetry transformations themselves, rather than on the objects transformed.

In order to show the logical structure very clearly I will concentrate on the simple examples of the perfect cubes, mass versus density, and water mixed with the wine. Let us begin by analysing the intuitive reaction to the cube factory, which led us into paradox. Focusing first on the parameter of length, we used the natural length measure for intervals:

$$m(a, b) = b - a$$

This is the *underlying measure*[12] that gave us our probabilities for cases inside the range [0,2]:

$$P(a, b) = m(a,b)/m(\text{range}) \text{ for } a, b \text{ inside the range.}$$

Now this underlying measure has a very special feature, from the point of view of symmetry:

Translation invariance: if $x' = x + k$
then $m(a, b) = m(a', b')$.

Up to multiplication by a scalar, m is the unique measure to have this feature. That is easy to see, because one interval can be moved into another by a translation exactly if they have the same length (and are of the same type: open, half-open, etc.).[13]

The number K represents the scale, if $m' = Km$, because for example the length in inches is numerically 12 times that in feet. It will not affect the probability at all, because it will cancel out (being present in both numerator and denominator in the equation for P in terms of m). We have therefore the following result:

> *Translation invariant measure.* The probability distribution on a real valued parameter x is uniquely determined, if we are given its range and the requirement that it derive from an underlying measure which is translation invariant.

In what sort of example would the given be exactly as required? Suppose I tell you that Peter is a marksman with no skill whatever, and an unknown target. Now I ask you the probability that his bullet will land between 10 and 20 feet from my heart, given that it lands within 20 feet. Treating this formally, I choose a line that falls on both my heart and the impact point of the bullet, coordinatize this line by choosing a point to call zero, and one foot away from it a point to call + 1. I choose a measure m' on this line, call my heart's coordinate X, and calculate

$$m'(X + 10, X + 20)/m'(X, X + 20)$$

and give you the resulting number as answer. If my procedure was properly in tune with the problem, this answer should better not depend on how I chose the points to call zero and + 1 (which two choices together determined the coordinate X). That entails that m' must be translation invariant, and is therefore now uniquely identified. We note with pleasure that the answer is also not affected by the choice of the foot as unit of measurement—as indeed it should not, because nothing in the problem hinged on its Anglo-Saxon peculiarities.

Now, in what sort of problem is the 'given' so different that this procedure is inappropriate? Obviously, when translation invariance is the wrong symmetry. This happens when the range of the physical quantity in question is not closed under addition and subtraction, for example, if the quantity has an infimum, which acts as natural zero point. For example, no classical object has negative or zero volume, mass, or absolute temperature.

In such a case, the scale or unit may still be irrelevant. For the transformation of the scaling unit consists simply in multiplication by a positive number, which operation does not take us out of this range. Consider now von Kries's problem, which concerns the positive quantities mass and density. With the units of measurement essentially irrelevant we look for an underlying measure

$$M(a, b) = M(ka, kb)$$

for any positive number k (invariance under *dilations*).

There is indeed such a measure, and it is unique in the same sense.[14] That is the *log uniform* distribution:

$$M(a, b) = \log b - \log a$$

where log is the natural logarithm. This function has the nice properties:

$$\log(xy) = \log x + \log y$$
$$\log(x^n) = n \log x$$
$$\log(1) = 0$$

but should be used only for positive quantities, because it moves zero to minus infinity. The first of these equations shows already that M is dilation invariant. The second shows us what is now regarded as equiprobable:

> The intervals (b^n, b^{n+k}) all receive the same value $k\log b$, so if within the appropriate range, the following are series of equiprobable cases:
>
> $(0.1, 1), (1, 10), (10, 100), \ldots, (10^n, 10^{n+1}), \ldots$
>
> $(0.2, 1), (1, 5), (5, 25), \ldots, (5^n, 5^{n+1}), \ldots$

and so forth. A probability measure derived from the log uniform distribution will therefore always give higher probabilities 'closer' to zero, by our usual reckoning.

For example, in the case of temperature we have since Kelvin accepted that this is essentially a positive quantity. Of course we are at liberty to give the name -273 to absolute zero. But this does not remove the infimum; subtraction eventually takes one outside the range. The presence of this infimum creates, or rather is, an asymmetry: it obstructs translation invariance. But it is no obstacle to dilation invariance, so the log uniform distribution is right—it is dictated by the symmetries of the problem.

This reasoning, being rather abstract, may not get us over our initial feeling of surprise. But as Roger Rosencrantz pointed out, we can test all this on the von Kries problem.[15] Our argument implies that von Kries's puzzle is due to focusing on the wrong transformation group. Attention to the right one dictates use of the log uniform distribution. To our delight this removes the conflict:

$$\begin{aligned} M(1/b \leqslant 1/x \leqslant 1/a) &= K(\log a^{-1} - \log \ b^{-1}) \\ &= K(\log b - \log a) \\ &= M(a \leqslant x \leqslant b). \end{aligned}$$

This is certainly a success for this approach to Indifference.

Consider next the perfect cube factory. Suppose that again we regard the unit of measurement as essentially irrelevant to this problem, conceived in true generality, but observe that length, area, volume are positive quantities. The uniqueness of the log uniform measure for dilation invariance, forces us then to use it as underlying the correct probabilities. This will not help us, unless we ask all our questions about intervals that exclude *zero*; but for them it works wonderfully well:

> What is the probability that the length is $\leqslant 2$, given that it is between 1 and 3 inclusive?
> What is the probability that the area is $\leqslant 4$, given that it is between 1 and 9 inclusive?
> The probability that the length is $\leqslant 2$, given that it is between 1 and 3 inclusive, equals $M(1, 2)/M(1, 3) = \log 2/\log 3 = 0.631$.
> The probability that the area is $\leqslant 4$, given that it is between 1 and 9 inclusive, equals $M(1, 4)/M(1, 9) = 2\log 2/2\log 3 = 0.631$.

Thus the two equivalent questions do receive the same answer. The point is perfectly general, because the exponent becomes a multiplier, which appears in both numerator and denominator, and so cancels out. This is again a real success. By showing us how to reformulate the problem, and then using its symmetries to determine a unique solution, this approach has as it were taught us how to understand our puzzled but insistent intuitions.

There is therefore good prima-facie reason to take this approach seriously. In the *Proofs and Illustrations* I shall show how this approach does give us a neat solution for the puzzle of Buffon's needle, construed as a Bertrand problem. But in subsequent sections we'll see that the approach does not generalize sufficiently to save the Principle of Indifference.

Proofs and illustrations

I shall here explain this rescue by geometric symmetries with another illustration. For this purpose I choose Buffon's needle problem again, for properly understood it can itself be described as a rudimentary Bertrand paradox.[16].

Buffon assumes no marksmanship—the location of the lines on the floor does not, as far as we know, affect the location of the fallen needle. So our description of the situation utilizes a frame of reference chosen for convenience, in which we treat as X-axis the line through needle point A which is parallel to the drawn lines, as in Fig. 12.4. Here d is the Y-coordinate of line L, θ the inclination of line AB to the X-axis, and y is Y-coordinate of B, and A is the origin.

Why not assume that y and d are independent and uniformly distributed? We must be careful to describe y so that it does not depend on d. But it is just sine θ, and θ does not depend on d, so that is fine. Thus y ranges from -1 to $+1$ (being measured from the X-axis, chosen so that the line L has equation $Y = 1$). The possible and favourable cases are depicted in Fig. 12.7, and we see that the probability of $y \leqslant d$ equals $\frac{1}{4}$. Hence by applying the Principle of Indifference to Buffon's problem *differently but equivalently described*, we have arrived at a different solution.

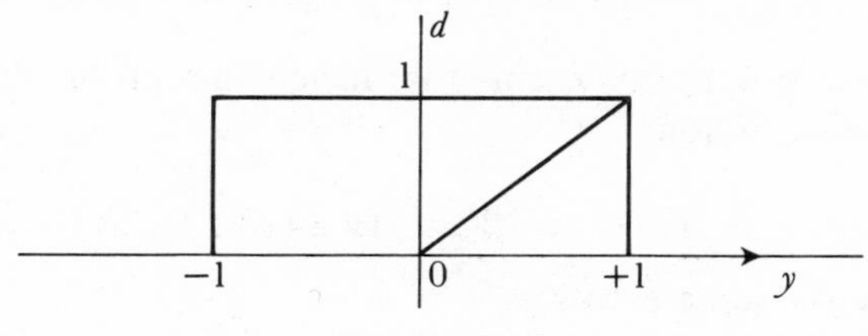

FIG. 12.7

But our description—or rather the solution that utilizes this description in the Principle of Indifference—may itself be faulted for failing to respect geometric symmetry. Consider what happens if the axes are rotated through some angle around point A—that is, the orientation of the lines drawn on the floor is changed. Whatever method of solution we propose, should not make the answer—probability of a cut—depend on this orientation, for the problem remains essentially unchanged. (The aspect varied did not appear at all in the statement of the problem.) How do the two rival solutions vary with respect to this criterion?

Buffon's solution fares very well. For the initial parameter (angle which the needle makes with the X-axis) is changed by adding something (the angle of rotation), modulo 360°. A uniform distribution on that initial parameter induces automatically a uniform distribution also on its transform—equal angular intervals continue to receive equal probability.

But, and here is the rub, if we assume that y is uniformly distributed, it follows that y' (the corresponding coordinate in the rotated frame) is not. The easy way to see this is to look at equal increments in y and notice that they do not correspond to equal increments in y'.

To see this it is necessary to use the formula that transforms coordinates, when the frame is rotated. If the original coordinates of a point are (x, y) they become, upon rotation through angle α around the origin

$$t(x) = x \cos \alpha - y \sin \alpha$$
$$t(y) = x \sin \alpha + y \cos \alpha$$

In our case, point B has coordinates (x, y) but because $AB = 1$ we know that $x^2 + y^2 = 1$. Hence $x = \sqrt{(1 - y^2)}$ and we have

$$t(y) = \sqrt{(1 - y^2)} \sin \alpha + y \cos \alpha$$

Let us now look at two events that have equal probability if y has uniform distribution:

$$[y \,\varepsilon\, (0, 1/2)] \qquad [y \,\varepsilon\, (1/2, 1)]$$

These are the same events as

$$[t(y) \,\varepsilon\, (t(0), t(1/2))] \qquad [t(y) \,\varepsilon\, (t(1/2), t(1))]$$

If the variable $t(y)$ has uniform distribution, these events will have equal probabilities only if

$$t(1/2) - t(0) = t(1) - t(1/2)$$

so that is what we need to check. A single counter-example will do, so let us choose the angle of 30° (i.e. $\pi/6$ radians) which has sine $\frac{1}{2}$ and cosine $(\sqrt{3})/2$. Therefore:

$$t(0) = \sin(\tfrac{1}{6}\pi) = \tfrac{1}{2}$$
$$t(\tfrac{1}{2}) = (\sqrt{3})/2 \sin \tfrac{1}{6}\pi + \tfrac{1}{2} \cos \tfrac{1}{6}\pi$$
$$= (\sqrt{3})/2\,(\tfrac{1}{2}) + \tfrac{1}{2}((\sqrt{3})/2) = (\sqrt{3})/2$$
$$t(1) = \cos \tfrac{1}{6}\pi = (\sqrt{3})/2$$

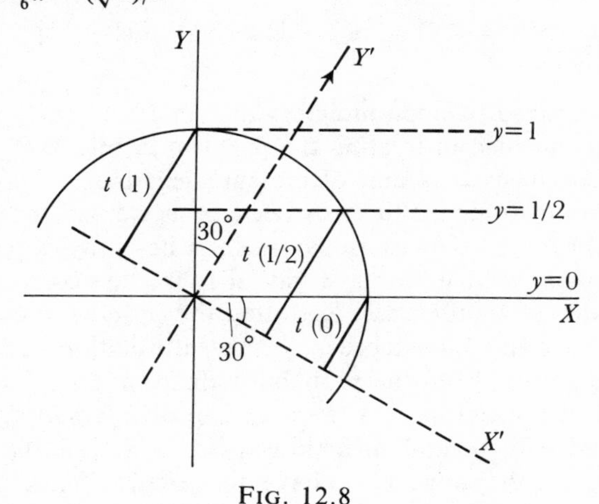

FIG. 12.8

It is very obvious that our desired equation does not hold. Figure 12.8 shows the different cases.

7. PYRRHIC VICTORY AND ULTIMATE DEFEAT

The successes we found in the preceding section, even together with their more sophisticated variants (to be discussed in the *Proofs and Illustrations*), constitute only a Pyrrhic victory. Again we can see this in simple examples, just because of the power of the uniqueness results utilized. Recall that invariance under translations and invariance under dilations each dictate an essentially *unique* answer to all probability questions. What happens when the examples take on more structure?

Peter Milne, writing about Rosencrantz's solution to von Kries's problem, has shown exactly how things go wrong.[17] To show this, he asked how the above results are to be applied to von Mises's water and wine problem. Let us again ask the same question in two different ways, referring back to the notation we had before.

What is the probability that at least 5 cc is water?
b = proportion of water to the total
x = proportion of wine to water.
$P(b \geqslant 0.5 \mid 0.1 \leqslant b \leqslant 0.9 = (\log 0.9 - \log 0.5)/(\log 0.9 - \log 0.1) =$ 0.267
$P(x \leqslant 1 \mid 1/9 < x \leqslant 9) = \log 1 - \log(1/9))/(\log 9 - \log(1/9)) =$ $\log 9/2\log 9 = 0.5$

We have received two contradictory answers.

Were we justified in treating the problem in this way? Well, the problem specified cc as unit of measurement but we have just as much warrant to regard this as irrelevant as we had for cm in the cubes problem. If we focus on parameter x here, say, we must treat it in the same way, if we have indeed found the correct form of the Principle of Indifference. Restating the problem then in terms of b, we have not introduced any new information—so we must derive the answer from the probability distribution on x, plus the definition of b in terms of x. Exactly the same would apply if we had started with b, and then moved on to x. But the two end results are not the same, so we have our paradox back.

It is also rather easy to see the pattern that will produce such paradoxes. A translation invariant measure will be well behaved with respect to addition and multiplication, while a dilation invariant measure will be equally good with respect to multiplication and exponentiation. But the relation between b and x uses both sorts of operations:

b = water/(water + wine) = water/10
x = wine/water = (10 − water)/water
hence, water = $10b$ and $x = (100 - b)/b$.

Now neither sort of measure will do. If the required dilation invariance did not dictate an essentially unique measure, we would perhaps have had some leeway to look for something other than logarithms—but we do not.

The history of the Principle of Indifference is instructive. If its mention in scientific sermons serves to remind us to look for symmetries, then it serves well. But a rule to determine initial probabilities a priori it is not. It violates a higher symmetry requirement when it is conceived of in that way.

Even if the Principle were unambiguous, the question whether its results would be probability functions with a good fit to actual frequencies in nature, would anyway be a purely contingent one. To imagine that it would not be—that empirical predictions could be made a priori, by 'pure thought' analysis—is feasible only on the assumption of some metaphysical scheme such as Leibniz's, in which the symmetries of the problems which God selects for attention, determine the structure of reality.[18] But because it is not unambiguous, even that assumption would leave us stranded, unless we knew how God selected his problems.

When E. T. Jaynes[19] discussed Bertrand's chord paradox, although noting that most writers had regarded it as an ill-posed problem, he responded:

> But do we really believe that it is beyond our power to predict by 'pure thought' the results of such a simple experiment? The point at issue is far more important than merely resolving a geometric puzzle; for . . . applications of probability theory to physical experiments usually lead to problems of just this type. . . . (p. 478)

Jaynes's analysis of the Bertrand chord problem is along the lines of the preceding section. He shows that there is only one solution which derives from a measure which is invariant under Euclidean transformations.

But when we look more carefully at other parts of Jaynes's paper we see that his more general conclusions nullify the radical tone. Jaynes says of von Mises's water and wine problem, that the fatal ambiguities of the Principle of Indifference remain. More important: the strongest conclusion Jaynes manages to reach is merely one of *advice*, to regard a problem as having a definite solution until the contrary has been proved. The method he advises us to follow is that of symmetry arguments:

> To summarize the above results: if we merely specify complete ignorance, we cannot hope to obtain any definite prior distribution, because such a statement is too vague to define any mathematically well-defined problem. We are defining what we mean by complete ignorance far more precisely if we can specify a set of operations which we recognize as transforming

the problem into an equivalent one, and the desideratum of consistency then places non-trivial restrictions on the form of the prior.[20]

But as we know, this method always rests on assumptions which may or may not fit the physical situation in reality. Hence it cannot lead to a priori predictions. Success, when achieved, must be attributed to the good fortune that nature fits and continues to fit the general model with which the solution begins.

Proofs and illustrations

Harold Jeffreys introduced the search for invariant priors into the foundations of statistics; there has been much subsequent work along these lines by others.[21] We must conclude with Dawid, however, that the programme 'produces a whole range of choices in some problems, and no prior free from all objections in others' ('Invariant Prior Distributions', 235). I just wish to take up here the elegant logical analysis that Jaynes introduced to generalize the approach which we have been studying in these last two sections.[22]

Here some powerful mathematical theorems come to our aid. For under certain conditions, there exists indeed only one possible probability assignment to a group, so there is no ambiguity.

The general pattern of the approach I have been outlining is as follows. First one selects the correct group of transformations of our sets K which should leave the probability measure invariant. Call the group G. Then one finds the correct probability measure p on this group. Next define

(1) $P(A) = p(\{g \text{ in } G\colon g(x_0) \,\varepsilon A\})$

where x_0 is a chosen reference point in the set K on which we want our probability defined. If everything has gone well, P is the probability measure 'demanded' by the group.

What is required at the very least is that (*a*) p is a privileged measure on the group; (*b*) P is invariant under the action of the group; and (*c*) P is independent of the choice of x_0. Mathematics allows these desiderata to be satisfied: if the group G has some 'nice' properties, and we require p to be a left Haar measure (which means that $p(S) = p(\{gg' : g' \,\varepsilon\, S\})$ for any part S of G and any member g) then these desired consequences follow, and p, P are essentially unique.[23]

This is a very tight situation, and the required niceties can be expected in geometric models such as are used to define Bertrand's chord problem. But other sorts of models will not be equally nice; and even if they are, different models of the same situation could fairly bring us diverse answers. In any case there is a no a priori reason why all phenomena should fit models with such 'nice' properties only.

8. THE ETHICS OF AMBIGUITY

From the initial example, of a traveller on a treacherous shore, to the partial but impressive successes in the search for invariant priors, I have tried to emphasize how much symmetry considerations tell us. That is the positive side of this definitive dissolution of the idea of unique logical probability. Yet the story is far from complete, and its tactical and strategic suggestions for model construction far from exhausted.

But throughout the history of this subject, there wafts the siren melody of empirical probabilities determined a priori on the basis of pure symmetry considerations. The correct appreciation leads us to exactly the same conclusion as in Chapter 10. Once a problem is modelled, the symmetry requirement may give it a unique, or at least greatly constrained solution. The modelling, however, involves substantive assumptions: an implicit selection of certain parameters as alone relevant, and a tacit assumption of structure in the parameter space. Whenever the consequent limitations are ignored, paradoxes bring us back to our senses—symmetries respected in one modelling of the problem *entail* symmetries broken in another model. As soon as we took the first step, symmetries swept us along in a powerful current—but nature might have demanded a different first step, or embarkation in a different stream.

Facts are ambiguous. It is vain to desire prescience: which resolution of the present ambiguities will later facts vindicate? Our models of the facts, on the other hand, are not ambiguous; they had better not be. To choose one, is therefore a risk. To eliminate the risk is to cease theorizing altogether. That is one message of these paradoxes.

13.
Symmetries of Probability Kinematics

MOST of probability theory is resilient under shifts of interpretation: the same theorems are proved for personal probability as for objective chance. The question of how probabilities change with time, however, is approached quite differently. In quantum theory for example, one can see a veritable dynamics of objective probabilities, which evolve in time, constrained by symmetries which induce conservation laws. This topic I shall leave aside here, to concentrate on rational change of opinion, probabilistically conceived. Here will be found also demonstrations to make good assertions relied on in Part II.

Recent literature has often given the appearance of strong opposition between those who do, and those who do not, look to Simple Conditionalization as the alpha and omega of this subject. I shall argue that this is only superficial appearance. It is indeed true that no admissible rule can rival Conditionalization on its own ground, and also (trivially) true that every rational opinion changer can be simulated by a pure Conditionalizer—but those truths place no severe limit on general probability kinematics, nor answer many of its questions.

I. A GENERAL APPROACH TO OPINION CHANGE

Epistemology is the theory of knowledge and opinion. For empiricism, fascinated with the erosion of certainty in our view of nature, opinion is the more important topic.

Personal probability is one model of opinion. Or rather, to be more accurate, it is the main ingredient in several competing, but related models. I will rely on the introduction to this subject in Chapter 7 for the basic schema of representation. Opinion is expressed in judgements. All such judgements can be expressed as judgements of personal probability, assigned to the factual propositions involved. We represent those propositions by areas on a Venn diagram, and their proportional probability by mud heaped on them. This Muddy Venn Diagram model is a better guide than any axioms or rules for calculation, and it does equally well. (There are in fact deep theorems to show that we have here the most general model of probability theory, provided the mud is so fine as to be continuous; see section 3 below.)

Let us now turn to change of opinion. Here I want to concentrate solely on change in response to experience, and its rationality. (In other words, I leave out at this point deliberation, theoretical innovation, conjecture, conceptual change, or whatever else there be.) *There is a general obstacle*. Suppose I were to write a recipe, or book of recipes, that would tell you how to amend your opinion rationally in view of your experience. The recipe would begin with a description of a state of opinion, and then a description of experience, or of the deliverances of experience, and then prescribe new opinions. Now you could use this recipe to evaluate how well someone else was doing, if you thought you knew what his experience and opinion were. But if you tried to use it on yourself, you would first have to describe your experience. And that is already a response to experience, and would be expressed in terms of a whole new set of judgements that are yours. So you would already have done a great deal of the job that the recipe is meant to guide. You cannot really step out of yourself and compare the representation with what it represents.

I described this problem not because I think there could be a serious use for recipes here, nor because I want to discuss foundationalism of any sort. I wanted to bring out instead the need for a model of experience, if we want to continue the epistemological story. And I do not mean a physiological or psychological model, because the focus of interest is not so much on how things happen, as on the rationality of the response that it must somehow be possible to evaluate. The model must pertain to phenomena presumably reported in such utterances as: 'I saw a flying saucer last year; and ever since I have been a firm believer in reincarnation.'

There are three models. The first was inherited I think from the main tradition. It says that in experience, some proposition E is received as evidence (or, it is taken as evidence). The subject becomes immediately totally certain that E, and adjusts his opinions accordingly. The rule for adjustment is *Simple Conditionalization*, to which I will return below. Perhaps also this model is perfectly suitable to another source for the Bayesian tradition: the working statistician, after all, is paid to accept certain propositions as data, not to question them but to use them as input for his calculations. But as a model for experience it is a bit simple-minded. I call it the *Revelation Model* because in it, experience speaks with the voice of an angel and gives you new total certainties.

A second model was described by Hartry Field in his article about Jeffrey.[1] He takes it as an article of faith that any epistemology must be compatible with materialism. So the input is a physical stimulus; the opinion is in physical storage, and is physically modified. I call this the *Robot Model.* Here the input is not a proposition, obviously, and need have nothing to do with propositions directly. Nor is there any question of an evaluation of the rationality of the response. It may be possible to speculate about general features of the mechanism, perhaps with an eye to survival value.

The third model, which I propose and endorse, also does not take the deliverances of experiences to be propositions, though still intimately connected with them. There is undoubtedly some level of response which we cannot criticize effectively in ourselves except retrospectively, at some later point. Let me call this the primary response. But I do not think that this primary response must already be a judgement, let alone a new total certainty. It is instead, I think, *the acceptance of some constraint* on what your opinion (henceforth) should be. Thus the deliverances of experience are not propositions, but commands (to oneself). A limiting case is possible: the command to become totally certain that *E*. If you accept that as a constraint, what happens next must be what happens next in the revelation model. But many different sorts of constraints are possible—for example, to raise a subjective probability a little, or accept new odds, or a new conditional probability, or indeed anything that could constrain opinion. Because of the crucial roles of the terms 'accept', 'constraint', 'command', it seems natural to call this model a *Voluntarist* one.[2]

Here questions of evaluation can and do arise. Suppose you accept a constraint on your actions by saying you will post a letter for me. I can later evaluate (*a*) whether your actions satisfied the accepted constraint, and (*b*) if so, how well you did. If for example you dropped the letter in the mud and got it to the mail box several days later, you satisfied the constraint, but not optimally so, in some interesting respects. Similarly for adjustments to opinion, made as secondary response, when certain constraints have been accepted.

In the remainder of this chapter I will discuss examples of primary response, and putative rules to govern the corresponding secondary response. Let us begin with the very simplest sort: the limiting case in which the constraint is simply a new certainty. In this case it is indeed as if experience has simply handed us some proposition E on a platter—our 'total new evidence'—and has spoken as if with the voice of an angel.

Suppose I go for a walk in the garden and come away absolutely convinced that a flying saucer has landed there. I reconstruct this as follows: I had originally a certain state of opinion, but accepted the constraint to become certain of this new proposition E, and adjusted my opinion accordingly. There is, as I mentioned, a rule for this adjustment, *Simple Conditionalization*. It is easily explained in terms of the Muddy Venn Diagram (see Fig. 7.1).

You simply wipe away all the mud on the area representing not-E. This has two effects: it raises the probability of E to 1 (for all the mud remaining is on E) and keeps the odds between propositions that entail E the same (for the mud 'inside' E was not disturbed). This is a complete description of the rule. If probability function P is changed in this way we say it is *conditionalized* on E and we call the result P_E or $P(-|E)$.

It is also possible to state this rule in algebraic form (which follows logically from its description in terms of the mud model): $P(X|E) = P(X \text{ and } E)/P(E)$. In either form it is obvious that the rule cannot apply if the prior probability $P(E)$ equals zero. In that case wiping away the mud on (not-E) would remove all mud, thus destroying the model.

So much for the geometric and algebraic descriptions of the rule—*but is it right*? This question of justification is a very fair one. Ian Hacking, writing in 1967, noted that Bayesians took this rule for granted and he called it the *Bayesian Dynamic Assumption*.[3] Textbook presentations tend to darken counsel, as usual, by suggesting that the meaning of 'conditional probability' is 'the probability you would have if you had learned that'. This cannot help because even if it were the meaning (which it is not), that meaning has logically nothing to do with my present odds for (X and E) to (not X and E)—which is the information conveyed by whatever number $P(X \text{ and } E)/P(E)$ *is*.[4] But the rule has an all-but-completely a priori justification, namely, by means of a symmetry argument.

2. SYMMETRY ARGUMENT FOR CONDITIONALIZATION

In Part II I presented the view that if we look for a rule to follow when changing opinion, then Simple Conditionalization is the only rule. Logic—specifically, coherence—requires it. In this section I shall give a proof of that assertion, not through the idea of coherence itself, but by a simple symmetry argument. The proof given in section 4 will be more rigorous, more general and have the present result as corollary. That will require the technical precision to be introduced in section 3, while here the reasoning will be intuitive only.

Three preliminary remarks first. Each rule has a domain of application, which may be more or less wide. Simple Conditionalization is the only admissible rule, *when it is applicable*. This leaves room for other, or more general rules if it is not always applicable—a question we shall take up later. Secondly, this discussion is about *rules*. Whether rationality requires rule-following is a separate question. Thirdly, there is a trivial sense in which Conditionalization is supreme, and this triviality tends to crop up in rhetoric. We had better discuss it first.[5]

Suppose someone never violates the probability calculus in describing his opinions at any one time. Imagine however that he changes his opinion *apparently* by leaps and bounds. Then we can still always claim consistently that his opinion never changes except by Conditionalization. This sounds spectacular, but it is a pure triviality. For we can simply postulate a hidden (unconscious) event $f(t)$ for each time t, which we call this person's (unconscious) insight. Then we can embed his opinion, represented as a function $P(t)$—which is a probability function for each time t—in one with larger domain, call it $P'(t)$ such that $P'(t + 1)$ is the conditionalization of $P'(t)$ on a proposition $F(t)$. The latter proposition is the conjunction of the evidence $E(t)$ which this person consciously acknowledges and the 'insight' proposition $f(t)$ of which he is not aware.[6]

There is a nice construction of $f(t)$ in higher-order probability theory (where propositions may describe the person's own states of opinion). That is interesting, but does not remove the fact that the above claim—'every rational person can be regarded as a Conditionalizer'—is already true on trivial grounds.

What substantive questions remain, after this trivial point is made? We should avert our eyes from the useless *post hoc* question of how all behaviour can be *retrospectively* simulated by machine. That is trivially so. No such trivial result can give us real insight. We should look instead to possible answers to the question posed by the conscious person: how *shall* I conduct myself, what *shall* I take to guide me, as I change my opinion in response to my experience?

Let us first of all narrow this problem so that Conditionalization becomes clearly applicable. This person's opinion is well defined for a certain family of propositions, call it F, which is closed under all logical operations. His opinion at a given time is represented by a probability function P, and the new deliverances of experience are summed up entirely in a proposition E which belongs to that family F. Call P his *prior* opinion and E his new *evidence*. What should be his new, *posterior* probability P', which accords certainty to the evidence E? Abstractly speaking there are many possibilities, that $P'(E) = 1$ can be satisfied in many ways.

So we narrow the problem to: what *rule* could define P', in terms of P and E? Now we have a problem of the typical sort for which a symmetry argument can be constructed.

Is anything tacitly assumed about P and E? We were narrowing the problem so that Conditionalization would become at least applicable—hence we require that $P(E)$ be positive. But apart from that the names F, P, E can stand for *any* probability function with domain F, and E in F. This is a very large class; what are its symmetries?

Structure in this class appears under two headings: logic and probability. There are the relations of logical implication among the propositions; and there are the numbers assigned as probabilities. So a transformation g which is applied to domain F_1 and probability function P_1, and sends them into a new domain F_2 and probability function P_2, is a symmetry if it preserves *that* structure.

The overall argument for Conditionalization must therefore be that this is the only rule which does not violate symmetry. Assume we have some rule, call it R, which yields a posterior probability function $R(P)$ when we specify prior P, its domain F, and the evidence E. An essentially similar problem will be one which has another prior P', domain F', and evidence E', which are connected to the former set-up by some symmetry transformation g. There the rule R will specify posterior $R(P')$. The symmetry requirement is then that also $R(P')$ must be connected to $R(P)$ by the same symmetry transformation. This requirement is just what we described in general at the end of Chapter 10.

To see intuitively how this symmetry argument can be carried through, three simple points need to be appreciated. The first is that if the rule R leaves ratios of probabilities invariant, for propositions which have no mud wiped off (i.e. propositions that imply E), then clearly R is just Conditionalization. For after all the posterior $R(P)$ must give 1 to E, so if A is any other proposition $R(P)$ assigns to A the same as to $A \cap E$. And secondly, the ratio between that number and what $R(P)$ assigns to E—i.e. 1—equals then the ratio $P(A \cap E)/P(E)$.

The second point is that a symmetry can be an embedding, that is, it can relate a probability function defined on a small domain of propositions to one defined on a large domain. If we think about this momentarily in terms of language, suppose that one person's language does not include the sentence 'Snow is white' while another's does. Then the second person's probabilities can still be the same as the first with respect to all propositions that both understand. This point becomes mathematically important when we think of how much mathematics deals with continua. That a rule should have essentially the same effect when we embed a problem set-up into a similar but continuous one, is very informative.

The third point is that the symmetry requirement explained entails that, in intuitive terms, the rule will depend solely on two parameters: the prior probability function and the evidence proposition. From this we can deduce that there is a functional relationship between the prior and posterior probabilities assigned to propositions which imply the evidence. The argument for this, as well as the argument that ties all these threads together will be given in section 4. But for now I'll just add this: what ties them together is a lemma that in the continuous cases (see point 2), the functionality (point 3) implies the invariance of probability ratios (point 1). This lemma is due to Paul Teller and Arthur Fine (see n. 12 below).

3. PROBABILITY AS MEASURE: THE HISTORY

It is time to become very precise. Before we go on, we must examine the mathematical foundations of probability theory. To have a well-defined probability, you must specify to what object the probability is assigned. This may be a family of events that may or may not occur, of propositions that may or may not be true, or of (purported) facts which may or may not be the case. The family must have at least the simple sort of structure that allows representation by means of sets.

The Muddy Venn Diagram described in Chapter 7 is the perfect intuitive guide to probability. From that model we can immediately derive (where Λ is the empty set):

I. $P(\Lambda) = 0 \quad P(K) = 1 \quad 0 \leqslant P(A) \leqslant 1$
II. $P(A \cap B) + P(A \cup B) = P(A) + P(B)$

To understand probability properly, though, one needs to know why it is such a good guide. I will first state the formal theory in its current form, and then describe how it came to have this form. After that I will state the results that convey such a privileged status on the muddy diagram models. It will be seen that the geometric probabilities (which Buffon rightly claimed to have introduced) are still, also in the abstract expressionism of our day, the core of the subject.

Let us distinguish between a *probability function* and a *probability measure*. The first is the subject of probability theory when we do not impose any continuity requirements or other concerns about infinity. Bruno De Finetti insisted this should remain the complete subject, and we will look at his reasons. In any case, a probability measure is a special kind of probability function.

A *field* of sets on a set K is a class F of subsets of K such that:
K and Λ are in F
If A, B are in F, so are $A \cap B$, $A \cup B$, $A - B$
A field F on K is a *Borel field* or *sigma-field* on K if in addition F contains the union of any countable class $A_1, \ldots, A_n, \ldots$ of its members.

It follows automatically that if $A_1, \ldots, A_n, \ldots$ are in a Borel field, so is their intersection.

P is a *probability function* on set K, defined in field F, exactly if

1. $P(A) \in [0, 1]$ for each member A of F
2. $P(\Lambda) = 0$, $P(K) = 1$
3. $P(A \cap B) + P(A \cup B) = P(A) + P(B)$

for any members A, B of F

P is a *probability measure* on K, defined on F, exactly if F is a Borel field and

4. $P(A) = \lim_{n\to\infty} P(A_i)$ if

A is the union of the series $A_1 \subseteq \ldots \subseteq A_n \subseteq \ldots$ of members of F.

Here 3 and 4 have equivalents some of which may look more familiar:

3*a*. $P(A \cup B) = P(A) + P(B)$ if $A \cap B = \Lambda$
4*a*. $P(\cup A_i) = \Sigma P(A_i)$ if $A_i \cap A_j = \Lambda$ for all $i \neq j$
4*b*. $P(\cap A_i) = \lim_{n\to\infty} P(A_i)$ if $A_1 \supseteq \ldots \supseteq A_n \supseteq \ldots$.

The property described by 3a is *finite additivity* and that described by 4*a* is *sigma-additivity* or *countable additivity*. It is clear from formulations 4 and 4*b* that the additional property that makes a probability function a measure is that it satisfies a continuity requirement.

So defined, probability theory is a part of measure theory. For in mathematics, a *measure* is a function defined exactly like a probability measure except that it need not have 1 as upper bound; indeed some sets may have infinite measure. Thus the condition $P(K) = 1$ is omitted, and 1 is replaced by

1′. $P(A) \in [0, \infty]$

If 4 is not satisfied, I shall call the function a *measure function*.

By these definitions, probability measures are probability functions which are also measures; and all three types are also kinds of measure functions. My use of 'measure' and 'probability measure' are standard, but the rest is a bit of verbal regimentation. I will describe a little of the history that introduced these subjects, in part to answer the question which must surely have occurred to you: why not have the probabilities defined for *every* subset of the total set K? Why this fiddling around with fields and such?

Measure theory began with some rather tentative and sceptical attempts to use Cantor's set theory in analysis.[7] These surface in the second edition of Camille Jordan's *Cours d'analyse* (Paris: Gauthier-Villars, 1893), where only finite additivity is noted as a defining requirement for measure. Countable additivity is made part of the definition in Émile Borel's monograph *Leçons sur la théorie des fonctions* (Paris: Gauthier-Villars, 1898). The measure which Borel defined on the unit interval [0, 1]—which we now call Lebesgue measure—is not defined for all sets of real numbers in that interval. The definition runs as follows: the measure of an interval of length s has measure s; a countable union of disjoint sets with measures $s_1,..., s_n,...$ has measure Σs_i; and if $E \supseteq E'$ have measures s_1 and s_2 then $E' - E$ has measure $s_2 - s_1$. The sets encompassed by these clauses he called *measurable*. What Borel calls the measurable subsets of [0, 1] we now call the *Borel sets* on that interval. It is clear that they form a Borel field, so here we see the origin of our terminology.

It was Henri Lebesgue who posed the explicit problem of defining a measure on all the subsets. That some such measures exist is trivial: assign 1 to any subset of [0, 1] that includes the number zero, and assign 0 to every other subset. Then you have a probability measure, but a trivial one. The question is not whether it can be done at all, but whether the requirement of including every subset in the domain would eliminate important or interesting functions. Here is the passage in which Lebesgue introduces his *Measure Problem*:

> We propose to assign to each set [in n-dimensional space] a non-negative number, which we shall call its measure, satisfying the following conditions:
> (i) There is a set whose measure is not zero.
> (ii) Congruent sets have equal measure.
> (iii) The measure of the sum [union] of a finite or denumerable infinity of disjoint sets is the sum of the measures of these sets. (Lebesque (1902), p. 236)

Congruence is the relation between sets which can be transformed into each other by symmetries of the space. In the Euclidean case, one of the symmetry transformations is translation, and so we see at once that the measure asked for is not a probability measure. For suppose a given cube has measure 1. By translation we turn it into infinitely many disjoint cubes congruent to it. The measure of their union—and hence of the whole space—is therefore ∞. But more than this we can also see that the measure Lebesgue calls for is already uniquely determined on all the Borel sets. In an n-dimensional space, not the intervals but the generalized rectangles—such as $\{(x_1,..., x_n): a_1 \leqslant x_1 \leqslant b_1,..., a_n \leqslant x_n \leqslant b_n\}$—are the beginning. The class of Borel sets is the smallest class that contains all these and is closed under countable union and set-difference. All the Borel sets can be approximated by choosing as beginning instead all the generalized rectangles congruent to a single very small cube R. The approximation gets progressively better as we take R smaller and smaller. But it is obvious that if the unit cube has measure 1, it can be divided into n disjoint cubes of dimensions $1/n^3$ which must (by finite additivity) all receive measure $1/n^3$. Thus the measure of all little cubes is uniquely determined, and hence by continuity, the measure of all the Borel sets.

It is well worth emphasizing this little point: *only* this natural generalization of the usual (length-area-volume) measure is invariant under translation. This measure, so defined on the Borel sets, we now call *Lebesgue measure*. The problem Lebesgue posed, and did not solve, in his dissertation, is whether this measure can be extended to all the other subsets of the space as well.

Enter the Axiom of Choice. Borel was one of its staunchest and most vocal opponents. Lebesgue was more moderate in his philosophical opposition, but rejected it as well. This was in the years 1904–5. In his own proofs, Lebesgue had apparently tacitly relied on the Axiom of Choice. But if the Axiom of Choice is true, it turns out that Lebesgue's 'Measure Problem' has no solution.

In 1905 Giuseppe Vitali, already relying on the Axiom of Choice, constructed a set not measurable in Lebesgue's sense. He used this to show that Lebesgue's Measure Problem has no solution for the real line. Lebesgue himself gave a further such example, but expressed doubts about the Axiom. Felix Hausdorff, who accepted the Axiom, used it to show in addition that even if the requirement of countable additivity is weakened to that of finite additivity, the Measure Problem has no solution for Euclidean spaces of dimension greater than 2.

Vitali had used the requirement of translation invariance which, as we saw, entails that the whole space has infinite measure. However Hausdorff's proof that the Measure Problem has no solution for Euclidean spaces of dimension greater than 2, used rotational invariance. He discussed measure on a sphere and showed that a measure defined on every subset could not be rotation invariant. This goes just as well for a measure which assigns 1 to the sphere itself, and thus tells us at once there cannot be such a probability measure on the sphere.[8]

It will now be quite clear, therefore, that the requirement to have probability defined everywhere, would be unacceptable. We must accept as genuine probability measures also those which cannot be extended to measures on all subsets of their domain. Fields and Borel fields are their natural habitat.

Are geometric probabilities really as logically central as this story made it look? The answer is that the geometry-oriented intuitions of Lebesgue, Borel, and Hausdorff were quite right. This is what I meant when I said that the Muddy Venn Diagram is in a certain sense a general model.

The results that establish this are the deepest in the foundations of probability theory, for they lead us to a classification of all possible probability measures, and an understanding of all of them in terms of Lebesgue measure.[9]

Therefore the focus will be on the paradigm probability space, consisting of the unit interval [0, 1] of real numbers, the family B_0 of Borel sets on this interval, and Lebesgue measure m defined on these Borel sets. The 'points' are the real numbers $0 \leqslant x \leqslant 1$. These are funny things: in some sense—namely by uncountable union—the unit sets $\{x\}$ generate the space, but each of them has measure zero, so their measures do not add up to the measure 1 of the whole space. We switch therefore to a more algebraic viewpoint, by identifying sets which differ only by measure zero. To be precise:

> Let $S = \langle K, F, P \rangle$ be a probability space with P a probability measure defined on the Borel field F of subsets of K. Call A and B *equivalent* if $P((A - B) \cup (B - A)) = 0$ and define the quotient F/P to be the following algebra:
>
> its elements are the sets $[A] = \{B\colon B \varepsilon F$ and B is equivalent to $A\}$, for A in F
>
> $[A] \wedge [B] = [A \cap B]$, $[A] - [B] = [A - B]$
>
> $[A] \leqslant [B]$ exactly if $A \subseteq C$ for some C in $[B]$, $p([A]) = P(A)$
>
> Then $\langle F/P, p \rangle$ is a *probability algebra* (the probability algebra generated by measure P on Borel field F).

The special case in which S is the paradigmatic probability space, yields in this way the probability algebra B_0-m with Lebesgue measure transposed to it as in the above definition.

Define an *atom* of such an algebra to be an element y such that there is nothing between it and the zero element $0 = [\Lambda]$. That is:

y is an *atom* exactly if for any x, if $0 \leqslant x \leqslant y$ then $x = 0$ or $x = y$

Clearly 0 is the *only* element that receives zero measure! For all sets A such that $P(A) = 0$ belong to $[\Lambda]$. So these atoms all have positive measures. We also see that the atoms cannot overlap: if y and z are distinct atoms, $y \wedge z = 0$. So we deduce at once that there can only be at most countably many atoms. For suppose we have any family of events with positive measure. There are at most 2 with measure $\geqslant \frac{1}{2}$ because their sum cannot be greater than 1. Similarly there are at most 3 with measure $\geqslant \frac{1}{3}$, at most 4 with measure $\geqslant \frac{1}{4}$, and so forth. But every one of them has a measure $\geqslant 1/n$ for some n, so there are at most $2 + 3 + 4 + \ldots$ of them, which is countable infinity. Therefore we now distinguish three sorts of probability algebra:

(1) the algebra has no atoms (*atomless*)
(2) the algebra has finitely or countably many atoms, whose measures add up to 1 (*atomistic*)
(3) the algebra has finitely or countably many atoms, whose measure add up to some number $x < 1$ (*mixed*)

The representation theory for mixed cases is of course complicated, but consists in showing that they are all constructions out of pure (atomless and atomistic) cases.

The second, atomistic case is quite easily represented in this algebra by a Muddy Venn Diagram: divide the square into countably many distinct parts, each of which represents one atom, and put a mass of mud on each proportional to the measure of that atom. In terms of our paradigmatic probability space we can do the same thing: to represent the algebra $\langle F/P, p \rangle$ divide the unit interval into a countable partition $\{A(y)$: y an atom of $F/P\}$. Do it so that each set $A(y)$ has positive Lebesgue measure. Define the function m': $m'(A(y)) = p(y)$. This function m' represents the probability. Of course, it is itself the beginning so to say of a probability measure. We can extend m' to the whole Borel field just by insisting on countable additivity. Obviously then we have reached our goal here: in the atomistic case, the probability can be represented by means of a measure defined in terms of Lebesgue measure.

The paradigmatic probability algebra B_0/m is itself an example of the atomless case. For suppose A is a Borel subset of [0, 1] and $m(A) = x > 0$. Could $[A]$ be an atom? No, because there will be a smaller Borel set B with $m(B) = x/2$; and so forth. It will not come as a surprise that in set theory it is possible to find probability spaces whose Borel fields have cardinalities incredibly much higher than the relatively small infinities we deal with in the case of real numbers or Euclidean spaces. But what about the algebras they give rise to?

On the probability algebra $\langle F/P, p\rangle$ we can define a metric: $d(x, y) = p(x \vee y) - p(x \wedge y)$, where $\vee$ is defined from $\wedge$ by DeMorgan's law. Thus we can apply metric concepts, and we have the theorem:

> (Birkhoff) Any atomless probability algebra which has a countable dense subset is isometrically isomorphic to the probability algebra generated by Lebesgue measure on the unit interval.

I shall end this exposition here; it will be clear now that even from a strictly logical point of view, probability theory is a subject which stays close to the earth of its geometric history.

4. SYMMETRY: AN ARGUMENT FOR JEFFREY CONDITIONALIZATION

In 1965, Richard Jeffrey created the new subject of *probability kinematics*. He did this by describing different sorts of changes in opinion, asking for a theoretical description, and proposing a rule that generalizes Simple Conditionalization. This new rule has since been generally known as *Jeffrey Conditionalization.* It has simple Conditionalization as a special case (and the result of this section has that in section 2 as corollary).

Here is an example: I walk through a room, in which I glimpse roses on a table lit by candlelight. I had expected the scene, had a prior opinion that the roses would, as likely as not, be red. Now I am more inclined to think they are red . . . it now seems twice as likely to me as not that they are red. How should I adjust my total state of opinion 'accordingly'?

A bit later or at roughly the same time as Jeffrey was thinking about this, I have been told, Wade Edwards and his collaborators were discussing a prima-facie different problem. The rule they came up with is formally the same as Jeffrey's. Their problem was this: we have a spy in Iran, and he sends us regular reports, but he is non-reliable, he tends to lie a little. Reading his reports, what constraints shall we place on our posterior opinion? There is an obvious analogy with Jeffrey's case, a suggestion that we think of our senses as Plato did, as lying spies in the garden of earthly delights. Perhaps my eyes were saying 'red' when I walked through that room, but they did not speak with the voice of an angel, and my response was only to become more inclined toward, *not* certain of, the proposition they spoke.

Let us restate the problem in an even more schizophrenic mode. There are two spies, Cain and Abel; they are twin brothers, and their reports are always each other's negation. But we consider Cain twice as likely to speak the truth as Abel. Now we can say this: if I accept Cain's reports totally, I shall conditionalize on them; if I accept Abel's totally, I shall conditionalize on the negations of Cain's.

Neither of these brothers speaking like an angel, I shall make my own opinion a proportional mixture of theirs (see Fig. 13.1).

This idea of a *mixture* of two states of opinion works perfectly in the alternative epistemology. In the dogmatist oversimplification, the best one could do would be to take what is implied by both Cain's and Abel's reports—i.e. tautologies only.

This is the complete description of the rule of Jeffrey Conditionalization. Can we justify this rule?

We can regard it as concerning a special case, among the following possible constraints on the posterior probability:

C_1. the new probability for B equals 1
C_2. the new probability for B equals 0.7
C_3. the new conditional probability for A given B equals 0.6
C_4. the new odds of A to B equal 6:3
C_5. the new expectation value of parameter x equals 4.5.

(Expectation is probability mean value: $\Sigma\, rP(x \text{ has value } r)$ with the sum over all possible values of r.)

Each of these sorts of constraint requires that the probabilities—initial and final—be defined at least on the events mentioned (B

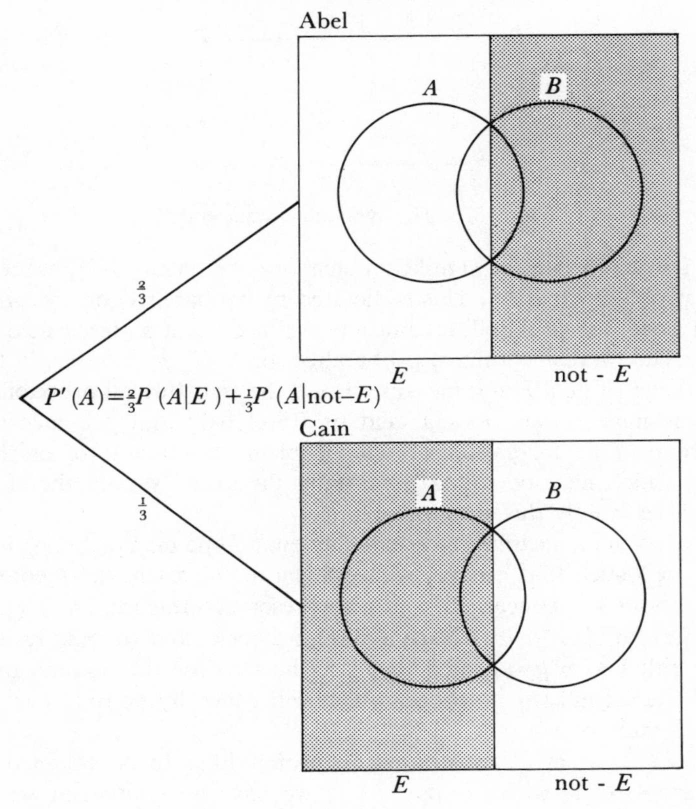

FIG. 13.1. The lying spies

for C_1 and C_2, A and B for C_3 and C_4 and the events [x has value r] for C_5). Thus if we consider the problem starting with initial probability P_1, and transform it into a different, essentially similar, problem, with initial probability P_2, we have here already one aspect which must be left intact. For the problem to make sense, P_1 had to be defined on certain events—the corresponding P_2 must be defined on corresponding events, and subject to a corresponding constraint. And of course, our solution must transform the corresponding *priors* into corresponding *posteriors* (final probability functions). We will have the usual closed diagram, as in Fig. 13.2. We must now implement these intuitive ideas about symmetry requirements, in the formal framework established in the preceding section.[10]

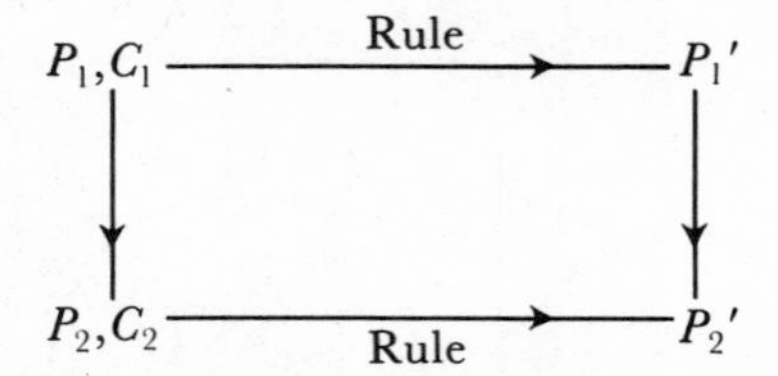

FIG. 13.2. The symmetry requirement

First we isolate the significant structure which can be present in the problem situation. This is dictated by probability theory: prior and posterior probabilities must be defined on a suitable field of sets. The total structure of probability space $\langle F, p\rangle$ consists in the character of field F and measure p, and the structure all probability spaces have is that F is a field or Borel field and p a measure thereon. Transformation of one problem situation into another one, which may be called 'essentially the same' should therefore preserve exactly this structure.

Let us call *a measure embedding* any one-to-one map of $\langle F, p\rangle$ into $\langle F', p'\rangle$ such that g is an isomorphism as far as the set-theoretic operations are concerned, and also preserves measure, i.e. $p'(gA) = p(A)$ for all A in F. Clearly g has an inverse, and we may restate this either as $p'g = p$ or $p' = pg^{-1}$ (the first for the domain of g and the second for its range, which will generally be only part of the domain of p').

If CC is a set of constraints on probabilities to be assigned to elements of X, which is part of F, we have an equivalent set of constraints CC' imposed on the images $g(A){:}A \varepsilon X$, applicable to probabilities defined on the target field F'. The requirement upon our general solution is therefore that cc', the function we present for imposing CC' on priors defined on F', should be related to cc in the following way:

> Symmetry: If g is a measure embedding of $\langle F, p\rangle$ into $\langle F', p'\rangle$ then $ccp(A) = cc'p'(gA)$ for all A in F.

In other words, when p is just $p'g$ then ccp should be $cc'p'g$.

The practical effect of following this principle is that, when we try to identify cc, we are always allowed to switch our attention to a more tractable 'equivalent' probability space (either the domain or the range of a measure embedding relating it to the one we had).

The constraints CC are stated as conditions on the posterior probabilities to be assigned to a certain set X of elements of F. Without loss of generality, we can take X to be a subfield; alternatively so long as X is countable we can, without loss, take it to be a partition. I shall henceforth assume the latter.

As a working hypothesis, let us assume cc to exist, and when applied to p, to give the posterior probabilities $ccp(B) = r_B$ to members B of partition X.

It is clear that $p'(A) = \Sigma \{p'(A \cap B)\colon B \varepsilon X\}$ for any function p', so we will have identified ccp if we can determine what values it gives to subsets of members B of X. Therefore we begin with:

Lemma 1: Let E, E' be subsets of member B of X and $p(E) = p(E')$. Then $ccp(E) = ccp(E')$.

(See *Proofs and illustrations*)

Because of this lemma, we now know that for a subset E of a member of the partition, the sole relevant factor is its prior probability. So there exists for a given member B of X a function f such that $ccp(E) = f(p(E))$ when $E \supseteq B$. What is this function like? Here it is more convenient to embed our problem in a context where real analysis can apply.

A probability space $\langle F, p\rangle$ is *full* when for each element A of F, p takes every value in $[0, p(A)]$ on the elements of F which are subsets of A.

Lemma 2: There exists a measure embedding * of $\langle F, p\rangle$ in a full space $\langle F^*, p^*\rangle$.

This embedding is easily constructed with $\langle F^*, p^*\rangle$ the product of $\langle F, p\rangle$ with $\langle [0, 1], m\rangle$ where m is Lebesque measure. Then F^* is the family $\{AxQ : A \text{ in } F \text{ in } F \text{ and } Q \text{ a Borel set on the unit interval}\}$, $p^*(AxQ) = p(A)m(Q)$, and $A^* = Ax[0,1]$.

To continue the main argument, we now apply Lemma 1 to our thinking about this full space (see *Proofs and illustrations*), and arrive at:

Symmetry Theorem: If there exists a function cc corresponding to constraints CC on the posterior probabilities for members of partition X, then for each probability function p to which cc is applicable and such that p is positive on all members of X, we have

$$ccp(-\,|B) = p(-\,|B) \text{ for all } B \text{ in } X$$

It was assumed of course that any such function must satisfy the *Symmetry* principle we had above. As an immediate corollary we have the theorem:

1st Corollary (Simple Conditionalization Rule). If $p(B) \neq 0$ and CC is the constraint that the posterior probability for B equal one, then $ccp(-) = p(- | B)$

Richard Jeffrey proposed for the constraint that the posterior probability for B equal r, the rule $p'(-) = rp(- | B) + (1 - r)p(- | B)$. We deduce similarly the uniqueness of his rule for this constraint, in the general form:

2nd Corollary (Jeffrey Conditionalization Rule): If X is a countable partition with $p(B) \geqslant 0$ for all B in X, and CC the constraint than the posterior probability for B equal r_B, for all B in X, then $ccp(-) = \Sigma\{r_B p(- | B) : B \varepsilon X\}$

In these cases therefore, our first theorem already singles out a unique way of imposing the constraint.

We may sum up our results as follows: whatever the constraints CC (stated with reference to partition X) are, the effect of the function cc is equivalent to an operation that determines posterior probabilities on the partition X, *followed by* the operation of Jeffrey Conditionalization with those posterior probabilities. The task that remains is to investigate the first operation with greater generality.

There are no equally satisfying results that go beyond this point. Jaynes and his collaborators have explored the rule to minimize relative information (equivalently, to maximize relative entropy) as a solution to this general problem. Of course that rule agrees with the Simple and Jeffrey Conditionalization rules—otherwise it would violate the basic symmetries of the general problem. But I shall now leave this topic, except for references, and a brief discussion of Jaynes's rule below.[11]

Proofs and illustrations

To prove Lemma 1, we look at the subfield of F generated by $X \cup \{E, E'\}$; call it F_o. Let p_o be p restricted to F_o. The identity function is then a measure embedding of $\langle F_o, p_o \rangle$ into $\langle F, p \rangle$. We now construct a measure embedding of the former onto itself as follows:

$$g(E - E') = E' - E$$

$g(E' - E) = E - E'$
g is the identity function on $X \cup \{E \cap E', B - (E \cup E')\}$
$g(A \cup A') = g(A) \cup g(A')$ when A, A' are disjoint.

A Venn diagram suffices to depict this subfield and this automorphism; because $p(E) = p(E')$ it follows that $p(E - E') = p(E' - E)$ so it is a measure embedding.

We can now apply Symmetry to this measure embedding g of $\langle F_o, p_o \rangle$ onto itself to deduce that $cc_o p_o(E) = cc_o p_o(gE) = cc_o p_o(E')$. But secondly, the identity map is a measure embedding of $\langle F_o, p_o \rangle$ into $\langle F, p \rangle$ and so we also deduce that $ccp(E') = cc_o p_o(E') = cc_o p_o(E) = ccp(E)$ as required.

To prove the *Symmetry Theorem*, we apply Lemma 1 to the full space described in Lemma 2. In that full space $\langle F^*, p^* \rangle$, we see now that for any element B^* there exists a numerical function f such that $cc^*p^*(E) = f(p^*(E))$ when $E \subseteq B^*$. There the function is perceived to be a map of $[0, p(B)] = [0, p^*(B^*)]$ onto $[0, r_B]$. It must be additive, for if E, E' are disjoint parts of B^* we have

$$\begin{aligned} f(p^*(E)) + f(p^*(E')) &= cc^*p^*(E) + cc^*p^*(E') \\ &= cc^*p^*(E \cup E') \\ &= f(p^*(E \cup E')) \\ &= f(p^*(E) + p^*(E')) \end{aligned}$$

and because the space is full we have disjoint parts E, E' with probabilities r, s respectively whenever $r + s \leqslant p(B)$.

A theorem of calculus[12] implies that such an additive function has a constant derivative, thus

$$f(x) = kx + m$$

Looking at $x = p(\Lambda)$ and $x = p(B)$ respectively we deduce that $f(0) = 0$ so $m = 0$, and $k = r_B/p(B)$

$$f(x) = r_B x/p(B).$$

Hence by *Symmetry* the function cc, like cc^*, is given by the equation:

$$\begin{aligned} ccp(E) &= r_B \frac{p(E)}{p(B)} \text{ for } E \subseteq B \in X \\ &= r_B p(E|B). \end{aligned}$$

5. LEVI'S OBJECTION: A SIMULATED HORSE-RACE

Also writing in 1967, in a review of Jeffrey's book, Isaac Levi raised certain objections to the idea of probability kinematics. One of these has variants for all continuations of the programme as well. Surely one's opinion, formed in response to a succession of experiences, should not depend on the *order* in which these experiences occur? That is obvious if the experience fits the Revelation Model. Simple Conditionalization successively on E_1 and E_2 is the same as doing it on (E_1 and E_2)—so there order does not matter. But Jeffrey Conditionalizing, with successive constraints [$E_1 = 0.7$]!, [$E_2 = 0.8$]! for example, clearly does depend on the order. (Think of the limiting case $1 = 2$.) And so it should; but if we have here a correct description of adjustment of opinion in response to relevant experience, we have a bit of a paradox. Two persons, who have the same relevant experiences on the same day, but in a different order, will not agree in the evening even if they had exactly the same opinions in the morning. Does this not make nonsense of the idea of learning from experience?

I shall approach this challenge in two ways. First, let us ask whether experience can contradict itself? Surely not: what my experience is at two different times, is a priori unrestricted—what happens to me is up to nature, so to say, and there are no a priori bounds on nature. (This is an empiricist speaking.) But what about responses to successive experiences? Do we have any reason to suspect irrationality if those are incompatible? I think not; else why ever look twice? It is along these lines that I understand Jeffrey's own response. That is that a person who wishes to raise the probability of E_1 to 0.7 and a bit later that of E_2 to 0.8, has the choice of imposing the joint constraint [$E_1 = 0.7$ and $E_2 = 0.8$]! at the later time, provided the two constraints are logically compatible. And if they are not compatible, he clearly *wants* to discard the previous judgement.

This response has two problems. First, Jeffrey's rule does not tell you how to impose the joint constraint. This requires a rule going beyond his; and such rules raise questions of their own, which I shall discuss below. The second problem is that the incompatibility is not a matter of simply discarding past judgements. Suppose E_1 and E_2 are mutually incompatible. Then they cannot have 0.7 and 0.8 at the same time; and which you impose later makes a difference to what everything gets.

TABLE 13.1. *Successive Jeffrey Conditionalizations*

	E_1	E_2	not (E_1 or E_2)	
P	0.1	0.8	0.1	
				impose [$E_1 = 0.7$]!
P'	0.7	$\frac{8}{9}$	$\frac{1}{9}$	
				impose [$E_2 = 0.8$]!
P''	$\frac{21}{22} \cdot \frac{2}{10}$	0.8	$\frac{1}{22} \cdot \frac{2}{10}$	

Table 13.1 shows clearly that going back to the original value for E_2, for instance, does not get you back to the original state of opinion. If the constraints had been imposed in opposite order, the first would have effected no change at all, and P' would have been the end result.

So we need a more sophisticated approach. I think we should momentarily forget the technicalities, and ask: What does happen to learning from experience? Has this become impossible? Suppose a person *learns*, in the sense that he successively accepts constraints, each of which correctly represent the real, objective probabilities. Will his opinion become more correct, or not?

This question could be investigated experimentally, to some extent. I decided to simulate an experiment on a computer. A man wishes to bet on a horse race; on each day preceding the race he is allowed to study the horses. Suppose that on each day D_n he updates his probability that a certain horse H_n will win to P_n, using Jeffrey's rule. Suppose also that the correct or objective probabilities are indeed $P_1, \ldots, P_n, \ldots$ Should we expect his opinions on the day of the race to be closer to the correct probabilities than they were initially? For simplicity I used three horses, and asked the question in this form: how many days does it take for his subjective probabilities to come within a preset amount of the objective ones? Table 13.2 shows some results (stated in terms of odds rather than probabilities).

TABLE 13.2. *The horse-race: learning by Jeffrey Conditionlization*

	Prior	Target	Degree of Approximation			
			1/100	$1/10^4$	$1/10^8$	$10/10^{10}$
(*A*)	1, 1, 1	1, 10^3, 10^6	2	2	2	3
(*A*)	1, 1, 1	1, 10, 100	2	4	6	8
(*C*)	1, 1, 1	1, 1, 20	2	5	8	11
(*D*)	3, 3, 4	1, 2, 7	3	5	11	13
(*E*)	7, 2, 1	1, 2, 7	5	7	11	14
(*F*)	5, 6, 7	6, 7, 8	1	5	11	15
(*G*)	5, 6, 7	7, 8, 9	1	5	12	15
(*H*)	1, 5, 25	1, 2, 4	3	6	11	17

There is no doubt that this person is learning, and quickly. He gets to within 1 per cent of the right probabilities in a few days, and to within an astronomical degree of approximation within about two weeks. Not only that, there is a definite convergence, and it is not true that successive experiences 'undo' the preceding learning process to a radical extent. Of course I assumed that the *input*—the imposed constraints—were exactly correct. But while such acute perceptions may be lacking in reality, any lack of overall success will not be due to the rule being followed.[13]

6. GENERAL PROBABILITY KINEMATICS AND ENTROPY

In 1965 Jeffrey had formulated a simple, new problem: in the roses-by-candelight case, where Simply Conditionalization is not applicable, what should one do? And as we saw, his proffered solution has a very special status, for it can be justified a priori by a symmetry argument. There may be cases in which this result still gives us no guidance. This can be for one of two reasons:

(*a*) We may want our posterior opinion to depend on other factors besides the prior opinion and the given, simple constraint

(*b*) We may wish to impose another constraint, of more complex form.

Reason (*a*) is always with us, and rightly keeps us from conceiving epistemology as in principle a special sort of arithmetic or other mechanical procedure. But we should also consider the case—presumably 'normal' in the sense that deliberation and theoretical innovation and creativity must be the exception rather than the norm in our daily life—in which reason (*a*) is absent.

This brings us to the problem of *general probability kinematics*: find rules that transform priors, subject to given constraints, and for which the prior and constraint are the sole relevant factors. The symmetry argument of section 4 really demonstrated the following general result. Suppose that we are given constraints on the posterior probabilities $P'(A_1),..., P'(A_n)$ defined with reference to a partition $(A_1,..., A_n)$ of the space of possibilities. Then any rule that transforms a prior P into such a posterior P' must in effect proceed in two steps:

(i) Assign exact new probabilities to $A_1,..., A_n$
(ii) Treat the outcome of step (i) as input for a Jeffrey Conditionalization.

This result is very helpful, because it means that if we are given a constraint of a new and strange sort, like:

> Change your opinion so that the probabilities for rain and snow become equal.

we know that we need not worry about finer subdivisions like (rain and cereal for breakfast), (rain and scrambled eggs for breakfast), etc. For step (ii) must always be carried out in the same way. With this in hand, let us begin to investigate constraints of more sophisticated form.

If we want to go on piecemeal, here are three special instances of the general problem, which we have not yet covered:

Impose constraints of form:

(*a*) change simultaneously both the probabilities of A and of B, to 0.3 and 0.7 respectively,
(*b*) change the odds of A to B, to 3:1,
(*c*) change the expectation values of quantities $X, Y, Z,...$, to 4.1, 4.2, 4.3,... respectively.

Let us look at each of these in turn. In (*a*), if $A = \bar{B}$ we have the special case of Jeffrey Conditionalization. But the general case with A distinct from B was not handled just because, as we saw above, the order of successive Jeffrey Conditionalizations matters.

In (*b*) there is also a special case which is at least somewhat more tractable: when *A* implies *B*. Then the odds is a ratio of type $P(A \cap B)/P(B)$, and so the constraint is to change a conditional probability. The problem of finding a rule for this case I have elsewhere called the Judy Benjamin problem.[14] It derives from the movie *Private Benjamin*, in which Goldie Hawn, playing the title character, joins the army. She and her platoon, participating in war games on the side of the 'Blue Army', are dropped in the wilderness, to scout the opposition ('Red Army'). They are soon lost. Leaving the movie script now, suppose the area to be divided in two halves, Blue and Red territory, while each territory is divided into Headquarters Company area and Second Company area. They were dropped more or less at the centre, and therefore feel it is equally likely that they are now located in one area as in another. This gives us the Muddy Venn diagram, drawn as a map of the area, shown in Fig. 13.3. They have some difficulty contacting their own HQ by radio, but finally succeed and describe what they can see around them. After a while, the officer at HQ radios: 'I can't be sure where you are. If you are in Red Territory, the odds are 3:1 that you are in HQ Coy. area...' At this point the radio gives out. As in the movie, the platoon goes on to capture enemy headquarters.

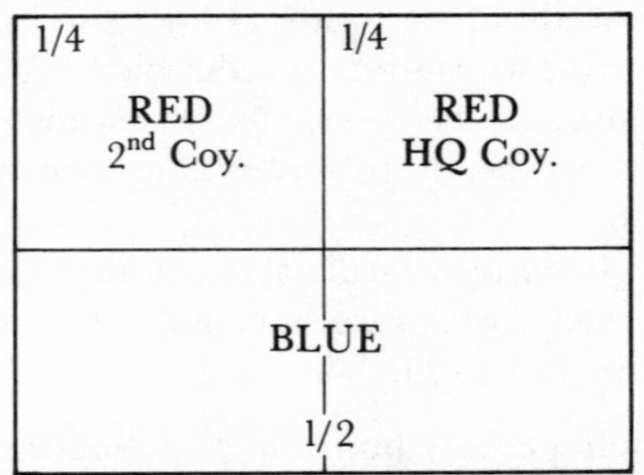

FIG. 13.3. The Judy Benjamin problem

We must now consider how Judy Benjamin should adjust her opinions, if she accepts this radio message as the correct and sole constraint to impose. The question on which we should focus is: what does it do to the probability that they are in friendly Blue Territory? Does it increase, or decrease, or stay at its present level of $\frac{1}{2}$? I originally proposed this question in a seminar in the autumn of 1980, and have raised it in various seminars, classes, and conference audiences since. The 'intuitive' or unreflective response has always been, overwhelmingly, that after the radio message about Red Territory, it should seem neither more nor less likely to Judy Benjamin that she is still in Blue.

Peter Williams brought out the main flaw in that sort of response. Suppose the radio officer had not said '3:1' but, in more general form:

> The probability that you are in HQ area, given that you are in Red Territory, equals q

for some value q—$\frac{3}{4}$ in our other example—which must lie between 0 and 1. Our task is to find a general rule to cover all. But if the radio officer had said '1'—if q had equalled 1—he would have told her, in effect, 'You are *not* in Red-2 Coy. area'. In that case she could have used Simple Conditionalization, and wiped the mud off the top-left square in the diagram. This would have the result of giving Blue Territory $\frac{2}{3}$ of the remaining mud. Hence the probability of Blue would go up; and we cannot make it a general rule that it stays the same.

In the cited papers, three distinct rules are found which appear to satisfy all symmetry requirements, as well as some others, equally well. This raises the possibility that the uniqueness result for Jeffrey Conditionalization will not extend to more broadly applicable rules in general probability kinematics. In that case rationality will not dictate epistemic procedure even when we decide that it shall be rule governed.

Problem (*c*) is the most general we have touched on yet. Indeed, (*a*) and (*b*) are special cases of (*c*). So a rule that handles (*c*) handles all the problems we have come across so far.

Does this mean we should simply concentrate on (*c*), i.e. Expectation constraints? There is a good deal of literature on this in connection with the rule *Infomin*, which I shall explain below; it is indeed applicable to such constraints. The physicist Edwin T. Jaynes proposed this rule in statistical mechanics, and had considerable success with it. The rule has also had successful engineering applications in data analysis, specifically photographic image enhancement. Lately there have appeared new deductions of the rule from much weaker premisses, including interesting symmetry arguments. All of this suggests an affirmative answer to our question.[15]

But the prominence of *Infomin* is due in great part to its being the only rule that is known and has been investigated at length, which can handle expectation value constraints. Mathematically speaking, other such rules exist, but we don't know them in the same way.[16] The symmetry arguments mentioned above should rule out these others from our consideration, but symmetry arguments are often based in part on desiderata or assumptions which are not totally incontrovertible, as we know.

Apart from that, should we rule out from consideration, say, any rule that solves the Judy Benjamin problem, but cannot be extended to Expectation constraints? The general principle behind such a decision—*require such rules to be extendible to other sorts of constraints*—would make us discard *Infomin* as well. For it cannot be used to impose such a constraint as, for example,

(*d*) change your probabilities so as to make A and B statistically independent

i.e. so that the posterior P' has $P'(A \text{ and } B) = P'(A).P'(B)$.

If the Judy Benjamin problem has no rationally compelled *unique* answer, then—mathematically and logically speaking—it can still be a special instance of a more general problem, which does have a unique solution. (After all, solutions take the form 'here is a rule that covers all the cases in your target class'.) But that does not mean automatically that the Judy Benjamin problem has a unique solution. And for someone who asks, 'Am I rationally compelled to change my opinions in such and such a fashion?', it is not a complete answer to be told 'If you decide to adopt a rule for solving all such problems with constraints definable in terms of posterior expectation values, then *yes*.'

But let us go on to consider a constraint on the so-called expectation value, for tosses yielding an even number, for example

1. $E(x|x \text{ is even}) = 2\text{Prob}(x = 2|x \text{ is even}) + 4\text{Prob}(x = 4|x \text{ is even}) + 6\text{Prob}(x = 6|x \text{ is even})$

that it should equal a certain number, say 3. That implies that the number 2 is more likely to come up than the number 6; but how much more likely?

If the Principle of Indifference could do its job, it would suffice to imply a unique answer here. Since that is definitely not so, one could make a proposal. A model was indeed proposed which recommends itself by its simplicity, plausibility, and advantages in application. The proposal made by Jaynes, and since explored in many contexts, was in effect as follows.[17] We need a measure of *information*, or *informativeness*, and choose the probability function which is the least informative among those which satisfy the imposed constraints. Such measures of information are available; the best known is Shannon's, also called *negative entropy*. It is defined by:

2. $I(P) = -\Sigma P(x) \log P(x)$

where log is again the natural logarithm, and x ranges over the set of basic alternatives—in our case the six faces of the die. This function has the following properties:

3. $I(P) \geqslant 0$
4. $I(P) = \infty$ if $P(x) = 0$ for any x
5. If x ranges over N alternatives, $I(P)$ has its minimum value, namely $\log N$, for the measure P which assigns $1/N$ to each alternative.

This information measure also behaves very well when various probability measures are combined. In equation 1, we really see four probability measures altogether. Besides the defined P' on set $\{1, 2, 3, 4, 5, 6\}$ we have:

P_1: $P_1(\text{Even}) = 0.75$
$P_1(\text{Odd}) = 0.25$
P_2: P_2 is defined on the set Even $= \{2, 4, 6\}$
P_3: P_3 is defined on the set Odd $= \{1,3,5\}$
P': $P'(x) = P_1(\text{Even})P_2(x)$ if x is even
$P'(x) = P_1(\text{Odd})P_3(x)$ if x is odd

and a little calculation shows us that:

6. $I(P') = 0.75I(P_2) + 0.25I(P_3) + I(P_1)$.

The constraint that Even was to be three times as likely as Odd fixes the last term; therefore to minimize quantity 6, we must minimize the other two terms on the right, i.e. choose P_2 and P_3 to be uniform on their domains. This gives us back the equation 2 for the correct choice of P'.

These are nice consequences. But isn't the choice of information measure I, as defined by 2, itself arbitrary? There do exist elegant deductions of the uniqueness of I, as defined, from desirable properties such as 3–5. Of course the justification for requiring such features as the additivity in equation 6 is only something like mathematical convenience or ease of calculation. Perhaps we should be happy with this, however, once we give up the false ideal of Indifference.

What about Bertrand's paradoxes? Proposition 5 above shows clearly how the division into basic alternatives is the initial step that determines the model. When different such divisions are available, as in our paradoxical examples, these will lead to different calculations of informativeness, and contradictory solutions. But Jaynes recognizes this clearly, and considers measure I as appropriate only for finitary and countable cases.[18] When the set of alternatives is defined, in terms of a continuous parameter, we need to choose an 'informationless' prior in some other way. Then we can find the right solution for a problem that imposes further constraints by shifting from that prior to some other measure, in a way that minimizes the *relative information*. In the countable case this is defined by

7. $I(P';\ P) = \Sigma P'(x)\log(P'(x)/P(x))$

which in the limit of a continuous parameter x becomes

8. $I(P';\ P) = \int P'(x)\log(P'(x)/P(x))\mathrm{d}x.$

In a Bertrand paradox we encounter a function g which transforms x into $y = g(x)$. The expression 8, unlike 2, is invariant under this transformation, provided it is monotone.

The rule to choose that posterior probability, among those which satisfy the given constraint, which has the least information relative to the prior, i.e. the rule to minimize relative information (or maximize cross-entropy), I call *Infomin*. It is not universally applicable, but there are nice theorems to delineate when it is.[19]

Proofs and illustrations

The relative information $I(P';\ P)$ and its companion $I(P;\ P')$ are often called the *directed divergences*, and their sum the divergence between P' and P. The family of their linear combinations was uniquely characterized most elegantly by Rodney Johnson.[20] Let p, q be strictly positive probability densities on a parameter x, and F the functional defined by

$$F(p, q) = \int f(p(x), q(x))\mathrm{d}x$$

for some measurable function f on the real numbers. The following requirements then entail that F belongs to that family:

Finiteness: $F(p, p)$ is finite
Additivity: $F(p, q) = F(p', q') + F(p'', q'')$ when $p(x', x'') = p'(x')p''(x'')$ and $q(x'x'') = q'(x')q''(x'')$
Positivity: $F(p, q)$ is non-negative, equalling zero only when $p = q$.

Johnson proves along the way that the first two properties already entail that the functional remains invariant when the domain of p or q is transformed by a non-singular linear transformation.

7. CONCLUSION: NORMAL RULE-FOLLOWING

In Chapter 7 I argued that rationality does not require rule-following. Rational opinion change need not be change in accordance with some rule or recipe. It does require that we do not sabotage ourselves by our own lights—for example by committing ourselves to a rule for opinion change which makes us incoherent. Besides the conclusion that it is rational not to follow rules, there was the other point that, if one does follow a rule, it must be Conditionalization.

In this chapter we have seen why that is. But we have also seen the limits of that point: Conditionalization is the sole rule if the input (deliverances of experience) is propositions, and the sole relevant factors are that input and the prior opinion.

The exploration of probability kinematics subject to other constraints was based on the premiss that not all responses to experience (need) consist in taking propositions as newly certain. That seems a weak enough premiss (Richard Jeffrey has urged that a rational agent give probability *one* to tautologies only), and yet it has been strongly resisted in the literature. Friedman and Shimony began the resistance in an article directed at Jaynes, to the effect that *Infomin* conflicts with conditionalization.[21] There has been a good deal of literature since. In my opinion, the criticism of *Infomin* and other such rules are based on the assumption that what I called the Revelation model is the correct model of experience. In addition, there are assumptions involved about exactly what the propositions constituting the new certainties must be. None of these assumptions are demonstrated. Nor do I see how they could be.[22]

One example may perhaps suffice. Suppose I respond to my experience by accepting as (total) constraint that my posterior opinion must involve a probability of 90 per cent that it will rain tomorrow. Then perhaps, if I am very consciously introspective, I shall also be aware that this is so, i.e. that I'm accepting this constraint. Could it be that I am simply conditionalizing on the latter, autobiographical proposition? Well it is *possible*. But personally I am so aware of the unreliability of introspection that I would not take that proposition to be certainly true. Perhaps I would give it 90 per cent probability with 10 per cent for its opposite. And as a result my probability for rain tomorrow would become 81 per cent instead of 90 per cent. However, I doubt that such autobiographical commentary is normally involved. It may be postulated of course. More *recherché* evidence taking may also be postulated. This brings us back to the argument in section 1, that it is trivially possible to reconstruct everyone as a conditionalizer. But not fruitful.

We have now come to the end of this exploration of symmetry. It is rather gratifying to note that even this section and the last contain a number of disputed points and unsolved problems. In this Part as well as in the preceding there are, as far as our discussion is concerned, large uncharted areas that remain. To an empiricist it must necessarily be so, for whatever insight symmetry brings us for theory and model construction, scientific progress must always rest on contingent theoretical assumptions. Any a priori certainty it can enjoy is at best conditional.

NOTES

Chapter 1

1. J. E. Ruby, 'The Origins of Scientific "Law" ', *Journal of the History of Ideas*, 47 (1986), 341–59.
2. Ibid. 341.
3. My historical speculation will inevitably be biased by antirealist sympathies. For a different way of telling the story, compare E. McMullen's 1984 presidential address to the American Philosophical Association, 'The Goal of Natural Science', *Proc. APA* 58 (1984), 37–64. I want to thank Prof. McMullen for very helpful discussions and correspondence.
4. For 'occult' see E. J. Dijksterhuis, *The Mechanization of the World Picture* (Oxford, 1961), 157 (II-87). The locality involved is not spatial; teleological causes may operate over spatial distances (McMullen). The locality principle is rather that everything that happens must be due to the action of specific individual substances on each other.
5. *The Works of the Honourable Robert Boyle*, ed. T. Birch (London, 1672), iii. 13.
6. Sir R. Blackmore, *The Creation*, bk. ii, ll. 295–6; 313–18; 321–30, in J. Heath-Stubbs and P. Salman (eds.), *Poems of Science* (Harmondsworth, 1984), 120–1.
7. There is a famous argument by Einstein that (special) relativity entails that every conservation law is local (cf. R. Feynman, *The Character of Physical Law* (Cambridge, Mass., 1965), 63–4). In classical physics, as I shall note below, the conservation laws are global. It is possible to derive them for any mechanical system, subject to certain conditions; but that any given system, or even the whole world, satisfies these conditions, is not derivable, but needs to be assumed. In general relativity we find scope for global principles again, because the global geometry of space–time is not determined by its local geometry. Quantum mechanics, where the whole is notoriously richer than the sum of its parts, furnishes many examples.
8. *Dante's Inferno*, tr. T. Philips (London, 1985), 10.
9. Cf. Job 38: 25–9; Clement of Rome, First Epistle to the Corinthians, 20; the anonymous Epistle to Diognetus, 7. For the latter two, see

M. Staniforth (tr.), *Early Christian Writings* (New York, 1968), 33–4 and 178.

10. The view in question is generally known as occasionalism (sometimes, voluntarism). Although Aquinas's reaction is quite clear from the passages cited in the next note, the tension between the Aristotelian notion of nature and the occasionalist view did not become fully clear until the fourteenth century, during the realist–nominalist debates. See A. J. Freddoso, 'Medieval Aristotelianism and the Case against Secondary Causes in Nature', in T. Morr (ed.), *Divine and Human Action: Essays in the Metaphysics of Theism* (Ithaca, NY, 1988), and E. McMullen, 'The Development of Philosophy of Science 1600–1850' in J. Hodge *et al.* (eds.), *Encyclopedia of the History of Science*, forthcoming.
11. *Summa Contra Gentiles*, bk. 2, especially sects. 4–5, 7–9, 309.
12. Ibid., sect. 7.
13. For recent discussion of Descartes's concept of law, see G. Loeck, 'Auskunft über die Gesetzesartigkeit aus ihrer Konstruktion', *Osnabrücker Philosophische Schriften*, Universität Osnabrück, 1985, and *Der cartesische Materialismus: Maschine, Gesetz und Simulation* (Frankfurt, 1986).
14. Newton himself is not exactly clear on this. Discussing forces in *Principia*, bk. 1, def. viii, he characterizes them as the causes of the alteration of motion. But then he writes they are to be taken 'mathematically, not physically', and tells us 'not to imagine that by those words I anywhere take upon me to define the kind or the manner of any action, the causes or the physical reason thereof'. Cf. McMullen, 'The Development of Philosophy of Science 1600–1850', who comments on this vacillation (or equivocation?) that this separation of the mathematical and the physical enables Newton 'to bracket all questions about how forces function as *causes*, while retaining enough of a causal overtone to suggest he has somehow explained *why* the motion occurs'.
15. I want to thank R. Cooke for making this clear to me. Instructive also are the examples Earman gives in his *A Primer on Determinism* (Dordrecht, 1986), 32, 34, 36, and the possible reactions to them listed on 38. See also ch. 10, sect. 4 below.
16. The metaphor of 'source' should alert us to the danger that an empiricist may also presuppose some such metaphysics. Historical empiricism did indeed succumb to this and fell to Kant's critique as well. See I. Kant, *The Critique of Pure Reason*, tr. N. Kemp Smith (London, 1980), B127, B498–9, B792–6, B882.
17. R. Descartes, *Principles of Philosophy*, author's letter prefaced to the French translation; *Philosophical Works of Descartes*, tr. E. S. Haldane and R. T. Ross (Cambridge, 1985), i. 211.

18. A. Wickham, *The Contemplative Quarry* (New York, 1921), Envoi. This was quoted by H. Weyl in his seminal lecture 'Symmetry', *Journal of the Washington Academy of Sciences*, 28 (1938), 253–71; reprinted in his *Gesammelte Abhhandlungen* (Berlin, 1968), iii. 592–610.
19. Cf. the analysis of such reasoning by E. Mach, *The Science of Mechanics* (LaSalle, Ill. 1974), 516–20, 549 (see also 456–9).
20. See further my *An Introduction to the Philosophy of Time and Space*, 2nd edn. (New York, 1985).

Chapter 2

1. D. Lewis, 'New Work for a Theory of Universals', *Australasian Journal of Philosophy*, 61 (1983), 343–77; citations from 356–357, 364, 365.
2. For Peirce's discussion of the history of this idea, see his essay on Hume's 'Of Miracles'; *Collected Paper of Charles Sanders Peirce*, ed. C. Hartshorne and P. Weiss (Cambridge, Mass., 1931–5, vol. VI, bk. ii, ch. 5, especially pp. 364–6. See also the section 'Letters to Samuel P. Langley, and "Hume on Miracles and Laws of Nature" ' in P. Wiener (ed.), *Values in a World of Chance: Selected Writings of C. S. Peirce* (New York, 1966).
3. *Collected Papers*, vol. V, sect. 5.93–5.119 (pp. 64–76). That first section bears the title 'Scholastic Realism'.
4. This may be compared with Reichenbach's conception of indeterminism which I discussed in my 'The Charybdis of Realism: Epistemological Implications of Bell's Inequalities', *Synthese*, 5 (1982), 25–38; see also ch. 5, sect. 6 below and *The Scientific Image*, ch. 2, sect. 5.
5. Reprinted Hartshorne and Weiss vi. 75–85; see esp. 76–9.
6. *A System of Logic* (London, 1846), vol. I, i; bk. iii, ch. 3; J. S. Mill, *Collected Works* (Toronto; 1963), 306.
7. D. Davidson, *Essays on Actions and Events* (Oxford, 1980), 213–14. I want to thank Mark Johnston for helpful conversations; I have also drawn on his 'Why Having a Mind Matters', in E. LePore and B. McLaughlin (eds.), *Actions and Events* (Oxford, 1985), 408–26.
8. The statement is general provided only that the universe satisfies the principle of identity of indicernibles: for in that case, whatever class of events we describe will have some uniquely identifying description.
9. The same can be said of Davidson's analysis of causation, relied on in this article. See 'Causal Relations' in his *Essays on Actions and Events* (Oxford, 1980), 149–62.
10. B. Willey, *The Eighteenth Century Background* (London, 1940), 126–7. This is not an accurate assessment if taken as a description of Hume's argument; as Peirce pointed out, that concerned probability

alone. But that Hume expressed this faith in uniformity is certainly evident.

11. Some writers do not balk; see e.g. N. Swartz, *The Concept of Physical Law* (Cambridge, 1985).
12. David Lewis tells me that such parallel pairs were discussed at UCLA in the 1960s, and I speculate this was the heritage of Reichenbach's discussion of such examples.
13. See my 'Essence and Existence', in *American Philosophical Quarterly* Monograph Series No. 12 (1978); J. Earman, 'The Universality of Laws', *Philosophy of Science*, 45 (1978), 173–81.
14. M. Tooley, 'The Nature of Law', *Canadian Journal of Philosophy*, 7 (1977), 667–98 (see esp. 686).
15. D. Armstrong, *What is a Law of Nature?* (Cambridge, 1983), 26.
16. See Earman, 'The Universality of Laws'; he does add that laws in the sense of Lewis are 'likely' to be universal.
17. My own attempts, utilizing formal pragmatics, are found in 'The only Necessity is Verbal Necessity', *Journal of Philosophy*, 74 (1977), 71–85, and 'Essences and Laws of Nature', in R. Healey (ed.), *Reduction, Time and Reality* (Cambridge, 1981).
18. Cf. Armstrong, *What is a Law of Nature?*, ch. 11; C. Swoyer, 'The Natural Laws', *Australasian Journal of Philosophy*, 60 (1982), 203–23.
19. For example, it is not in accord with my own, cf. *The Scientific Image*, ch. 5. An entirely different, but also contrary, point is made by McMullen, 1984: 'In the new sciences, lawlikeness is not an explainer, it is what has to be explained' (p. 51).
20. Perhaps the resemblance goes further: according to the *Posterior Analytics*, science deals only with the truly universal; and in the *Poetics*, Aristotle writes 'poetry is something more philosophic and of graver import than history, since its statements are of the nature of universals' (1451^7-9). The parallel can not be pushed too far, but remains striking nevertheless, if we are at all puzzled by our insistent craving for reasons why.
21. My own account of explanation, in *The Scientific Image*, implies that explanation is not a good in itself, and is worth pursuing only as a tactical aim, to bring us to empirical strength and adequacy if it can.
22. Armstrong, *What is a Law of Nature?*, 5; see P. Forrest, 'What Reasons do we have for Believing There are Laws of Nature?', *Philosophical Inquiry International Quarterly*, 7 (1985), 1–12.
23. N. Goodman, *Fact, Fiction, and Forecast* (Atlantic Highlands, NJ, 1954).
24. See *The Scientific Image*, p. 118, and 'Essences and Laws of Nature'.
25. R. Stalnaker, *Inquiry* (Cambridge, Mass., 1984), 149–50.
26. Cf. F. Dretske, 'Laws of Nature', *Philosophy of Science*, 44 (1977) 248–68; see esp. 251–2 and 254–5.

27. H. Margenau and R. B. Lindsay, *Foundations of Physics* (New York, 1957), 20.
28. That is certainly in accordance with one use of the term, which may indeed be quite common in some idiolects. Thus P. Abbot and H. Marshall, *National Certificate Mathematics*, 2nd edn. (London, 1960) designed 'for students taking mechanical or electrical engineering courses' has a chapter entitled 'Determination of Laws' which teaches how to find equations that fit given sets of data.
29. N. Cartwright, *How the Laws of Physics Lie* (Oxford, 1983).
30. Ed. M. Gardner (New York, 1966).

Chapter 3

1. I wish to thank D. Lewis for his comments on an early draft.
2. This is just the one metaphysical element which P. Duhem, in other respects an empiricist hero, introduced in his theory of science. In a later chapter we shall see attempts to elaborate much more extensive anti-nominalist theories into rival accounts of law and necessity.
3. For this chapter I am indebted to previous critical discussions by J. Earman, P. Forrest, and M. Tooley. Lewis has recently (1986) proposed an amendment to his account. I shall discuss it only in a note appended to this chapter.
4. *Counterfactuals* (Cambridge, Mass., 1973), 73–5.
5. J. Earman, 'Laws of Nature: The Empiricist Challenge', in R. J. Bogdan (ed.), *D. H. Armstrong* (Dordrecht, 1984), 191–224; quotes from 229–30.
6. For example, it may be that simplicity exists only in the eye of the beholder. If that is so, it may or may not matter to Lewis's analysis. After all he does *not* say that the best theories about a given world are those which are *there regarded* as optimally simple and strong. He assumes instead that the classification of theories as simple or strong is independent of questions of truth, and of the historical features of the world under consideration. I will leave this assumption unchallenged.
7. 'New Work for a Theory of Universals', *Australasian Journal of Philosophy*, 61 (1983), 343–77; see esp. 367–8.
8. See R. S. Woolhouse, *Locke's Philosophy of Language and Science* (New York, 1971) or ch. 3 of his *Locke* (Minneapolis, 1983); M. H. Carré, *Realists and Nominalists* (Oxford, 1950).
9. I do not mean that what we regard as virtuous simplicity is historically unconditioned; but given what we so regard, it is a feature that theories can have even in e.g. uninhabited worlds.

10. These assumptions are that the best theories, by the presently used criteria for good and better, are *also* the best by the criteria of (actual, historical) science, and *in addition* the best by the criteria for evaluating explanations.
11. This criticism is perhaps related to the charge in P. Forrest's challenging article 'What Reasons Do We Have for Believing That There Are Laws of Nature?', *Philosophical Inquiry*, 7 (1985), 1–12, that if laws are simply what is common to all the best 'summaries' of what actually happens, then the fact that something is a law, is just not the sort of fact which can explain anything. But this charge also carries the complaint that Lewis does not do justice to the connotation of necessity, I think.
12. I want to thank Mark Johnston for discussions of this subject.
13. See the incisive critique of natural kinds, in the context of similar issues, by P. Churchland, 'Conceptual Progress and World/Word Relations: In Search of the Essence of Natural Kinds', *Canadian Journal of Philosophy*, 15 (1985), 1–17. I recognize that this reasoning is a plausibility argument only. In a later chapter I will briefly examine A. Plantinga's argument that, given different factual premisses—e.g. that we are created in God's image—a different conclusion follows. I would contest that argument unless the premisses entail highly specific (and correspondingly less plausible) evaluations of our theoretical activity.
14. *Philosophy of Science*, 45 (1978), 173–81; quote from p. 180.
15. Kuhn, *Structure of Scientific Revolutions* 2nd edn. (Chicago, 1970), 107–8.
16. Since I am no friend of real necessity, as conceived in pre-Kantian metaphysics, this criticism may sound inappropriate on my tongue. But it seems to me that the misgivings of the metaphysician translate also into criticisms of Lewis's account, regarded as an analysis of modal discourse. For such as me, there is little relation between ontology and semantics—but not for Lewis.

Chapter 4

1. This is a large issue, which no analytic philosopher can safely ignore. My own view is that modal discourse is very strongly but tacitly context-dependent, and when we supply all the tacit relevant parameters, we are left with purely 'verbal' necessity and possibility only. This has of course been, in one form or other the traditional nominalist-empiricist line on modality. See my *The Scientific Image*,

ch. 6, and references therein; or, more closely connected with our present subject, my 'Essences and Laws of Nature'.

2. H. Reichenbach, *Elements of Symbolic Logic* (New York, 1947), see esp. 360, 368; F. B. Fitch, *Symbolic Logic* (New York, 1952), ch. 3, sect. 11.19; R. Montague, ch. 1 in R. H. Thomason (ed.), *Formal Philosophy: Papers by R. Montague* (New Haven, 1974).
3. W. Sellars, 'Concepts as Involving Laws and Inconceivable without Them', *Philosophy of Science*, 15 (1948), 287–315. At first sight, Sellars's account looks like a combined possible-worlds and universals account. But the relations between universals are eventually defined in terms between corresponding classes in possible worlds, exactly in the way that later became the standard modal account of properties.
4. 'Laws and Modal Realism', *Philosophical Studies*, 46 (1984), 335–47.
5. 'Time and the Physical Modalities', *Monist*, 53 (1969), 426–46; 'Counterfactuals Based on Real Possible Worlds', *Nous*, 18 (1984), 463–77; 'Laws of Nature and Nomic Necessity', MS, 1987.
6. 'Explicating Lawhood', to appear in *Philosophy of Science*, 55 (1988).
7. Walter Benjamin, 'Theses on the Philosophy of History, IX', in his *Illuminations* (New York, 1969), 267.
8. Compare Benjamin's notes on the angelic figure SPES of Andrea de Spisano: the *Florence Baptistry* entry in 'One-Way Street' in his *One-Way Street and Other Writings* (London, 1979). (Thanks to Glen Most.)
9. See articles cited above; the idea appeared already in Montague's early articles on the uses of possible-world semantics for philosophy. See further R. H. Thomason, 'Indeterminist Time and Truth-Value Gaps', *Theoria*, 36 (1970), 264–81; 'Combinations of Tense and Modality', in D. Gabbay and F. Guenther (eds.), *Handbook of Philosophical Logic* (Dordrecht, 1984).
10. The reader may have guessed at another difficulty: how should we distinguish a world which shares our history exactly until, say, 1900, from another world in which the initial history is exactly the same for equally long, but all events happen exactly four minutes later than in ours? This is not an insurmountable difficulty, but is better discussed in another context; see the section on determinism in Part III.
11. For details on the logic of branching time, see R. H. Thomason, op. cit. and my 'A Temporal Framework for Conditionals and Chance', *Philosophical Review*, 89 (1980), 91–108.
12. The main advocate of the views discussed here is P. Vallentyne, 'Explicating Lawhood'. In this section and the next, I am indebted to D. Lewis's lucid discussions of his own attempts to combine his conceptions of law and objective chance. See especially his *Philosophical Papers*, ii (Oxford, 1986), xiv–xvii and 126–31.

13. For ease of exposition I shall leave this and similar examples a little imprecise. The most stable isotope of radium has a half-life of 1620 years. The immediate disintegration product of radium is the radioactive gas radon; its most stable isotope has a half-life of 3.825 days.
14. This question is closely related to what Putnam calls 'Peirce's puzzle'. See H. Putnam, *The Many Faces of Realism* (LaSalle, Ill., 1987), 80–6.
15. A fact long emphasized by I. Hacking; see his *The Emergence of Probability* (London, 1975).
16. See D. Miller, 'A Paradox of Information', *British Journal for the Philosophy of Science*, 17 (1966); R. C. Jeffrey, Review of Miller *et al.*, *Journal of Symbolic Logic*, 35 (1970), 124–7; D. Lewis 'A Subjectivist's Guide to Objective Chance', in W. Harper *et al.* (eds), *Ifs* (Dordrecht, 1981); and my 'A Temporal Framework for Conditionals and Chance'.
17. See *The Scientific Image*, ch. 6. This chapter contains a modal frequency interpretation of physical probability; even a brief look, however, will show that it is not of a sort that could help the friends of laws here.
18. I say 'roughly': the precise form of the problem, and hence the solution, relates a probability measure on the possible states to a measure of duration, in a fashion which is theoretically contingent. It would be impossible to say therefore that the probabilities as such constrain or explain the proportional times of sojourn. Their connection is simply contained in the definition of ergodicity. In addition, it must be added that the agreement between probability and duration is deduced to occur in a class of trajectories of measure *one*. The measure is derived from the probability measure in question. Thus we have the same problem as for the Law of Large Numbers: even if ergodicity is postulated, we must ask why our subjective probability should follow suit upon *that* probability measure rather than some other one, or none at all.
19. The 'modal interpretation' of quantum mechanics has some *formal* similarity to the many-worlds interpretation, but without this realism about possible worlds. See my 'Semantic Analysis of Quantum Logic', in C. A. Hooker (ed.), *Contemporary Research in the Foundations and Philosophy of Quantum Theory* (Dordrecht, 1973), 80–113, and 'A Modal Interpretation of Quantum Mechanics', in E. G. Beltrametti and B. C. van Fraassen (eds.), *Current Issues in Quantum Logic* (New York, 1981).
20. L. E. Ballentine, 'Can the Statistical Postulate of Quantum Theory Be Derived? A Critique of the Many-Universe Interpretation', *Foundations of Physics*, 3 (1973), 229–40. See also S. J. Norman, *Subsystem States*

in Quantum Theory and Their Relation to the Measurement Problem, Ph.D. Dissertation, Stanford University, 1981.

21. R. Jeffrey and B. Skyrms have discussed this in different ways; see further ch. 8.
22. This point was forcefully made by R. Foley in his commentary on my paper 'What Are Laws of Nature?' at the New Jersey Regional Philosophy Conference, Nov. 1985.
23. I do agree that common-cause models are pre-eminently good responses to prevalent sorts of explanation request—but that is all.
25. Pargetter, like D. Lewis, denies this, strictly speaking (personal communication).
26. I gave the example which follows, near enough, in my 'Essences and Laws of Nature'.
27. Thus, given such a crystal and its history, we can never be sure that it does *not* belong to the large 'randomizing' class. If it has responded differently in two tests, then of course we know that it does.

Chapter 5

1. Parts of this chapter appeared in earlier form as my 'Armstrong on laws and Probabilities', *Australasian Journal of Philosophy*, 65 (1987), 253–60. See also J. Carroll, 'Ontology and the Laws of Nature', which appeared in the same issue (261–76) and contains similar criticisms but suggests a different conclusion.
2. See my 'Theory Comparison and Relevant Evidence', in J. Earman (ed.), *Testing Scientific Theories*, *Minnesota Studies in the Philosophy of Science*, x (Minneapolis, 1984), and 'Glymour on Evidence and Explanation', ibid.
3. These problems are implicit also in D. Lewis's critique of Armstrong (Lewis, 'New Work for a Theory of Universals', *Australasian Journal of Philosophy*, 61 (1983), 366; *Philosophical Papers*, ii (Oxford, 1986), xii).
4. M. Tooley, 'The Nature of Law', *Canadian Journal of Philosophy*, 7 (1977), 673.
5. D. Armstrong, *What is a Law of Nature?* (Cambridge, 1983), 92.
6. F. Dretske 'Laws of Nature', *Philosophy of Science*, 44 (1977), 264.
7. The inference problem for laws conceived as relations among universals, is exactly parallel to that which Schlick urgued concerning objective values.
8. Dretske, 'Laws of Nature', 265.
9. Tooley, 'The Nature of Law', 678 ff.

10. Such examples must always, in an anti-nominalist context, have the caveat that the described properties may not exist.
11. My definition here is not exactly Tooley's of nomological relations, but very close if his definition incorporates the discussion that follows it ('The Nature of Law', 679–80); I have tried to do this by my choice of definitions for 'purely' and 'irreducibly'.
12. Sects. 3 and 4 are based on parts of my 'Armstrong on Laws and Probabilities', *Australasian Journal of Philosophy*, 65 (1987), 243–60. See also D. Armstrong, 'Reply to van Fraassen', *Australasian Journal of Philosophy*, 66 (1988), 224–9.
13. In ch. 6, sect. 5 of *What is a Law of Nature?* Armstrong explores the possibility of getting an 'independent fix' (p. 96) on necessitation *via* the notion of causality. I shall concentrate instead on sects. 3, 4, and 6 of that chapter.
14. Armstrong, *What is a Law of Nature?*, ch. 9.
15. By the generality of *now* and the continuity of time, three of the canvassed possiblities require an infinity of atoms of the power of the continuum, which is not even sensible, let alone possible.
16. A look back will also reveal the corollary, that remaining stable for interval *t*, rather than decaying within *t*, is the real universal. To my knowledge, this is the first deduction of the reality of a specific universal from a law of physics, and I respectfully offer it as a contribution to the theory.
17. 'Reply to van Fraassen', *Australasian Journal of Philosophy*, 66 (1988), 226.
18. The difficulty I raise here had been previously and independently raised by J. Collins in correspondence with Armstrong.
19. M. Tooley, *Causation: A Realist Approach* (Oxford, 1987), ch. 4.
20. There were certain other problems: the rules that could determine such probabilities were obviously sensitive to the structure of the language, and so the probabilities would change when the language was extended or translated into another language. Tooley correctly points out that the theory of universals—or anti-nominalism generally—can help here, in effect by determining which is the 'correct' language, in the weak sense of formulating a distinguishing difference in its own terms. But that does not help the main problem.
21. This refers to the early stage of Carnap's programme, in *The Logical Foundations of Probability*, 2nd edn. (Chicago, 1962). In later stages Carnap himself made no such claim of uniqueness.
22. University of Sydney, 1981. I must also thank Collins for his comments on my earlier correspondence with Armstrong and with Tooley.
23. The probability function which gives equal probabilities to all *TV*'s is Carnap's *m*† if restricted to the quantifier-free fragment. It has also

feature that $P(Fn|Fa \& ... \& Fm) = P(Fn)$, and Carnap rejected it for that reason.

24. If we switch to non-standard numbers, the conclusion is only that at most countably many can receive non-infinitesimal probability. This helps with some technical difficulties, but it makes the non-uniqueness problem worse.
25. *The Works of the Honourable Robert Boyle*, ed. T. Birch (London, 1673), iii. 13; see R. Woolhouse, *Locke's Philosophy of Science and Language* (Oxford, 1971) for an excellent discussion.

Chapter 6

1. London, 1985, p. 92.
2. See G. Harman, 'The Inference to the Best Explanation', *Philosophical Review*, 74 (1965), 88–95. In this chapter I am much indebted to discussions of this issue by A. Fine in his 'The Natural Ontological Attitude', in J. Leplin (ed.), *Scientific Realism* (Berkeley, Calif., 1984).
3. For other discussions of these failures, see my 'Empiricism in the Philosophy of Science', in P. M. Churchland and C. A. Hooker (eds.), *Images of Science: Essays on Realism and Empiricism, with A Reply by Bas C. van Fraassen* (Chicago, 1985) and J. Watkins, *Science and Scepticism* (Princeton, NJ, 1984).
4. Compare Putnam's short discussion of Mill, Reichenbach, and Carnap in *The Many Faces of Realism* (LaSalle, Ill., 1987), 73–4.
5. This is essentially how the rule of induction is presented in Russell's *The Problems of Philosophy* (New York, 1959).
6. The point is not defeated by the stipulation that all inductive rules must converge to the straight rule as the sample size goes to infinity, even if that be accepted. The wide divergences will appear at every finite size, however large.
7. For another telling line of criticism, see H. Putnam, 'Reflexive Reflections', *Erkenntnis*, 22 (1985), 143–54. For still another one, see A. P. Dawid, 'The Impossibility of Inductive Evidence', *Journal of the American Statistical Association*, 80 (1985), 340–1 and sect. 7.1 of his 'Calibration-Based Empirical Probability', *Annals of Statistics*, 13 (1985), 1251–73, both of which refer to the result of D. Oakes, 'Self-calibrating Priors do not Exist', *Journal of the American Statistical Association*, 80 (1985), 339.
8. This is directed against certain Bayesian ideas about confirmation; see e.g. P. Horwich, *Probability and Evidence* (Cambridge, 1982).
9. See D. Armstrong *What is a Law of Nature?* (Cambridge, 1983),

ch. 5, sect. 4; my 'Armstrong on Laws and Probabilities', *Australasian Journal of Philosophy*, 65 (1987), 243–60 and D. Armstrong 'Reply to van Fraassen', ibid. 66 (1988), 224–9.

10. I want to thank Y. Ben-Menachim for helpful discussion and correspondence.
11. This is not a subject without a history. Most of the arguments bandied about appeared already in the debates between Cartesian and Newtonian in the seventeenth century. These debates suffered however from a too easy equation of epistemology and methodology. In the terminology of Herschel's distinction, they tended to confuse the tactics of the context of discovery with rules proper to the context of justification. They also suffered from historical loyalties that kept the 'method of hypotheses' and 'method of induction' as the great and sole alternatives, allied to rival programmes in the natural sciences. We are therefore doomed to repeat this history, not through ignorance, but through loss of innocence. The opposition of induction and hypothesis reappeared in a new key at each subsequence stage, with Whewell's consilience, Peirce's abduction, Popper's bold conjecture, and so forth. See the illuminating essays by L. Laudan in his *Science and Hypothesis* (Dordrecht, 1981).
12. M. Friedman seemed reluctantly to come close to this position in 'Truth and Confirmation', *Journal of Philosophy*, 76 (1979), 361–82 (see esp. 370); R. Boyd appears to take it in P. Churchland and C. A. Hooker (eds.), *Images of Science* (Chicago, 1985); see also my reply to Boyd there.
13. For an effective critique of such evolutionary epistemology, from the vantage of current biology, see M. Piatelli-Palmarini, 'Not on Darwin's Shoulders; A Critique of Evolutionary Epistemology', Boston Colloquium for the Philosophy of Science, Jan. 1988.
14. Cf. the end of M. Tooley, 'The Nature of Law', *Canadian Journal of Philosophy*, 7 (1977), which suggests that IBE can furnish non-subjective prior probabilities for hypotheses.
15. 'Reply to van Fraassen', p. 228.
16. J. J. C. Smart, 'Laws of Nature and Cosmic Coincidences', *Philosophical Quarterly*, 35 (1985), 272–80; quote from p. 273.

Chapter 7

1. From R. Boyd, 'The Current Status of Scientific Realism', in J. Leplin (ed.), *Scientific Realism* (Berkeley, Calif., 1984), 67.
2. *The Port-Royal Logic*, 1662; cited R. C. Jeffrey, *The Logic of Decision*, 2nd edn. (Chicago, 1983), 1.
3. It is an avowal, not an autobiographical description. For discussion

of this difference, and the various functions of first-person declarative sentences, see for example P. M. S. Hacker, *Insight and Illusion* (Oxford, 1986), 297 ff.

4. The point of view here adopted, with its insistence on a strict separation between expression of opinion and statement of biographical fact about opinion, I call Voluntarist, for reasons more specifically explained in 'Belief and the Will', *Journal of Philosophy*, 81 (1984), 235–56. Concerning vagueness in personal probability, see R. C. Jeffrey 'Bayesianism with a Human Face', in J. Earman (ed.), *Minnesota Studies in the Philosophy of Science*, ix (Minneapolis, 1983), 133–56.
5. See N. Rescher, *The Logic of Commands*; C. Hamblin, *Imperatives* (Oxford, 1987).
6. The idea needs to be refined. Consider for example the number of days on which he speaks; it must be finite. Therefore he'd have no chance of perfect calibration if he made the x an irrational number. But it is not a point of logic that a probability cannot be an irrational number. So we have to think about how the announced number relates to the proposition in some more complex fashion. See my 'Calibration: Frequency Justification for Personal Probability', in R. S. Cohen and L. Laudan (eds.), *Physics, Philosophy, and Psychoanalysis* (Dordrecht, 1983), 295–319.
7. For a general introduction to probability theory, see B. Skyrms, *Choice and Chance* (Belmont, Calif., 1986).
8. B. De Finetti, 'Probability: Beware of Falsifications', in H. Kyburg, Jun. and H. Smokler (eds.), *Studies in Subjective Probability* (Huntington, NY, 1980).
9. As reported in P. Teller, 'Conditionalization and Observation', *Synthese*, 26 (1973), 218–58.
10. More examples of such bets, as well as the general strategy, are described in my 'Belief and the Will'. There is a danger, when these coherence arguments are written in terms of bets, that they will be perceived as being essentially about betting behaviour. That is not so; they are about consistency in judgement. See B. Skyrms, *Pragmatics and Empiricism* (New Haven, Conn., 1984), ch. 2.
11. What I mean by this will be explained and demonstrated in ch. 13.
12. B. De Finetti, 'Methods for Discriminating Levels of Partial Knowledge concerning a Test Item', *British Journal of Mathematical and Scientific Psychology*, 18 (1965), 87–123; R. C. Pickhardt and J. B. Wallace, 'A Study of the Performance of Subjective Probability Assessors', *Decision Sciences*, 5 (1974), 347–63 and references therein.
13. I have attempted a longer sketch, though with a somewhat different focus, in 'Empiricism in the Philosophy of Science'.
14. (Oxford, 1912; New York, 1959) 7.

15. That we are of the opinion that our opinions are reliable must, however, be construed very delicately; see 'Belief and the Will'.
16. W. James, 'The Will to Believe'. Page references are to his *Essays in Pragmatism* (New York, 1948). Clifford's lecture is found in W. K. Clifford, *Lectures and Essays* (London, 1879).
17. Unlike e.g. Shimony, Friedman, and Seidenfeld, but like R. Jeffrey, and P. Williams, I allow here for different sorts of deliverances of experience, not always equivalent to simply taking propositions as evidence. See further ch. 13.
18. See further my article 'Rationality does not Require Conditionalization', forthcoming.
19. See, for related reflections, Lecture IV: 'Reasonableness as a Fact and as a Value', esp. pp. 77–80, in Hilary Putnam, *The Many Faces of Realism* (LaSalle, Ill., 1987). Merely to deny the relevant dichotomies does not remove the problem; but that is one way to begin the task of elucidating the parallel in relations between practices and standards, for all sorts of enterprises.
20. In the form of R. Jeffrey's Bayesianism 'with a human face'. If I still resist the name 'Bayesian', which has been stretched far beyond orthodoxy, it is to distance myself from certain ideas concerning scientific methodology, held by some Bayesians.
21. See R. Carnap, *The Continuum of Inductive Methods* (Chicago, 1952), sect. 18.
22. (Cambridge, Mass., 1967.)
23. This subject of 'calibration' is explored in my 'Calibration' and in A. Shimoney, 'An Adamite Derivation of the Principles of the Calculus of Probability' (forthcoming), which had been presented as a lecture already in 1982. That rightness, i.e. calibration, is not enough, since opinion is subject to other criteria as well, is clearly shown in T. Seidenfeld, 'Calibration, Coherence, and Scoring Rules', *Philosophy of Science*, 52 (1985), 274–94.
24. D. Armstrong, 'Reply to van Fraassen', *Australasian Journal of Philosophy*, 66 (1988), 224–9.
25. In more traditional perspective, this is to say that I opt for a voluntarist rather than an idealist refutation of scepticism. See the discussion of St Augustine's *Against the Academics*, in my 'The Peculiar Effects of Love and Desire', forthcoming in A. Rorty and B. McLaughlin (eds.), *Perspectives on Self-Deception*, (Los Angeles, Calif.).

Chapter 8

1. My views on explanation are mainly presented in *The Scientific Image*, ch. 5; in 'Salmon on Explanation', *Journal of Philosophy*, 82 (1985),

639–51; and in '*Was ist Aufklärung?*' in G. Schurz (ed.), *Erklären und Verstehen* (Munich, 1988). For critique, see e.g. K. Lambert and G. Brittan, *An Introduction to the Philosophy of Science*, 3rd edn. (Atascadero, Calif., 1987), and P. Kitcher and W. Salmon, 'Van Fraassen on Explanation', *Journal of Philosophy*, 84 (1987), 315–30.

2. 'Rationalism and Empiricism: An Inquiry into the Roots of Philosophical Error', in H. Reichenbach, *Modern Philosophy of Science* (New York, 1959).
3. H. Weyl, 'The Ghost of Modality', in M. Farber (ed.), *Philosophical Essays in Memory of Edmund Husserl* (Cambridge, Mass., 1940), 278–303.
4. MS, circulated spring 1987; forthcoming with the University of Illinois Press.
5. Urbana, Ill., 1974.
6. New York, 1979.
7. Chicago, 1988.
8. New York, 1970; 2nd edn. with new preface and postscript, 1985.
9. In S. Morgenbesser (ed.), *Philosophy of Science Today* (New York, 1967).
10. Westport, Conn., 1988.
11. Albany, NY, 1988.
12. J. Beattie, 'What's Wrong with the Received View of Evolutionary Theory', in P. Asquith and R. Giere (eds.), *PSA 1980*, ii (East Lansing, Mich., 1981), 397–426.
13. See W. Stegmüller, *The Structuralist View of Theories* (Berlin, 1979); C. U. Moulines, 'Approximate Application of Empirical Theories', *Erkenntnis*, 10 (1976), 201–27. For comparisons see A. R. Perez Ransanz, 'El concepto de teoría empirica según van Fraassen' (with English tr.), *Critica*, 17 (1985), 3–20 and my reply, 'On the Question of Identification of Scientific Theory', ibid. 21–30. In the actual analysis of scientific theories, the structuralist and semantic approach proceed in much the same way, except that the former tends to be more formal. An interesting approach that shares characteristics with both semantic and structuralist views has been developed by Erhard Scheibe; see e.g. his 'On the Structure of Physical Theories', *Acta Philosophica Fennica*, 30 (1978), 205–23.
14. See my 'Empiricism in the Philosophy of Science', sect. I. 6.
15. For an excellent study of this option see J. Hanna, 'Empirical Adequacy', *Philosophy of Science*, 50 (1983), 1–34.
16. This last sentence is not just provocative, but rejects the reality of objective chance. It is not an account of chance for it does not respect the logic of that notion.

17. For a review of the literature and a careful diagnosis of the fallacy involved in the paradox, see R. C. Jeffrey, *Journal of Symbolic Logic*, 35 (1970), 124–7. See also D. Lewis, 'A subjectivist's Guide to Objective Chance', in W. Harper *et al.* (eds.), *Ifs* (Boston, Mass., 1981), 267–98 where several of the following points are clearly made.
18. See my 'A Temporal Framework for Conditionals and Chance'.
19. By 'fully believe' I mean that I give it subjective probability 1; in this sense I fully believe that the mass of the moon in kilograms is not a rational number. Some further distinctions are clearly necessary, and can be made e.g. in terms of Popper or Renyi functions—I'll leave this aside here.
20. It is not even easy to apply it to our own actions. Could I not be rational, though certain that I will do something if circumstances allow, and yet believe that it is not a physically settled fact that I shall (i.e. believe that I can do otherwise)?
21. See R. C. Jeffrey, *The Logic of Decision*, section 12.7, and 'Choice, Chance, and Credence', in G. Floistad and G. H. von Wright (eds.), *Philosophy of Language/Philosophical Logic* (The Hague, 1981), 367–86. This reflection is at the heart of De Finetti's and cognate reconstructions of talk about objective chance.
22. See ch. 6 of *The Scientific Image*.
23. C. Gaifman, 'A theory of higher order probability', in B. Skyrms *et al.* (eds.), *Causality, Chance, and Choice* (Dordrecht, 1988).
24. See the discussion of calibration in my 'Belief and the Will'.
25. My 'voluntarist' resolution here of the problem of why the probabilities in accepted scientific theories constrain personal expectation in this way, seems to me at least similar, and perhaps the same, as that sketched by Putnam in the closing passages of *The Many Faces of Realism*.
26. For this topic and applications, see further B. Skyrms, 'Conditional chance', in J. Fetzer (ed.), *Probabilistic Causation: Essays in Honor of Wesley C. Salmon* (Dordrecht, 1988).
27. We can replace the simplifying assumption 2 by a much more general one, allowing mixtures to be made by integration instead of finite sum. There are still some limitations. In the case of a well-formulated physical theory, if not a human expert, we will have an exact mathematical description of the range of probability functions it allows.
28. See T. Seidenfeld, J. B. Kadane, M. J. Schervish, 'On the Shared Preferences of two Bayesian Decision Makers' MS, circulated 1987, and J. Broome, 'Bolker-Jeffrey Decision Theory and Axiomatic Utilitarianism', MS, 1988.
29. There is a voluminous literature by Skyrms, Cartwright, Lewis,

Harper and Gibbard, and others; D. Lewis, Causal Decision Theory', *Australian Journal of Philosophy*, 59 (1981), 5–30, and B. Armendt, 'A Foundation for Causal Decision Theory', *Topoi*, 5 (1986), 3–19, contain references to the rest of the literature.

30. J. Worrall, 'An Unreal Image', Review of van Fraassen (1980), *British Journal for the Philosophy of Science*, 35 (1984), 65–80.
31. This answers a question posed in another review, the one by M. Friedman, *Journal of Philosophy*, 79 (1982), 274–83; see also P. M. Churchland and C. A. Hooker (eds.), *Images of Science* (Chicago, 1985), 302–3.
32. See R. A. Rynasiewicz, 'Falsifiability and Semantic Eliminability', *Brit. J. Phil. Sci.*, 34 (1983), 225–41, and his Ph.D. dissertation, University of Minnesota, 1981; see further J. Earman, *A Primer on Determinism* (Dordrecht, 1986), 105–6.
33. See my 'The World We Speak of and the Language We Live in', in *Philosophy and Culture: Proc. of the XVII World Congress of Philosophy at Montreal*, 1983 (Montreal, 1986), 213–21.
34. It should be added however that I soon found it much more advantageous to concentrate on the propositions expressible by elementary statements, rather than on the statements themselves. This is how my emphasis changed progressively in my articles on logical aspects of quantum mechanics, from 1968 onward. At later points there is not even a bow in the direction of syntactic description.
35. This is a quick sketch of my attempts to make sense of modality without metaphysics. See ch. 6 of *The Scientific Image*, and my articles 'The Only Necessity is Verbal Necessity', Essence and Existence', and 'Essences and Laws of Nature'. See also R. Stalnaker, 'Anti-Essentialism', in P. French *et al.* (eds.), *Midwest Studies in Philosophy*, iv (Minneapolis, 1979), 343–55.

Chapter 9

1. C. B. Daniels and S. Todes, 'Beyond the Doubt of a Shadow: A Phenomenological and Linguistic Analysis of Shadows', in D. Ihde and R. M. Zaner (eds.), *Selected Studies in Phenomenology and Existential Philosophy* (The Hague, 1975), 203–16.
2. A model consists, formally speaking, of entities and relations among those entities. Not all parts are intended in this description of empirical adequacy. For example, a non-Euclidean space might be ismorphic to some part of a Euclidean space, if we allowed the introduction of new relations to single out this 'substructure'. That is not meant.
3. I use the word deliberately: it was a tragedy for philosophers of science

to go off on these logico-linguistic tangles, which contributed nothing to the understanding of either science or logic or language. It is still unfortunately necessary to speak polemically about this, because so much philosophy of science is still couched in terminology based on a mistake.

4. The impact of Suppes's innovation is lost if models are defined, as in many standard logic texts, to be partially linguistic entities, each yoked to a particular syntax. In my terminology here the models are mathematical structures, called models of a given theory only by virtue of belonging to the class defined to be the models of that theory.
5. Unlike perhaps Giere, I take it that normally the asserted relation of real systems to members of the defined class is not identity but some embedding or approximate embedding. See the Postscript to the 2nd edn. of my *An Introduction to the Philosophy of Time and Space* (New York, 1985).
6. For this topic, and for a sensitive analysis of the relations between state-spaces, parameters, and laws see E. Lloyd, *The Structure and Confirmation of Evolutionary Theory* (Westport, Conn., 1988).
7. See B. Ellis, 'The Origin and Nature of Newton's Laws of Motion', in R. Colodny, *Beyond the Edge of Certainty* (Englewood Cliffs, NJ, 1965), 29–68, and J. Earman and M. Friedman, 'The Nature and Status of Newton's Laws of Inertia', *Philosophy of Science*, 40 (1973), 329–59.
8. See M. Przelewski, *The Logic of Empirical Theories* (London, 1969); R. Wojcicki, 'Set Theoretic Representations of Empirical Phenomena', *Journal of Philosophical Logic*, 3 (1974), 337–43; M. L. Dalla Chiara, and G. Toraldo di Francia, 'A Logical Analysis of Physical Theories', *Rivista de Nuovo Cimento*, Serie 2, 3 (1973), 1–20; and 'Formal Analysis of Physical Theories', in G. Toraldo di Francia (ed.), *Problems in the Foundations of Physics* (Amsterdam, 1979); F. Suppe, 'Theories, The Formulations and the Operational Imperative', *Synthese*, 25 (1972), 129–59; P. Suppes, 'What is a Scientific Theory?', S. Morgenbesser (ed.) in *Philosophy of Science Today* (New York, 1967), 55–67; and 'The Structure of Theories and the Analysis of Data', in F. Suppe (ed.), *The Structure of Scientific Theories* (Urbana, Ill. 1974) 266–83.
9. I have discussed this further, with examples, in 'Theory Construction and Experiment: an Empiricist View', in P. Asquith and R. Giere (eds.), *PSA 1980*, ii (East Lansing, Mich., 1981), 663–78.
10. These reflections clearly bear e.g. on Glymour's theory of testing and relevant evidence, and his use of this important and original theory in arguments concerning scientific realism; see C. Glymour, *Theory and Evidence* (Princeton, NJ, 1980) and J. Earman (ed.), *Testing Scientific Theories*, *Minnesota Studies in the Philosophy of Science*, x

(Minneapolis, 1983). See also D. Baird, 'Tests of Significance Violate the Role of Implication', in P. Kitcher and P. Asquith (eds.), *PSA 1984*, (East Lansing, Mich., 1985), 81–92, esp. sect. 4.

Chapter 10

1. Taken from G. E. Martin, *Transformation Geometry* (New York, 1982), ch. 4. It is related to the argument, already given in ancient times, for the optical law of reflection; see ch. 1 sect. 4. The two alternative ways of solving the problem, by differentiation and by symmetry, and also its relation to Fermat's reasoning about optical reflection, are presented fully in A. Ostrowski, *Differential and Integral Calculus*, i. 318–19 (Glenview, Ill., 1968).
2. For general discussions see H. Weyl, *Symmetry* (Princeton, NJ, 1952) and J. Rosen, *Symmetry Discovered* (Cambridge, 1975).
3. See my *An Introduction to the Philosophy of Time and Space*, ch. 4 sect. 4b.
4. *Collected Works of Charles Sanders Peirce*, v, (Cambridge, Mass., 1964), 45–6.
5. See my *An Introduction to the Philosophy of Time and Space*, ch. 3 sect. 3.
6. Celsius originally made 100° the freezing point and 0° the boiling point; see P. van der Star (ed.), *Fahrenheit's Letters to Leibniz and van Boerhaave* (Amsterdam, 1983), p. 28 n. 1. I want to thank James Lenard for this reference.
7. Lucretius, *On the Nature of the Universe*, tr. R. E. Latham (New York, 1985), bk. ii, p. 66.
8. B. Russell, 'On the Notion of Cause with Applications to the Free Will Problem', in H. Feigl and M. Brodbeck, *Readings in the Philosophy of Science* (New York, 1953).
9. 'Deterministic Theories', in R. H. Thomason (ed.), *Formal Philosophy: Papers by R. Montague* (New York, 1974).
10. Dordrecht, 1986. I want to thank Roger Cooke for a helpful discussion.
11. Is the imagined world deterministic or indeterministic by our account? The question is elliptical: it applies only to the world classified as a certain kind of system. The kind of system described by classical physics minus conservation of mass and energy, this example shows us, is indeterministic.
12. The example also violates what Poincaré called the hypothesis of central forces, that is, the eighteenth-century idea that every force can be regarded as being exerted by some body or bodies. Here the

deceleration corresponds to a force not apparently covered by that hypothesis.

Chapter 11

1. I want to thank my student James Lenard for helpful comments on this chapter.
2. See further G. E. Martin, op. cit.
3. See the exposition and criticism by Ernst Mach, *The Science of Mechanics*, tr. T. McCormack (LaSalle, Ill. 1942), ch. 1 sect. 3.
4. G. D. Birkhoff, *Collected Mathematical Papers* (New York, 1950) ii. 890–9; iii. 788–804.
5. See E. Mach, *The Science of Mechanics*, ch. 3 sect. 3 for a full account.
6. See also J. C. C. McKinsey and P. Suppes, 'On the Notion of Invariance in Classical Mechanics', *British Journal for the Philosophy of Science*, 5 (1955), 290–302.
7. *Theory of Groups and Quantum Mechanics* (New York, 1931), ch. 3 sect. 14.
8. *Concepts of Mass* (New York, 1961), ch. 12.
9. In retrospect it is easy to see that some of Carnap's attempts to reformulate logical theory were inspired by the use of these notions in physics. His *logically determinate* corresponds to our *covariance*; and he had two explications for it. The first and main one was semantic: such a statement is either true for all interpretations or true for none. The second was partly syntactic: the result of syntactically transforming the statement by a uniform substitution (i.e. not just x for y but simultaneously also y for x) has always the same truth-value of the original. A third criterion which is much closer to our present usage would be obtained if we characterized transformations of interpretations (logical models) and defined the character in question as preservation of the truth value under all such transformations. The notion of covariance is certainly essentially a logical one, if applied to propositions; it betokens a certain kind of generality which amounts, in the extreme case, to the character of being either tautologous or self-contradictory. However, this extreme case is reached only if the group of transformations is so large as to preserve only logical structure.
10. See J. C. C. McKinsey, A. C. Sugar, and P. Suppes, 'Axiomatic Foundations of Classical Particle Mechanics', *Journal of Rational Mechanics and Analysis*, 2 (1953), 253–72.
11. Compare J. Aharoni, *Lectures on Mechanics* (Oxford, 1972), 290–304.

12. $\delta H/\delta t$ is also discarded at this point. See Aharoni, *Lectures on Mechanics*, 295.
13. *Symmetry* (Princeton, NJ, 1952), 26–7.
14. The concept of generality has a logical fascination all its own; in 'Essence and Existence' I have attempted to show how permutation symmetry helps to explicate it (and its contrary, the relation of being peculiarly about something specific), in modal semantics.

Chapter 12

1. I want to thank Mr Moore for allowing use of this example, and Dorothy Edgington for telling me about it.
2. See I. Todhunter, *A History of the Mathematical Theory of Probability* (London, 1865), 222–3.
3. See L. E. Maistrov, *Probability Theory: A Historical Sketch* (New York, 1974), 118–19.
4. See I. Todhunter, op. cit., 491–4.
5. In the Euclidean plane, a hyperbola is described by an equation of form $(x^2/a^2) - (y^2/b^2) = 1$, an ellipse by $(x^2/a^2) + (y^2/b^2) = 1$, and a parabola by $y^2 = 2px$.
6. I. Hacking, 'Equipossibility Theories of Probability', *British Journal for the Philosophy of Science*, 22 (1971), 339–55; K.-R. Bierman and M. Falk, 'G. W. Leibniz' *De incerti aestimatione*', *Forschungen und Fortschritte*, 31 (1957), 168–73; Leibniz, *Opuscules et fragments inédits*, ed. L. Couturat (Paris, 1903), 569–71.
7. Buffon's needle problem is discussed in many probability texts (e.g. J. V. Uspensky, *Introduction to Mathematical Probability* (New York, 1937) and in the standard histories of probability. It appeared in G. Buffon's supplement to his Natural History, *Essai d'arithmétique morale*. See further E. F. Schuster, 'Buffon's Needle Experiment', *American Mathematical Monthly*, 81 (1974), 26–9 and for a survey, H. Solomon, *Geometric Probability* (Philadelphia, 1978), ch. 1.
8. Results of the experiment are described in M. G. Kendall and P. A. P. Moran, *Geometrical Probability* (London, 1963). For serious doubts as to the reliability of the actual experiments, see N. T. Gridgeman, 'Geometric Probability and the Number π', *Scripta Mathematica*, 25 (1960), 183–95. The number of trials required according to Gridgeman is of the order of 90.10^{2n} for precision to n decimal places.
9. *Calcul des probabilités* (Paris, 1889), 4–5; 2nd edn. 1907, 4–7 (reprinted as 3rd edn., New York, 1972). See further the discussion of Bertrand's book in section 12.6 below.

10. Bertrand himself stated the problem in roughly this form: the problem of choosing a number at random from [0, 100] is the same as that of choosing its square (*Calcul des probabilités*, 2nd edn., 4). He adds that these contradictions can be multiplied to infinity. His own conclusion is that when the sample space is infinite, the notion of choosing at random '*n'est pas une indication suffisante*'—presumably not sufficient to create a well-posed problem.
11. See E. T. Jaynes, 'The Well-Posed Problem', *Foundations of Physics*, 3 (1973), 477–92, which has references to preceding discussions.
12. The concept of measure will be discussed more formally in the next chapter. Note here that a measure assigns non-negative numbers and is additive.
13. See ch. 13 sect. 3.
14. The probabilities must be the same for the events ($a \leqslant x \leqslant b$) and ($ka \leqslant y \leqslant kb$), so we deduce:

$$\int_0^b f(x)\,\mathrm{d}x = \int_0^{kb} f(kx)\,\mathrm{d}(kx) \text{ for all } b$$

hence $f(x) = kf(kx)$, for any positive constant k. This equation has a unique solution up to a constant multiplier:

$$f(x) = (1/x)$$

This gives us the basic measure:

$$M(a \leqslant x \leqslant b) = \mathrm{K}(\log b - \log a)$$

because $(1/x)$ is the derivative of $\log x$ (natural logarithm).

15. R. D. Rosenkranz, *Inference, Method and Decision* (Dordrecht, 1977), 63–8. See also R. D. Rosencrantz, *Foundations and Applications of Inductive Probability* (Atascadero, Calif., 1981), sects. 4.2 and 4.1.
16. A discussion of Buffon's needle along these lines is provided by M. Kac, E. R. van Kampen, and A. Winter, 'On Buffon's Problem and its Generalizations', *American Journal of Mathematics*, 61 (1939), 672–4.
17. P. Milne, 'A Note on Scale Invariance', *British Journal for the Philosophy of Science*, 34 (1983), 49–55.
18. Despite some rhetoric that seems to express the wish it were not so, Jaynes's article really agrees. Specifically, he implies that to treat a problem as solvable by symmetry considerations is to assume—what might be empirically false—that all relevant factors have been indicated in the statement of the problem ('The Well-Posed Problem', 489). Thus to treat a specific problem that way can itself not be justified a priori; the solution is correct for reality only conditional on that substantial assumption.
19. Ibid. 477–92.

20. E. T. Jaynes, *Papers on Probability, Statistics and Statistical Physics*, ed. R. Rosencrantz (Dordrecht, 1983), 128.
21. See the concise, perspicuous exposition in A. P. Dawid, 'Invariant Prior Distributions', in S. Kotz and N. L. Johnson, *Encyclopedia of Statistical Sciences* (New York, 1983), 228–36. The main figures in the search for 'invariant priors' besides Jaynes were H. Jeffreys's classic text *Theory of Probability* (Oxford, 1939), and D. Fraser (see e.g. his 'The Fiducial Method and Invariance', *Biometrica*, 48 (1961), 261–80).
22. See E. T. Jaynes, 'Prior Probabilities', *IEEE Transactions of the Society of Systems Sciences Cybernetics* SSC-4 (1968), 227–41; C. Villegas, 'On Haar Priors', in V. P. Godambe *et al.* (eds.), *Foundations of Statistical Inference* (Toronto, 1971), 409–14; 'Inner Statistical Inference', *Journal of the American Statistical Association*, 72 (1977), 453–8 and *Annals of Statistics*, 9 (1981), 768–76. I want to thank Dr F. G. Perey, of the Oak Ridge National Laboratory, for letting me have a copy of his excellent and insightful presentation of this approach, 'Application of Group Theory to Data Reduction', Report ORNL-5908 (Sept. 1982).
23. The 'nice properties' referred to in the text are topological properties of the group; if it is locally compact and transitive (for any y and z in the set there is a member g of G such that $g(y) = z$) then the left Haar measure is unique up to a multiplicative constant. However, in order for P to be also independent of the choice of reference point x_0, the left and right Haar measure must be the same; this is guaranteed if the group is compact.

Chapter 13

1. 'A Note on Jeffrey Conditionalization', *Philosophy of Science*, 45 (1978), 361–7.
2. See my 'Rational Belief and Probability Kinematics', *Philosophy of Science*, 47 (1980), 165–87.
3. I. Hacking, 'Slightly More Realistic Personal Probability', *Philosophy of Science*, 34 (1967), 311–25.
4. Cf. F. P. Ramsey, 'Truth and Probability', repr. in H. E. Kyburg Jun. and H. E. Smokler (eds.), *Studies in Subjective Probability* (Huntingdon, NY, 1980), 23–52; 3 p. 40.
5. See my 'Rationality does not Require Conditionalization', forthcoming.
6. I think of time here as discrete, but the unit can of course be chosen as small as you like. The proposition $E(t)$ can be identified as follows: it is the logically strongest proposition X in the domain of P such that $P(t)(X) = 1$. Note that we must take into account also the case of someone who gives positive probability to a proposition which he

gave probability *zero* before. That case is left aside here but discussed in 'Rationality does not Require Conditionalization', where it is shown that nothing very advanced is needed to substantiate the assertion that such a person as here described can always be simulated by a perfect Conditionalizer.

7. For this history, I am especially indebted to G. H. Moore, 'The Origins of Zermelo's Axiomatization of Set Theory', *Journal of Philosophical Logic*, 7 (1978), 307–29; and 'Lebesgue's Measure Problem and Zermelo's Axiom of Choice: the Mathematical Effects of a Philosophical Dispute', *Annals of the New York Academy of* Sciences (1983), 129–54. The Measure Problem is stated in H. Lebesgue, 'Intégrale, longeur, aire', *Annali di Mathematica Pura ed Applicata*, 3 (1902), 231–359.
8. We may note in passing that Banach proved in 1923 that the Measure Problem does have solutions for dimensions 1 and 2 provided we settle for finite additivity (measure functions, as I called them). But he also proved a much stronger negative result for measures properly speaking: quite aside from requirements of congruence, there cannot be a measure defined on all subsets of [0, 1] which (like Lebesgue measure) gives zero to each point (i.e. to each unit set $\{x\}$).
9. The crucial results appealed to in the following are Theorems of Kuratowski, Birkhoff, and Horn, Tarski, and Maharam; see G. Birkhoff, *Lattice Theory*, 3rd edn. (Providence, RI, 1967); and D. A. Kappos, *Probability Algebras and Stochastic Spaces* (New York, 1969), chs. 2. 4 and 3. 3. For a more extensive discussion focusing on the relation between probabilities and frequencies, see my 'Foundations of Probability: A Modal Frequency Interpretation', in G. Toraldo di Francia (ed.), *Problems in the Foundations of Physics* (Amsterdam; 1979), esp. 345–65.
10. The following argument was first given in a different setting, see my 'A Demonstration of the Jeffrey Conditionalization Rule', *Erkenntnis*, 24 (1986), 17–24, 'Symmetry Arguments in Probability Kinematics' (with R. I. G. Hughes), in P. Kitcher and P. Asquith (eds.), *PSA 1984*, 851–69 (East Lansing, Mich., 1985), and 'Symmetries in Personal Probability Kinematics', in N. Rescher (ed.), *Scientific Inquiry in Philosophical Perspective* (Lanham, Md., 1987).
11. See further my papers cited earlier in this chapter, and also my 'Discussion: A Problem for Relative Information Minimizers', *British Journal for the Philosophy of Science*, 32 (1981), 375–9, and 'A Problem for Relative Information Minimizers in Probability Kinematics, Continued' (with R. I. G. Hughes and G. Harman), *British Journal for the Philosophy of Science*, 37 (1986), 453–75.
12. At this point the argument follows that of P. Teller and A. Fine, 'A

Characterization of Conditional Probability', *Mathematical Magazine*, 48 (1975), 267–70.

13. For the proof of convergence, see John Collins's Appendix to my 'Symmetries in Personal Probability Kinematics'.
14. See van Fraassen 'A Problem for Relative Information Minimizers'; van Fraassen, Hughes, and Harman, 'A Problem for Relative Information Minimizers, Continued'; van Fraassen 'Symmetries in Personal Probability Kinematics'. These articles also contain the calculations omitted below.
15. R. D. Levine, and M. Tribus, *The Maximum Entropy Formalism* (Cambridge, Mass., 1979); J. E. Shore and R. W. Johnson, 'Axiomatic Derivation of the Principle of Maximum Cross-Entropy', *IEEE Transactions Information Theory*, IT–26 (1980), 26–37; J. Skilling, 'The Maximum Entropy Method', *Nature*, 309 (28 June 1984), 748–9; Y. Tikochinsky, N. Z. Tishby, and R. D. Levine, 'Consistent Inference of Probabilities for Reproducible Experiments', *Physical Review Letters*, 52 (1984), 1357–60.
16. P. Diaconis and S. Zabell, 'Updating Subjective Probability', *Journal of the American Statistical Association*, 77 (1982), 822–30.
17. Levine and Tribus, op. cit.
18. For this problem, see A. Hobson, *Concepts in Statistical Mechanics* (New York, 1971), 36, 42, 49.
19. See P. M. Williams, 'Bayesian Conditionalization and the Principle of Minimum Information', *British Journal for Philosophy of Science*, 31 (1980), 131–144.
20. R. W. Johnson. 'Axiomatic Characterization of the Directed Divergences and Their Linear Combinations', *IEEE Transactions Information Theory*, IT–25 (1979), 709–16.

21 K. Friedman and A. Shimony, 'Jaynes' Maximum Entropy Prescription and Probability Theory', *Journal of Statistical Physics*, 3 (1971), 381–4. See also A. Shimony, 'Comment on the Interpretation of Inductive Probabilities', ibid. 9 (1973), 187–91.

22. For the most sensitive treatment so far, see B. Skyrms, 'Maximum Entropy as a Special Case of Conditionalization', *Synthese*, 636 (1985), 55–74, and 'Updating, Supposing and MAXENT', forthcoming.

BIBLIOGRAPHY

ABBOTT, P., and MARSHALL, H., *National Certificate Mathematics*, 2nd edn. (London: English Universities Press, 1960).

AHARONI, J., *Lectures on Mechanics* (Oxford: Oxford University Press, 1972).

ARMENDT, B., 'A Foundation for Causal Decision Theory', *Topoi*, 5 (1986), 3–19.

ARMSTRONG, D., 'Reply to van Fraassen', *Australasian Journal of Philosophy*, 66 (1988), 224–9.

—— *What is a Law of Nature?* (Cambridge: Cambridge University Press, 1983).

ASQUITH, P., and GIERE, R. (eds.), *PSA 1980* (East Lansing, Mich.: Philosophy of Science Assocation, 1981).

BALLENTINE, L. E., 'Can the Statistical Postulate of Quantum Theory Be Derived? A Critique of the Many-Universe Interpretation', *Foundations of Physics*, 3 (1973), 229–40.

BEATTIE, J., 'What's Wrong with the Received View of Evolutionary Theory', in P. Asquith and R. Giere (eds.), *PSA 1980*, 397–426.

BELTRAMETTI, E., and VAN FRAASSEN, B., *Current Issues in Quantum Logic* (New York: Plenum Press, 1981).

BENJAMIN, WALTER, *Illuminations* (New York: Schocken Books, 1969).

—— *One-Way Street and Other Writings* (London: NLB, 1979).

BERTRAND, J., *Calcul des probabilités* (1st edn., Paris: Gauthier-Villars, 1889; 2nd edn. 1907; repr. as 3rd edn., New York: Chelsea Pub. Co., 1972).

BIERMANN, K.-R., and FALK, M., 'G. W. Leibniz' *De incerti aestimatione*', *Forschungen und Fortschritte*, 31 (1957), 168–73.

BIRKHOFF, G., *Lattice Theory*, 3rd edn. (Providence, RI: American Mathematical Society, 1967).

BIRKHOFF, G. D., *Collected Mathematical Papers* (New York: American Mathematical Society, 1950).

BLACKMORE, SIR RICHARD, see Heath-Stubbs and Salman.

BOYD, R., 'The Current Status of Scientific Realism', in J. Leplin (ed.) *Scientific Realism*, 41–82.

BOYLE, R., *The Works of the Honourable Robert Boyle*, ed. T. Birch. (London, 1673).

BROOME, J., 'Bolker–Jeffrey Decision Theory and Axiomatic Utilitarianism', MS 1988.

CARNAP, R., *The Continuum of Inductive Methods* (Chicago: University of Chicago Press, 1952).

—— *Philosophical Foundations of Physics*, ed. Martin Gardner (New York: Basic Books, 1966).

CARRÉ, M. H., *Realists and Nominalists* (Oxford: Oxford University Press, 1950).

CARROLL, J., 'Ontology and the Laws of Nature', *Australasian Journal of Philosophy*, 65 (1987), 261–76.

CARTWRIGHT, N., *How the Laws of Physics Lie* (Oxford: Oxford University Press, 1983).

CHURCHLAND, P. M., 'Conceptual Progress and World/Word Relations: In Search of the Essence of Natural Kinds', *Canadian Journal of Philosophy*, 15 (1985), 1–17.

—— and HOOKER, C. A. (eds.), *Images of Science: Essays on Realism and Empiricism, with A Reply by Bas C. van Fraassen* (Chicago: University of Chicago Press, 1985).

CLIFFORD, W. K., *Lectures and Essays* (London: Macmillan and Co., 1879).

COHEN, R. S., and LAUDAN, L. (eds.), *Physics, Philosophy, and Psychoanalysis* (Dordrecht: Reidel, 1983).

COLODNY, R. (ed.), *Beyond the Edge of Certainty* (Englewood Cliffs, NJ: Prentice-Hall, 1965).

COUTURAT, L. (ed.), Leibniz, *Opuscules et fragments inédits* (Paris: 1903).

DALLA CHIARA, M. L., and TORALDO DI FRANCIA, G., 'A Formal Analysis of Physical Theories', in G. Toraldo di Francia (ed.), *Problems in the Foundations of Physics*.

—— and —— 'Logical Analysis of Physical Theories', *Rivista di Nuovo Cimento*, Serie 2,3 (1973), 1-20.

DANIELS, C. B., and TODES, S., 'Beyond the Doubt of a Shadow: A Phenomenological and Linguistic Analysis of Shadows', in D. Ihde and R. M. Zaner (eds.) *Selected Studies in Phenomenology and Existential Philosophy*, 203–16.

DANTE ALIGHIERI, *Dante's Inferno*, tr. T. Philips (London: Thames and Hudson, 1985).

DAVIDSON, D., *Essays on Actions and Events* (Oxford: Oxford University Press, 1980).

DAWID, A. P., 'Calibration-Based Empirical Probability', *Annals of Statistics*, 13 (1985), 1251–73.

—— 'The Impossibility of Inductive Evidence', *Journal of the American Statistical Association*, 80 (1985), 340–1.

—— 'Invariant Prior Distributions', in S. Kotz and N. L. Johnson, *Encyclopedia of Statistical Sciences* (New York: Wiley, 1983), 228–36.

DE FINETTI, B., 'Methods for Discriminating Levels of Partial Knowledge concerning a Test Item', *British Journal of Mathematical and Scientific Psychology*, 18 (1965), 87–123.

DE FINETTI, B., 'Probability: Beware of Falsifications', in H. Kyburg, Jun. and H. Smokler, *Studies in Subjective Probability*.

DESCARTES, R., *Philosophical Works of Descartes*, tr. E. S. Haldane and R. T. Ross (Cambridge: Cambridge University Press, 1985).

DIACONIS, P., and ZABELL, S., 'Updating Subjective Probability', *Journal of the American Statistical Association*, 77 (1982), 822–30.

DIJKSTERHUIS, E. J., *The Mechanization of the World Picture* (Oxford: Oxford University Press, 1961).

DRETSKE, F., 'Laws of Nature', *Philosophy of Science*, 44 (1977) 248–68.

EARMAN J., 'Laws of Nature: The Empiricist Challenge', in R. J. Bogdan (ed.), *D. H. Armstrong* (Dordrecht: Reidel, 1984), 191–224.

—— 'The Universality of Laws', *Philosophy of Science*, 45 (1978), 173–81.

—— *A Primer on Determinism* (Dordrecht: Reidel, 1986).

—— (ed.), *Testing Scientific Theories*, Minnesota Studies in the Philosophy of Science, x (Minneapolis: University of Minnesota Press, 1984).

—— and FRIEDMAN, M., 'The Nature and Status of Newton's Laws of Inertia', *Philosophy of Science*, 40 (1973), 329–59.

ELLIS, B., 'The Origin and Nature of Newton's Laws of Motion', in R. Colodny, *Beyond the Edge of Certainty*, 29–68.

FARBER, M. (ed.), *Philosophical Essays in Memory of Edmund Husserl* (Cambridge, Mass.: Harvard University Press, 1940).

FEIGL, H., and BRODBECK, M., *Readings in the Philosophy of Science* (New York: Appleton-Century-Crofts, 1953).

FETZER, J. H., (ed.), *Probability and Causality: Essays in Honor of Wesley C. Salmon* (Dordrecht: Kluwer, 1988).

FEYNMAN, R., *The Character of Physical Law* (Cambridge, Mass.: MIT Press, 1965).

FIELD, H., 'A Note on Jeffrey Conditionalization', *Philosophy of Science*, 45 (1978), 361–7.

FINE, A., 'The Natural Ontological Attitude', in J. Leplin (ed.), *Scientific Realism*, 83–107.

FITCH, F., *Symbolic Logic* (New York: Ronald, 1952).

FLOISTAD, G., and VON WRIGHT, G. H. (eds.), *Philosophy of Language/ Philosophical Logic* (The Hague: Nijhoff, 1981).

FORREST, P., 'What Reasons do We Have for Believing There Are Laws of Nature?', *Philosophical Inquiry International Quarterly*, 7 (1985), 1–12.

FRASER, D., 'The Fiducial Method and Invariance', *Biometrica*, 48 (1961), 261–80.

FREDDOSO, A. J., 'Medieval Aristotelianism and the Case against Secondary Causes in Nature', in T. Morr (ed.), *Divine and Human Action*.

FRENCH, P. (ed.), *Midwest Studies*, iv (Metaphysics) (Minneapolis: University of Minnesota Press, 1979).

FRIEDMAN, K., and SHIMONY, A., 'Jaynes' Maximum Entropy Prescription and Probability Theory', *Journal of Statistical Physics*, 3 (1971), 381–4.

FRIEDMAN, M., 'Truth and Confirmation', *Journal of Philosophy*, 76 (1979), 361–82.

—— Review of van Fraassen, *The Scientific Image*, *Journal of Philosophy*, 79 (1982), 274–83.

GAIFMAN, H., 'A Theory of Higher Order Probability', in B. Skyrms *et al.* (eds.), *Causality, Chance, and Choice* (Dordrecht: Reidel, 1988).

GIERE, R., *Understanding Scientific Reasoning* (New York: Holt, Rinehart, and Winston, 1979).

GLYMOUR, C., *Theory and Evidence* (Princeton, NJ: Princeton University Press, 1980).

GODAMBE, V. P., *et al.* (eds.) *Foundations of Statistical Inference* (Toronto: Holt, Rinehart, and Winston, 1971).

GOODMAN, N., *Fact, Fiction, and Forecast* (Atlantic Highlands, NJ: Athlone Press, 1954).

GRIDGEMAN, N. T., 'Geometric Probability and the Number π', *Scripta Mathematica*, 25 (1960), 183–95.

HACKER, P. M. S., *Insight and Illusion* (Oxford: Oxford University Press, 1986).

HACKING, I., *The Emergence of Probability* (London: Cambridge University Press, 1975).

—— 'Equipossibility Theories of Probability', *British Journal for the Philosophy of Science*, 22 (1971), 339–55.

—— 'A Slightly More Realistic Personal Probability', *Philosophy of Science*, 34 (1967), 311–25.

HALMOS, P. R., *Lectures on Ergodic Theory* (New York: Chelsea, 1956).

HAMBLIN, C., *Imperatives* (Oxford: Blackwell, 1987).

HANNA, J., 'Empirical Adequacy', *Philosophy of Science*, 50 (1983), 1–34.

HARMAN, G., 'The Inference to the Best Explanation', *Philosophical Review*, 74 (1965), 88–95.

—— *Change in View* (Cambridge, Mass.: MIT Press, 1986).

HARPER, W., *et al.* (eds.), *Ifs* (Dordrecht: Reidel, 1981).

HEATH-STUBBS, J., and SALMAN, P. (eds.), *Poems of Science* (Harmondsworth: Penguin, 1984).

HOBSON, A., *Concepts in Statistical Mechanics* (New York: Gordon and Breach, 1971).

HOOKER, C. A. (ed.), *Contemporary Research in the Foundations and Philosophy of Quantum Theory* (Dordrecht: Reidel, 1973).

HORWICH, P., *Probability and Evidence* (Cambridge: Cambridge University Press, 1982).

HUXLEY, A., *Ape and Essence* (London: Triad/Panther, 1985).

IHDE, D., and ZANER, R. M. (eds.), *Selected Studies in Phenomenology and Existential Philosophy* (The Hague: Nijhoff, 1975).

JAMES, W., *Essays in Pragmatism* (New York: Hafner, 1948).

JAMMER, M., *Concepts of Mass* (New York: Harper and Row, 1961).

JAYNES, E. T., 'Prior Probabilities', *IEEE Transactions of the Society of Systems Sciences Cybernetics* SSC-4 (1968), 227–41.

—— 'The Well-Posed Problem', *Foundations of Physics*, 3 (1973), 477–92.

—— *Papers on Probability, Statistics and Statistical Physics*, ed. R. Rosencrantz (Dordrecht: Reidel, 1983).

JEFFREY, R. C., 'Bayesianism with a Human Face', in J. Earman (ed.), *Testing Scientific Theories*, 133–56.

—— 'Choice, Chance, and Credence', in G. Floistad and G. H. von Wright (eds.), *Philosophy of Language/Philosophical Logic*, 367–86.

—— Review of D. Miller *et al.*, 'A Paradox of Information', in *Journal of Symbolic Logic*, 35 (1970), 124–7.

—— *The Logic of Decision*, 2nd edn. (Chicago: University of Chicago Press, 1983).

JEFFREYS, H., *Theory of Probability* (Oxford: Clarendon Press, 1939).

JOHNSON, R. W., 'Axiomatic Characterization of the Directed Divergences and Their Linear Combinations', *IEEE Transactions Information Theory*, IT-25 (1979), 709–16.

JOHNSTON, M., 'Why Having a Mind Matters', in E. LePore and B. McLaughlin (eds.), *Actions and Events* (Oxford: Blackwell, 1985), 408–26.

KAC, M., VAN KAMPEN, E. R., and WINTER, A., 'On Buffon's Problem and its Generalizations', *American Journal of Mathematics*, 61 (1939), 672–4.

KANT, I., *The Critique of Pure Reason*, tr. N. Kemp Smith (London: Macmillan, 1980).

KAPPOS, D. A., *Probability Algebras and Stochastic Space* (New York, 1969).

KENDALL, M. G., and MORAN, P. A. P., *Geometrical Probability* (London: Griffin, 1963).

KITCHER, P. and ASQUITH, P. (eds.), *PSA 1984* (East Lansing, Mich.: Philosophy of Science Association, 1985).

—— and SALMON, W., 'Van Fraassen on Explanation', *Journal of Philosophy*, 84 (1987), 315–30.

KUHN, T., *The Structure of Scientific Revolutions*, 2nd edn. (Chicago: University of Chicago Press, 1970).

KYBURG, Jun., H., and SMOKLER, H. (eds.), *Studies in Subjective Probability* (Huntingdon, NY: Krieger, 1980).

LAMBERT, K., and BRITTAN, G., *An Introduction to the Philosophy of Science*, 3rd edn. (Atascadero, Calif.: Ridgeview, 1987).

LAUDAN, L., *Science and Hypothesis* (Dordrecht: Reidel, 1981).

LEBESGUE, H., 'Intégrale, longueur, aire', *Annali di Mathematica Pura ed Applicata*, 3 (1902), 231–359.

LEIBNIZ, G. W., *Opuscules et fragments inédits*, ed. L. Couturat (Paris: 1903), 569–71.

LEPLIN, J. (ed.), *Scientific Realism* (Berkeley: University of California Press, 1984).

LEVINE, R., and TRIBUS, M., *The Maximum Entropy Formalism* (Cambridge, Mass.: MIT Press, 1979).

LEWIS, D. K., 'Causal Decision Theory', *Australasian Journal of Philosophy*, 59 (1981), 5–30.

—— 'New Work for a Theory of Universals', *Australasian Journal of Philosophy*, 61 (1983), 343–77.

—— 'A Subjectivist's Guide to Objective Chance', in W. Harper *et al.* (eds.), *Ifs*.

—— Counterfactuals (Cambridge, Mass.: Harvard University Press, 1973).

—— *Philosophical Papers* (Oxford: Oxford University Press, 1986).

LLOYD, E. A., *The Structure and Confirmation of Evolutionary Theory* (Westport, Conn.: Greenwood Press, 1988).

LOECK, G., 'Auskunft über die Gesetzesartigkeit aus ihrer Konstruktion', *Osnabrücker Philosophische Schriften* (Osnabrück: Universität Osnabrück, 1985).

—— *Der cartesische Materialismus: Maschine, Gesetz und Simulation* (Frankfurt: Peter Lang, 1986).

LUCRETIUS, *On the Nature of the Universe*, tr. R. E. Latham (New York: Viking Penguin, 1985).

MCCALL, S., 'Counterfactuals Based on Real Possible Worlds', *Nous*, 18 (1984), 463–77.

—— 'Time and the Physical Modalities', *Monist*, 53 (1969), 426–46.

MACH, E., *The Science of Mechanics*, tr. T. McCormack (LaSalle, Ill.: Open Court, 1942).

MCKINSEY, J. C. C., SUGAR, A. C., and SUPPES, P., 'Axiomatic Foundations of Classical Particle Mechanics', *Journal of Rational Mechanics and Analysis*, 2 (1953), 253–72.

MCKINSEY, J. C. C., and SUPPES, P., 'On the Notion of Invariance in Classical Mechanics', *British Journal for the Philosophy of Science*, 5 (1955), 290–302.

MCMULLEN, E., 'The Development of Philosophy of Science 1600–1850', in J. Hodge *et al.* (eds.), *Encyclopedia of the History of Science*, forthcoming.

—— 'The Goal of Natural Science', *Proceedings of the American Philosophical Association*, 58 (1984), 37–64.

MAISTROV, L. E., *Probability Theory: A Historical Sketch* (New York: Academic Press, 1974).

MARGENAU, H., and LINDSAY, R. B., *Foundations of Physics* (New York: Dover 1957).

MARTIN, G. E., *Transformation Geometry* (New York: Springer-Verlag, 1982).

MILL, J. S., *Collected Works* (Toronto: University of Toronto Press, 1963).

—— *A System of Logic* (London: Parker, 1846).

MILLER, D., 'A Paradox of Information', *British Journal for the Philosophy of Science*, 17 (1966), 59–61.

MILNE, P., 'A Note on Scale Invariance', *British Journal for the Philosophy of Science*, 34 (1983), 49–55.

MONTAGUE, R., 'Deterministic Theories', in R. H. Thomason (ed.), *Formal Philosophy: Papers by R. Montague.*

MOORE, G. H., 'Lebesque's Measure Problem and Zermelo's Axiom of Choice: The Mathematical Effects of a Philosophical Dispute', *Annals of the New York Academy of Sciences* (1983), 129–54.

—— 'Origins of Zermelo's Axiomatization of Set Theory', *Journal of Philosophical Logic*, 7 (1978), 307–29.

MORGENBESSER, S. (ed.), *Philosophy of Science Today* (New York: Basic Books, 1967).

MORR, T. (ed.), *Divine and Human Action: Essays in the Metaphysics of Theism* (Ithaca, NY: Cornell University Press, 1988).

MOULINES, C. U., 'Approximate Application of Empirical Theories', *Erkenntnis*, 10 (1976), 201–27.

NORMAN, S. J., 'Subsystem States in Quantum Theory and Their Relation to the Measurement Problem', Ph.D. dissertation (Stanford University, 1981).

OAKES, D., 'Self-calibrating Priors do not Exist', *Journal of the American Statistical Association*, 80 (1985), 339.

OSTROWSKI, A., *Differential and Integral Calculus*, i (Glenview, Ill.: Scott, Foresman, 1968).

PARGETER, R., 'Laws and Modal Realism', *Philosophical Studies*, 46 (1984), 335–47.

PEIRCE, C. S., *Collected Papers of Charles Sanders Peirce*, ed. C. Hartshorne and P. Weiss (Cambridge, Mass.: Harvard University Press, 1931–5 and 1964).

—— *Values in a World of Chance: Selected Writings of C. S. Peirce*, ed. P. Wiener (New York: Dover Books, 1966).

PEREY, F. G., 'Application of Group Theory to Data Reduction', Report ORNL-5908 (Oak Ridge National Laboratory, September 1982).

PEREZ RANSANZ, A. R., 'El concepto de teoría empirica según van Fraassen' (with English tr.) *Critica*, 17 (1985), 3–20.

PIATELLI-PALMARINI, M., 'Not on Darwin's Shoulders: A Critique of Evolutionary Epistemology', *Boston Colloquium for the Philosophy of Science* (January 1988).

PICKHARDT, R., and WALLACE, J. B., 'A Study of the Performance of Subjective Probability Assessors', *Decision Sciences*, 5 (1974), 347–63.

PRZELEWSKI, M., *The Logic of Empirical Theories* (London: Routledge and Kegan Paul, 1969).

PUTNAM, H., 'Reflexive Reflections', *Erkenntnis*, 22 (1985), 143–54.

—— *The Many Faces of Realism* (LaSalle, Ill.: Open Court, 1987).

RAMSEY, F. P., 'Truth and Probability', reprinted in H. E. Kyburg Jun. and H. E. Smokler (eds.), *Studies in Subjective Probability*, 23–52.

REICHENBACH, H., *Elements of Symbolic Logic* (New York: Macmillan, 1947).

—— *Modern Philosophy of Science* (New York: Humanities Press, 1959).

RESCHER, N., *The Logic of Commands* (London: Routledge, 1966).

—— (ed.), *Scientific Inquiry in Philosophical Perspective* (Lanham, Md.: University Press of America, 1987).

RORTY, A., and MCLAUGHLIN, B. (eds.), *Perspectives on Self-Deception* (Los Angeles, Calif.: University of California Press, 1988).

ROSEN, J., *Symmetry Discovered* (Cambridge: Cambridge University Press, 1975).

ROSENCRANTZ, R. D., *Foundations and Applications of Inductive Probability* (Atascadero, Calif.: Ridgeview, 1981).

—— *Inference, Method and Decision* (Dordrecht: Reidel, 1977).

—— (ed.), *E. T. Jaynes: Papers on Probability, Statistics and Statistical Physics* (Dordrecht: Reidel, 1983).

RUBY, J. E., 'The Origins of Scientific "Law" ', *Journal of the History of Ideas*, 47 (1986), 341–59.

RUSSELL, B., 'On the Notion of Cause with Applications to the Free Will Problem', in H. Feigl and M. Brodbeck, *Readings in the Philosophy of Science*.

—— *The Problems of Philosophy* (New York: Oxford University Press, 1959).

RYNASIEWICZ, R. A., 'Falsifiability and Semantic Eliminability', *British Journal for the Philosophy of Science*, 34 (1983), 225–41.

SCHEIBE, E., 'On the Structure of Physical Theories', *Acta Philosophica Fennica*, 30 (1978), 205–23.

SCHUSTER, E. F., 'Buffon's Needle Experiment', *American Mathematical Monthly*, 81 (1974), 26–9.

SEIDENFELD, T., 'Calibration, Coherence, and Scoring Rules', *Philosophy of Science*, 52 (1985), 274–94.

—— KADANE, J. B., and SCHERVISH, M. J., 'On the Shared Preferences of Two Bayesian Decision Makers', MS circulated 1987.

SELLARS, W., 'Concepts as Involving Laws and Inconceivable without Them', *Philosophy of Science*, 15 (1948), 287–315.

SHIMONY, A., 'An Adamite Derivation of the Principles of the Calculus of Probability', in J. H. Fetzer (ed.), *Probability and Causality: Essays in Honor of Wesley C. Salmon.*

SHIMONY, A., 'Comment on the Interpretation of Inductive Probabilities', *Journal of Statistical Physics*, 9 (1973), 187–91.

—— see also K. Friedman and A. Shimony.

SHORE, J. E., and JOHNSON, R. W., 'Axiomatic Derivation of the Principle of Maximum Cross-Entropy', *IEEE Transactions Information Theory*, IT-26 (1980), 26–37.

SKILLING, J., 'The Maximum Entropy Method', *Nature*, 309 (28 June 1984), 748–9.

SKYRMS, B., 'Conditional Chance', in J. Fetzer (ed.), *Probabilistic Causation: Essays in Honor of Wesley C. Salmon.*

—— 'Dynamic Coherence and Probability Kinematics', *Philosophy of Science*, 54 (1987), 1–20.

—— 'Maximum Entropy as a Special Case of Conditionalization', *Synthese*, 636 (1985), 55–74.

—— 'Updating, Supposing and MAXENT', forthcoming in *Theory and Decision.*

—— *Choice and Chance* (Belmont, Calif.: Wadsworth, 1986).

—— *Pragmatics and Empiricism* (New Haven, Conn.: Yale University Press, 1984).

—— *et al.* (eds.), *Causality, Chance, and Choice* (Dordrecht: Reidel, 1988).

SMART, J. J. C., 'Laws of Nature and Cosmic Coincidences', *Philosophical Quarterly*, 35 (1985), 272–80.

SOLOMON, H., *Geometric Probability* (Philadelphia, Pa.: Society for Industrial and Applied Mathematics, 1978).

STALNAKER, R., 'Anti-Essentialism', in P. French *et al.* (eds.), *Midwest Studies*, iv (Metaphysics).

—— *Inquiry* (Cambridge, Mass.: MIT Press, 1984).

STANIFORTH, M. (tr.), *Early Christian Writings* (New York: Viking Penguin, 1968).

STEGMÜLLER, W., *The Structuralist View of Theories* (Berlin: Springer-Verlag, 1979).

SUPPE, F., 'Theories, Their Formulations and the Operational Imperative', *Synthese*, 25 (1972), 129–59.

—— *The Semantic Conception of Theories and Scientific Realism* (Urbana, Ill.: University of Illinois Press, 1988).

—— (ed.), *The Structure of Scientific Theories* (Urbana, Ill.: University of Illinois Press, 1974).

SUPPES, P., 'The Structure of Theories and the Analysis of Data', in F. Suppe (ed.), *The Structure of Scientific Theories*, 266–83.

—— 'What is a Scientific Theory?', in S. Morgenbesser (ed.), *Philosophy of Science Today.*

SWARTZ, N., *The Concept of Physical Law* (Cambridge: Cambridge University Press, 1985).

SWOYER, C., 'The Nature of Natural Laws', *Australasian Journal of Philosophy*, 60 (1982), 203–23.

TELLER, P., 'Conditionalization and Observation', *Synthese*, 26 (1973), 218–58.

—— and Fine, A., 'A Characterization of Conditional Probability', *Mathematical Magazine*, 48 (1975), 267–70.

THOMASON, R. H., 'Combinations of Tense and Modality', in D. Gabbay and F. Guenther (eds.), *Handbook of Philosophical Logic* (Dordrecht: Reidel, 1984).

—— 'Indeterminist Time and Truth-Value Gaps', *Theoria*, 36 (1970), 264–81.

—— (ed.), *Formal Philosophy: Papers by R. Montague* (New Haven, Conn.: Yale University Press, 1974).

THOMPSON, P., *The Structure of Biological Theories* (Albany, NY: State University of New York Press, 1988).

TIKOCHINSKY, Y., TISHBY, N., and LEVINE, R. D., 'Consistent Inference of Probabilities for Reproducible Experiments', *Physical Review Letters*, 52 (1984), 1357–60.

TODHUNTER, I., *A History of the Mathematical Theory of Probability* (London: Macmillan, 1865).

TOOLEY, M., 'The Nature of Law', *Canadian Journal of Philosophy*, 7 (1977), 667–98.

—— *Causation: A Realist Approach* (Oxford: Oxford University Press, 1987).

TORALDO DI FRANCIA, G. (ed.), *Problems in the Foundations of Physics* (Amsterdam: North-Holland, 1979).

USPENSKY, J. V., *Introduction to Mathematical Probability* (New York: McGraw-Hill, 1937).

VALLENTYNE, P., 'Explicating Lawhood', forthcoming in *Philosophy of Science*, 55 (1988).

VAN DER STAR, P. (ed.), *Fahrenheit's Letters to Leibniz and van Boerhaave* (Amsterdam: Rodopi, 1983).

VAN FRAASSEN, B. C., 'Armstrong on Laws and Probabilities', *Australasian Journal of Philosophy*, 65 (1987), 243–60.

—— 'Belief and the Will', *Journal of Philosophy*, 81 (1984), 235–56.

—— 'Calibration: A Frequency Justification for Personal Probability', in R. S. Cohen and L. Laudan (eds.), *Physics, Philosophy, and Psychoanalysis*, 295–319.

—— 'A Demonstration of the Jeffrey Conditionalization Rule', *Erkenntnis*, 24 (1986), 17–24.

—— 'Discussion: A Problem for Relative Information Minimizers', *British Journal for the Philosophy of Science*, 32 (1981), 375–9.

van Fraassen, B. C., 'Empiricism in the Philosophy of Science', in P. M. Churchland and C. A. Hooker (eds.), *Images of Science*.

—— 'Essence and Existence', in N. Rescher (ed.), *Studies in Ontology*, American Philosophical Quarterly Monograph Series No. 12 (Oxford: Blackwell, 1978), 1–25.

—— 'Essences and Laws of Nature', in R. Healey (ed.), *Reduction, Time and Reality* (Cambridge: Cambridge University Press, 1981).

—— 'Foundations of Probability: A Modal Frequency Interpretation', in G. Toraldo di Francia (ed.), *Problems in the Foundations of Physics*.

—— 'Glymour on Evidence and Explanation', in J. Earman (ed.), *Testing Scientific Theories*.

—— 'A Modal Interpretation of Quantum Mechanics', in Beltrametti and van Fraassen.

—— 'On the Question of Identification of a Scientific Theory', *Critica*, 17 (1985), 21–30.

—— 'The Only Necessity is Verbal Necessity', *Journal of Philosophy*, 74 (1977), 71–85.

—— 'The Peculiar Effects of Love and Desire', in A. Rorty and B. McLaughlin (eds.), *Perspectives on Self-Deception*.

—— 'Rational Belief and Probability Kinematics', *Philosophy of Science*, 47 (1980), 165–87.

—— 'Rationality does not Require Conditionalization', forthcoming in E. Ullman-Margalit (ed.), *The Israel Colloquium Studies in the History, Philosophy, and Sociology of Science*, v (Dordrecht: Kluwer, 1989).

—— 'Salmon on Explanation', *Journal of Philosophy*, 82 (1985), 639–51.

—— 'Semantic Analysis of Quantum Logic', in C. A. Hooker (ed.), *Contemporary Research in the Foundations and Philosophy of Quantum Theory*, 80–113.

—— 'Symmetries in Personal Probability Kinematics', in N. Rescher (ed.), *Scientific Inquiry in Philosophical Perspective*.

—— 'A Temporal Framework for Conditionals and Chance', *Philosophical Review*, 89 (1980), 91–108.

—— 'Theory Comparison and Relevant Evidence', in J. Earman (ed.), *Testing Scientific Theories*.

—— 'Theory Construction and Experiment: An Empiricist View', in P. Asquith and R. Giere (eds.), *PSA 1980*, 663–78.

—— 'Was ist Aufklärung?', in G. Schurz (ed.), *Erklären und Verstehen* (Munich: Oldenbourg, 1988).

—— and Hughes, R. I. G., 'Symmetry Arguments in Probability Kinematics', in P. Kitcher and P. Asquith (eds.), *PSA 1984*, 851–69.

——, —— and Harman, G., 'A Problem for Relative Information Minimizers in Probability Kinematics, Continued', *British Journal for the Philosophy of Science*, 37 (1986), 453–75.

—— *An Introduction to the Philosophy of Time and Space* (1st edn., New York: Random House, 1970; 2nd edn. with new preface and postscript, New York: Columbia University Press, 1985).

—— *The Scientific Image* (Oxford: Oxford University Press, 1980).

VILLEGAS, C., 'Inner statistical inference', *Journal of the American Statistical Association*, 72 (1977), 453–8; and *Annals of Statistics*, 9 (1981), 768–76.

—— 'On Haar Priors', in V. P. Godambe *et al.* (eds.), *Foundations of Statistical Inference*, 409–14.

WATKINS, J., *Science and Scepticism* (Princeton, NJ: Princeton University Press, 1984).

WEYL, H., 'The Ghost of Modality', in M. Farber (ed.), *Philosophical Essays in Memory of Edmund Husserl*, 278–303.

—— 'Symmetry', *Journal of the Washington Academy of Sciences*, 28 (1938), 253–71.

—— *Gesammelte Abhhandlungen* (Berlin: Springer-Verlag, 1968).

—— *Symmetry* (Princeton, NJ: Princeton University Press, 1952).

—— *Theory of Groups and Quantum Mechanics* (New York: Dover, 1931).

WICKHAM, ANNA, *The Contemplative Quarry* (New York: Harcourt, Brace, and Co., 1921).

WILLEY, B., *The Eighteenth Century Background* (London: Chatto and Windus, 1940).

WOJCICKI, R., 'Set Theoretic Representations of Empirical Phenomena', *Journal of Philosophical Logic*, 3 (1974), 337–43.

WOOLHOUSE, R. S., *Locke* (Minneapolis: University of Minnesota Press, 1983).

—— *Locke's Philosophy of Science and Knowledge* (Oxford: Blackwell, 1971).

WORRALL, J., 'An Unreal Image', Review of van Fraassen, *The Scientific Image*, *British Journal for the Philosophy of Science*, 35 (1984), 65–80.

INDEX